DECISION ANALYSIS AND PORTFOLIO MANAGEMENT IN THE BIOPHARMA INDUSTRY

A PRACTITIONER'S GUIDE

Richard M. Bayney, Ph.D.
and Jose Orellana

J.ROSS PUBLISHING

ISBN: 978-1-60427-214-7
e-ISBN: 978-1-60427-869-9

Printed and bound in the U.S.A. Printed on acid-free paper.

10 9 8 7 6 5 4 3 2 1

Library of Congress Cataloging-in-Publication Data can be found in the WAV section of the publisher's website at www.jrosspub.com/wav.

Direct all inquiries to J. Ross Publishing, Inc., 151 N. Nob Hill Rd., Suite 476, Plantation, FL 33324. For EU safety/GPSR concerns, email: gpsr@mare-nostrum.co.uk or contact: Mare Nostrum Group B.V, Doelen 72, 4831 GR Breda, The Netherlands.

Phone: (954) 727-9333
Fax: (561) 892-0700
Web: www.jrosspub.com

CONTENTS

FOREWORD

Organizations that invest heavily in research and development (R&D) face a fundamental challenge. They must decide which ideas deserve scarce resources today in order to create value tomorrow. R&D portfolios unfold over long time horizons, operate under deep uncertainty, and require organizations to balance incremental improvements with transformational bets. Managing these investments well is not simply an operational concern. It is one of the most consequential strategic responsibilities any innovation-driven organization faces.

Over the past decade, the complexity of these decisions has increased significantly. Technology cycles move faster, development costs continue to rise, and leaders must weigh scientific promise against commercial opportunity while navigating uncertain regulatory, technical, and competitive landscapes. In this environment, the ability to manage innovation as a portfolio rather than as a collection of individual projects becomes essential.

In my work with strategic portfolio management teams across industries, including pharmaceuticals, manufacturing, chemicals, and high technology, I see these challenges play out every day. Organizations are increasingly applying advanced analytics and artificial intelligence to improve how they evaluate initiatives, identify emerging risks, and allocate resources across complex portfolios of work. These technologies can surface patterns, quantify uncertainty, and accelerate insight. Yet their value ultimately depends on the decision frameworks within which they operate. Without disciplined portfolio thinking, even the most sophisticated analytics do not translate into sound investment judgment.

Few people understand this discipline better than Dr. Richard Bayney. His career spans the full arc of R&D decision analysis and portfolio management. Trained as a molecular biologist, he began on the scientific front lines of drug discovery before moving into leadership roles across the pharmaceutical industry at Merck, Bayer, Bristol Myers Squibb, and Johnson & Johnson, where I first met him as a pragmatic and innovation-seeking leader. At Johnson & Johnson, he ultimately led portfolio management and decision analysis for all pharmaceutical R&D, helping guide investment decisions across complex, global development pipelines. In the years since, he has become a respected advisor, educator, and author, helping organizations around the world strengthen how they evaluate and govern innovation investments. This book reflects the perspective of someone who has spent decades navigating the practical realities of R&D decision making.

The chapters that follow guide the reader through a foundational methodology required to manage R&D portfolios effectively. Richard and his coauthor, Jose Orellana, begin by examining the structural realities of drug development, including the economics of the preclinical and clinical funnel and the necessary process of initiative attrition. They then introduce the principles of decision analysis and probabilistic thinking that allow leaders to evaluate opportunities under uncertainty.

Building on this foundation, the book presents a practical framework (**CREOPM**: asset **c**ategorization, **r**isk analysis, integrated **e**valuation, portfolio **o**ptimization, project **p**rioritization, **m**anagement best practices) for portfolio construction and governance that integrates strategic alignment, value assessment, and resource allocation. The result is a disciplined approach to managing innovation investments that brings greater transparency, rigor, and strategic clarity to some of the most difficult decisions organizations must make.

What further distinguishes this book from many treatments of portfolio management is the depth of its quantitative and decision analytic foundation. The authors introduce readers to the probabilistic and analytical tools required to evaluate R&D investments rigorously. Concepts such as probability of technical success, expected value modeling, scenario analysis, and risk-adjusted valuation are complemented by linear programming and stochastic optimization techniques that help organizations allocate constrained resources across portfolios of uncertain value. These methods translate scientific and technical risk into structured decision frameworks that allow leadership teams to compare opportunities across development stages, therapeutic areas, and time horizons while explicitly accounting for uncertainty.

At a time when scientific progress is accelerating and the stakes of innovation have never been higher, organizations must become better at deciding where to place their bets and be willing to take a long view on building lasting value. The work of Richard and Jose offers both the conceptual grounding and the practical guidance needed to meet that challenge. Leaders responsible for innovation, strategy, and portfolio governance will find in these pages a thoughtful, practical guide to making better R&D decisions about the future in a continuously challenging and competitive biopharma environment.

—Dr. Richard Sonnenblick, Chief Data Scientist, Planview

PREFACE

THE CURRENT STATE OF PORTFOLIO MANAGEMENT IN BIOPHARMACEUTICAL R&D

Across the biopharma sector, organizations continue to experience sustained pressure to restructure operating models, rebalance investment priorities, and improve capital efficiency in the face of escalating research and development (R&D) costs, increasing regulatory complexity, and persistently high attrition rates. Within this context, a commonly invoked internal mantra has emerged: *To be successful, we must do more with less.* While superficially compelling, this statement lacks the precision required to guide high-stakes, long-horizon investment decisions that define pharmaceutical R&D portfolios. From an executive perspective, the ambiguity embedded in this phrase is nontrivial. Notably absent is any explicit temporal framing without which the statement fails to provide the directional clarity required for scientific teams, portfolio managers, and senior leaders to align decisions across drug discovery, development, and commercialization.

In our experience, organizations that do not explicitly articulate their strategic and financial objectives—together with the metrics by which progress will be assessed over defined time frames—struggle to implement coherent portfolio management practices. A more precise articulation, albeit less succinct, would read:

> *To achieve sustained success—defined by measurable improvements in our primary financial and strategic performance indicators—over a defined multi-year horizon, we must maximize the value generated by our R&D investment portfolio, subject to finite financial, scientific, and organizational resources.*

Such clarity establishes the intellectual and operational foundation upon which disciplined portfolio management must rest. Without it, portfolio decisions become episodic, reactive, and susceptible to bias, rather than systematic and value maximizing.

FROM VALUE CREATION TO VALUE MAXIMIZATION IN R&D PORTFOLIOS

Value creation has become a ubiquitous concept in corporate and academic discourse. Yet, in practice, it is often invoked without reference to *optimality*. Assuming value is appropriately defined, the rational objective of portfolio management is not merely to create value, but to create *maximum* value given resource constraints under conditions of uncertainty.

In biopharma R&D, skepticism is sometimes expressed regarding the quantification of value, particularly in early discovery or high-uncertainty development stages. While it is true that not all dimensions of value can be perfectly monetized, all R&D investments ultimately exist to deliver benefit—whether to patients, healthcare systems, or society at large. At a minimum, such benefit can be expressed in terms of expected utility, clinical impact, or probabilistic economic outcomes. Framed in this way, *value maximization* may be more intuitively understood as *benefit maximization.*

RESTRUCTURING, COST PRESSURE, AND THE PORTFOLIO QUESTION

Over multiple decades, we have observed highly successful biopharma companies deliver transformative therapies, only to subsequently reconfigure their business models in response to market, pricing, or competitive pressures. These transitions are often accompanied by broad restructuring initiatives, cost containment programs, and workforce reductions. Yet a fundamental question frequently remains unanswered:

> *Given constrained resources, is the organization demonstrably maximizing the value it is capable of creating through its R&D portfolio?*

If portfolio-level value maximization were more explicitly assessed and transparently demonstrated, the rationale for many investment decisions—and disinvestments—would be clearer.

LIMITATIONS OF CURRENT R&D PORTFOLIO MANAGEMENT PRACTICE

While most biopharma organizations practice some form of R&D portfolio management, the rigor, transparency, and analytical maturity of these practices vary considerably. Furthermore, despite their natural complementarity, decision analysis is seldom integrated with portfolio management in the governance of biopharma R&D portfolios.

The two disciplines evolved to address different decision contexts: decision analysis emerged from operations research and applied statistics to support individual high-stakes decisions under uncertainty, while portfolio management developed largely as a managerial and financial control function focused on collections of projects competing for scarce resources. This divergence has led many organizations to apply decision analysis narrowly at the asset or milestone level, while reserving portfolio management for budgeting, prioritization, and governance cycles—without formally integrating the two.

Organizational structure further reinforces this separation. Decision analysis expertise in biopharma companies is often embedded within specialized quantitative groups (e.g., modeling, biostatistics, or advanced analytics), whereas portfolio decisions are typically owned by senior governance committees with heterogeneous analytical sophistication.

Methodological misalignment also plays a role. Portfolio management practices in R&D have historically relied largely on deterministic scoring models, ranking schemes, and stage-gate heuristics that are perceived as transparent and easy to communicate. In contrast, decision analysis explicitly embraces probability distributions, conditional dependencies, and nonlinear value functions—features that are essential for scientific realism.

Consequently, several recurring portfolio management deficiencies are evident:

- **Disassociation (or nonexistence) of decision analysis from portfolio management:** Many portfolio management decisions rest on fragile foundations, as the underlying data used to inform them often lack clear traceability, methodological rigor, and analytical defensibility.
- **Weak decision lineage:** It is often difficult to reconstruct why a given asset entered or progressed through the portfolio, what assumptions underpinned its approval, and how those assumptions have evolved over time.
- **Inadequate treatment of uncertainty:** Although risk is widely acknowledged, its quantification frequently lacks methodological rigor, traceability, and defensibility—particularly in early-stage programs.
- **Overreliance on point estimates:** Faced with uncertainty, organizations often default to single-point valuations to simplify decision making.
- **Conflation of prioritization with selection:** Asset ranking is frequently treated as synonymous with portfolio construction, even though prioritization alone is suboptimal to portfolio-level optimization.
- **Persistent execution underperformance:** Despite decades of focus on scope, schedule, and cost control, many R&D programs continue to experience delays, budget overruns, and scope erosion—indicating that broader portfolio-level drivers remain insufficiently addressed.

PURPOSE AND MOTIVATION OF THIS BOOK

This book is grounded in the conviction that decision analysis and portfolio management must be recognized and practiced as a continuous, enterprise-level discipline—not as a periodic budgeting or governance exercise. Absent such discipline, even commercially successful organizations leave substantial value unrealized. Worse, they may compensate through repeated restructuring cycles that erode institutional knowledge and destabilize scientific organizations. Embedding prescriptive decision analysis and portfolio management into the organizational fabric is therefore essential for long-term competitiveness in biopharma R&D.

Without this clarity, portfolio management becomes untethered from strategy and incapable of guiding meaningful trade-offs. Strategic planning is therefore a necessary precondition for effective R&D portfolio management.

THE EFFICIENT FRONTIER APPLIED TO R&D INVESTMENT PORTFOLIOS

Unlike financial portfolio management, which benefits from decades of theoretical and empirical development, portfolio management in R&D has struggled to achieve comparable disciplinary rigor. Central to financial portfolio theory is the *efficient frontier*, first articulated by Markowitz, which defines the set of portfolios that maximize expected return for a given level of risk.

This concept is directly applicable to R&D portfolios. For any given level of aggregate technical and financial risk, there exists a combination of R&D programs that maximizes expected portfolio value. Portfolios that lie below this frontier are, by definition, suboptimal—regardless of how attractive individual assets may appear in isolation.

THE CREOPM™ FRAMEWORK FOR R&D PORTFOLIO MANAGEMENT

To operationalize these principles, we introduce **CREOPM™**, a pragmatic framework for enterprise project portfolio management (EPPM) in R&D. The framework integrates six core components:

1. **Project categorization**
2. **Risk analysis**
3. **Integrated evaluation**
4. **Portfolio optimization**
5. **Project prioritization**
6. **Best practices in portfolio management**

CREOPM™ is intentionally modular and maturity-sensitive. While not presented as a universal prescription, its foundational principles—strategic clarity, analytical rigor, disciplined governance, and leadership accountability—are essential for any organization seeking to improve R&D portfolio performance systematically.

TARGET AUDIENCE AND BENEFICIARIES

This volume is designed to serve a diverse set of stakeholders engaged in the management, optimization, and strategic oversight of portfolios—whether of products, services, or organizational capabilities—within complex biopharma environments. The readership is expected to include the following categories:

1. **Portfolio management practitioners:** Professionals directly responsible for the execution and governance of project portfolios. For newcomers, this text provides a rigorous foundation in principles, methodologies, and analytical tools for robust project portfolio management (PPM). For experienced practitioners, it offers frameworks and case-based insights to elevate capability maturity and enhance decision rigor.
2. **Financial and resource managers:** Financial stewards who require transparency, defensibility, and traceability of project investment estimates, resource allocations, and expenditure patterns.
3. **Executive decision makers:** Senior leaders—including chief executive officers, chief financial officers, chief information officers, and heads of R&D—who may not be quantitative specialists but seek to evaluate the efficacy of their organization's PPM capabilities.
4. **Strategists and advisors:** Professionals charged with aligning organizational strategy with portfolio execution. The content provides actionable frameworks for integrating strategic objectives, R&D roadmaps, and business plans.
5. **Entrepreneurs and innovators:** Founders and business leaders navigating the complexities of both established and emerging markets. Those investing in novel technologies or platform-based innovations will find guidance on disciplined portfolio evaluation and risk-informed decision making.
6. **Graduate-level students and academics:** Master's-level students (e.g., MBA, EMBA, EMTM) with a serious interest in PPM and decision analysis. The book provides a bridge between the theoretical foundations of operations research/management science and the practical realities of organizational decision making in high-stakes R&D environments.

7. **Educators and thought leaders:** Instructors seeking to integrate quantitative rigor with organizational and strategic perspectives in portfolio management. The material offers a structured approach for teaching the intersection of analytical methods and executive decision making, emphasizing practical applicability in real-world contexts.
8. **Skeptics and pragmatists:** Individuals who question whether structured PPM frameworks add value beyond experiential judgment and intuitive decision making. The text addresses these perspectives by providing empirically grounded methodologies, illustrating how disciplined, evidence-based portfolio management complements, rather than replaces, managerial judgment.

ORGANIZATION OF THE BOOK

The book is structured into two sections containing a total of 10 chapters:

- **Section I: Decision Analysis and Portfolio Management in the Biopharma Industry**
  - **Chapter 1: Drug Research and Development Within an Evolving Industry—** This chapter argues that despite rising drug approvals and technological advances, biopharma R&D productivity remains constrained by persistent clinical attrition, high capitalized costs, and reliance on averages that obscure portfolio-level risk, making disciplined, probabilistic portfolio management essential for value creation.
  - **Chapter 2: Scenario Analysis and Strategic Planning—**This chapter positions scenario analysis as a decision-support discipline that helps biopharma leaders navigate deep uncertainty by clarifying causal logic, surfacing critical uncertainties, and stress-testing strategic robustness rather than forecasting outcomes.
  - **Chapter 3: Decision Analysis and Portfolio Management—**This chapter positions decision analysis as a rigorous, quantitative discipline for improving decision quality in biopharma R&D by explicitly framing uncertainty, trade-offs, and risk preferences, and by distinguishing sound decisions from stochastic outcomes.
- **Section II: CREOPM™—A Holistic Portfolio Management Framework**
  - **Chapter 4: Project Categorization—**This chapter explains project categorization as a core portfolio management discipline that enables consistent evaluation and governance by classifying initiatives into mutually exclusive categories based on obligation, risk, and strategic intent.
  - **Chapter 5: Risk Analysis—**This chapter presents a rigorous, probabilistic framework for quantifying and managing risk and uncertainty in biopharma R&D, distinguishing controllable operational risks from intrinsic scientific risks and extending analysis from projects to programs and portfolios.
  - **Chapter 6: Integrated Evaluation—**This chapter establishes a rigorous, decision-tree-based framework for valuing biopharma R&D investments using probabilistic financial metrics such as expected net present value, economic internal rate of return, and utility-adjusted value, explicitly modeling sequential risk resolution, timing, and interdependencies across projects and indications.
  - **Chapter 7: Portfolio Optimization—**This chapter presents portfolio optimization as a rigorous, quantitative framework for selecting the combination of biopharma R&D investments that maximizes value, balances risk, and satisfies strategic and resource constraints by identifying efficient-frontier portfolios.

- **Chapter 8: Project Prioritization**—This chapter describes project prioritization as a structured approach for ranking biopharma initiatives using financial metrics and multi-objective frameworks such as multiple objective decision analysis to integrate value, risk, cost, and strategic fit under resource constraints.
- **Chapter 9: Best Practices in Portfolio Management**—This chapter presents a comprehensive framework for achieving mature, value-driven EPPM in biopharma, emphasizing capability development across strategy, process, people, and technology domains.
- **Chapter 10: Epilogue**—This chapter summarizes CREOPM™ as a structured, enterprise-level framework that integrates categorization, risk analysis, evaluation, optimization, prioritization, and management to maximize value in complex biopharma R&D portfolios.

ACKNOWLEDGMENTS

The inherent complexity and sustained effort required to write a book are far from trivial, particularly when coauthors are working across different time zones while balancing professional responsibilities and family commitments. Under such conditions, successful collaboration depends critically on a shared mindset, aligned values, and a common passion for the subject matter. Fortunately, these elements were present from the very beginning of this project.

From the outset, we were fortunate to work with an exceptional editor, Stephen Buda of J. Ross Publishing, whose thoughtful refinement and steady guidance were instrumental in shepherding this manuscript from its earliest drafts through to final production. His contributions significantly improved both the clarity and coherence of the work.

From Richard Bayney:
To my wonderful wife, Pauls, who has been the bedrock of my existence and my constant companion through every chapter of life since Ma passed on—your strength, patience, and unwavering love have sustained me in ways words cannot adequately capture. You have stood beside me at every step of the journey, offering clarity in moments of uncertainty and warmth in times of challenge. To my precious children, Ma and Zico, whose presence has brought immeasurable joy, laughter, and purpose into my life—though I try, I can never fully express how profoundly fortunate I am to have you both. You are my greatest source of inspiration and my enduring reminder of what truly matters. And to my precious and loving granddaughters, Lexi and Savi—your boundless joy, infectious laughter, and unguarded curiosity have been a constant source of light in my life. Through you, I have learned anew the virtues of patience, presence, and gentle countenance that help define the true gift of grandparenthood.

I have been immensely fortunate to have formed friendships that have grown into bonds of lifelong brotherhood—Terry Banarsee, Robin Naraine, Martin White, and Subinay Ganguly, along with Rajendra Tiwari and Ruthven Thompson, who were both taken far too soon. Each of you enriched my life in ways that time cannot diminish, through shared experiences, loyalty, and unwavering support. Thank you all for welcoming me into your lives, for the memories we created together, and for the enduring imprint you have left on my heart.

As always, to my wonderful grandmother, Ma, who possessed a wisdom and humanity that surpassed that of anyone I have ever known. Her boundless and unconditional love, quiet strength, and deeply rooted values shaped the person I have become. I thank her every day for her sacrifices and her guidance, and for instilling in me a moral compass that has guided me throughout my life. Until we meet again, Ma.

From Jose Orellana:
To my wife, Sonia, and my daughters, Sonia Elizabeth and Emilia. I can't think of a better team to have in my corner or better people with whom to share our adventures around the world.

And to my parents, Hildegar and Jose, who made the difficult decision many years ago to let me pursue studies in a faraway place. That decision made me the person I am today, and I thank you for that.

I am honored to join Richard as a coauthor, and I thank him for his mentorship. I would also like to thank Lumivero for their support and for continuing to innovate and develop software in the risk analysis space.

ABOUT THE AUTHORS

RICHARD BAYNEY

Richard Bayney is the president and founder of Project & Portfolio Value Creation (PPVC), a specialized consulting boutique that delivers an integrated suite of services spanning executive education, decision-support analytics, asset valuation, and R&D portfolio strategy. PPVC's work is grounded in rigorous applications of decision analysis, risk analysis, and portfolio management, and is applied across multiple innovation-driven sectors, including biopharmaceuticals, medical devices and diagnostics, consumer care, and agriculture. Through this work, Richard advises organizations on aligning scientific uncertainty, regulatory risk, and capital allocation with long-term strategic and commercial objectives.

In parallel with his consulting practice, Richard has an extensive academic and professional speaking background. He has served as a lecturer at the University of Pennsylvania, where he taught graduate and executive courses in decision modeling and project portfolio management. He is also a frequently invited speaker at industry forums, where he addresses topics such as decision and risk analysis, portfolio optimization, and strategic planning in drug development.

Richard brings more than four decades of industry and consulting experience to his work. He is a 23-year veteran of the biopharma industry, having held roles at Merck, Bayer, Bristol-Myers Squibb, and Johnson & Johnson. In addition, he has spent 18 years in management consulting, working primarily with clients in the United States and Europe. Over the course of his career, he has built and led functions in strategic planning, decision analysis, and portfolio management, and has hands-on operational experience, including two years as an international project manager for a marketed cardiovascular therapy.

Earlier in his career, Richard spent a decade as a molecular biologist, conducting research on gene expression in drug detoxification systems and in the pathophysiology of Alzheimer's disease. He has an M.Sc. and Ph.D. from the University of London, an MBA from Columbia University, and professional certification as a Project Management Professional (PMP) from the Project Management Institute.

JOSE ORELLANA

Jose Orellana is a senior business consultant and quantitative solutions engineer at Lumivero with over 25 years of experience applying numerical modeling to high-stakes decision making. As a globally recognized expert in the DecisionTools Suite (@RISK, PrecisionTree, Evolver) and XLSTAT, Jose has advised organizations worldwide on navigating the inherent uncertainties of complex R&D cycles.

While his foundational expertise was forged in the demanding engineering environments of Schlumberger and P&G, his work now focuses on the *last mile* of decision analysis—the critical bridge between raw data and executive strategy. Leveraging a mastery of Python, machine learning, and traditional Monte Carlo simulation, Jose provides a robust framework for modeling the binary outcomes and long-tail risks characteristic of the pharmaceutical and life sciences sectors.

Jose holds a B.S. in Electrical Engineering from Cornell University, where he graduated Magna Cum Laude and was inducted into the Tau Beta Pi and Eta Kappa Nu honor societies. His interdisciplinary approach is further bolstered by a Data Analytics Certification from UC Berkeley and a Diploma of Management from the Australian Institute of Management. He currently resides in Perth, Western Australia, with his wife and two teenage daughters.

This book has free material available for download from the Web Added Value™ resource center at *www.jrosspub.com*

At J. Ross Publishing we are committed to providing today's professional with practical, hands-on tools that enhance the learning experience and give readers an opportunity to apply what they have learned. That is why we offer free ancillary materials available for download on this book and all participating Web Added Value™ publications. These online resources may include interactive versions of material that appears in the book or supplemental templates, worksheets, models, plans, case studies, proposals, spreadsheets and assessment tools, among other things. Whenever you see the WAV™ symbol in any of our publications, it means bonus materials accompany the book and are available from the Web Added Value Download Resource Center at www.jrosspub.com.

Downloads for *Decision Analysis and Portfolio Management in the Biopharma Industry* include figures and tables that were too large or complex to fit inside the printed book and a bonus chapter on attrition in drug development.

SECTION I

Decision Analysis and Portfolio Management in the Biopharma Industry

1

DRUG RESEARCH AND DEVELOPMENT WITHIN AN EVOLVING INDUSTRY

Some of the figures and tables in this chapter are too large or complex to properly format in printed form. Figures and tables marked with a "WAV" label can be downloaded from the Web Added Value Download Resource Center at www.jrosspub.com/wav for proper viewing.

1.1 THE RESEARCH AND DEVELOPMENT FUNNEL—THE IMPACT OF RISK, COST, AND TIME TO PRODUCTIVITY

Since the early 2000s, the biopharmaceutical industry (biopharma) has experienced sustained structural disruption, fundamentally reshaping operating models across leading organizations. A primary driver has been the widespread erosion of intellectual property protection for blockbuster products, undermining many companies' ability to sustain the double-digit revenue growth that characterized the late twentieth century.

This challenge has been compounded by persistent difficulty in replenishing late-stage and commercial portfolios fast enough to offset generic and biosimilar competition. Although global approvals of new molecular entities (NMEs) have increased—averaging ~51 per year from 2015 through 2019 and rising to ~76 annually from 2020 through 2024—the distribution of innovation remains highly concentrated. Of the 634 NMEs approved globally between 2015 and 2024, oncology (34.4%) and neurology (12.1%) accounted for nearly half of all launches (IQVIA 2025). In 2024 alone, emerging biopharma companies (EBPs), defined by research and development (R&D) spend below $200M and revenues under $500M, originated ~85% of the 48 NMEs approved in the U.S. market. That same year, small-molecule (new chemical entity) and biologic (new biologic entity) modalities contributed roughly equally, highlighting continued platform diversification (IQVIA 2025).

Despite major advances in discovery technologies—such as high-throughput screening, combinatorial chemistry, and genomics-enabled target identification—overall R&D conversion efficiency has improved only marginally. While discovery throughput has expanded dramatically, downstream clinical and regulatory attrition rates have remained largely unchanged. In response, the industry has increasingly externalized discovery and development through outsourcing and offshoring, while reducing internal R&D headcount. In parallel, mergers and acquisitions have become a dominant strategy to augment pipelines, diversify risk, and capture operational synergies, often followed by further workforce rationalization.

Although these measures may enhance near-term efficiency, their ability to deliver durable improvements in long-term R&D productivity is uncertain. As a result, sustainable value creation is increasingly viewed as dependent less on organizational scale and more on disciplined, data-driven portfolio management in an environment characterized by high technical failure rates and substantial commercial uncertainty. Effective portfolio governance requires decision frameworks that integrate strategic intent with probabilistic assessments of risk, value, and timing—addressing not only which assets to advance, but also when and why capital should be allocated.

A biopharma portfolio typically comprises two components: an R&D portfolio (*pipeline*) spanning research, clinical development, and regulatory submission, and a portfolio of marketed products (*base business*) generating current revenues. By convention, the pipeline is burdened primarily by technical and regulatory risk, while the commercial portfolio faces market uncertainty driven by competition and pricing pressure but minimal technical risk. Many organizations—particularly EBPs—may have robust pipelines without any marketed products. Managing the interaction between these portfolio components has therefore become a critical determinant of long-term resilience and shareholder value.

Industry attrition statistics underscore the magnitude of this challenge. Based on historical success rates, approximately 23–36 drug candidates of chemical or biological origin are required to achieve FDA approval of a single product (see Figure 1.1). Of these, roughly 12–18 advance to preclinical development, 8–14 enter clinical development, and only one or two reach the registrational phase, ultimately yielding one approved product. This funnel structure necessitates robust portfolio decision-making processes to maximize the value of constrained capital and human resources.

Drug *discovery*, typically spanning two to four years (Rang 2006), includes three core activities. *Target selection and validation* focus on identifying and confirming disease-relevant, druggable molecular targets using genomic, proteomic, computational, and translational approaches. *Lead identification* centers on developing and deploying biochemical, biophysical, and cell-based assays to screen

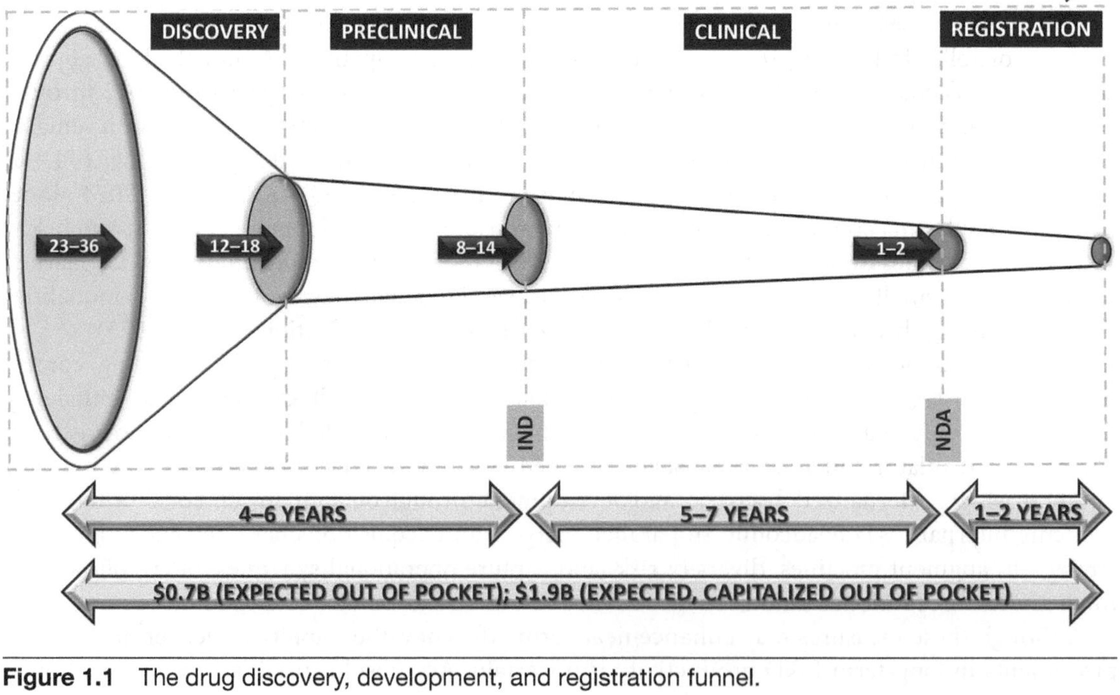

Figure 1.1 The drug discovery, development, and registration funnel.

chemical or biologic libraries, triaging hits based on potency, selectivity, and preliminary pharmacokinetics (PK); adsorption, distribution, metabolism, and excretion characteristics (ADME), and prioritizing lead series through structure–activity relationships (SAR). *Lead optimization* involves iterative refinement of selected leads through medicinal chemistry or modality-specific engineering, supported by in vitro and in vivo pharmacology, PK/PD (pharmacodynamic) characterization, developability assessments, and translational biomarker identification to enable candidate selection and readiness for Good Laboratory Practice (GLP) toxicology and first-in-human (FIH) studies.

Preclinical development typically spans 18–24 months and includes following Good Manufacturing Practices in the manufacture of drug substance and product, dose-ranging and tolerability studies, formulation and stability characterization, safety pharmacology, genotoxicity and repeat-dose toxicology, and establishment of validated analytical methods. These activities culminate in the preparation of the Investigational New Drug (IND) dossier and investigator's brochure from the Institutional Review Board, including Chemistry, Manufacturing, and Controls, for regulatory submission.

Following IND authorization, clinical development proceeds sequentially over ~5–8 years. Phase I studies assess safety, tolerability, and human PK/PD. Phase II trials refine dose–response relationships, establish proof-of-mechanism (PoM) or proof-of-concept (PoC), and select optimized dosing regimens. Phase III trials provide statistically powered confirmatory evidence of safety and efficacy to support regulatory approval, while Phase IV encompasses post-marketing commitments and pharmacovigilance. Upon completion, a New Drug Application is submitted, typically requiring 1–1.5 years for FDA review. Across the full discovery-to-approval continuum, the risk-adjusted, capitalized cost of a successful FDA approval is estimated at ~$1.7B–$3.4B, inclusive of the cost of failed projects (Paul et al. 2010; DiMasi et al. 2016; Sertkaya et al. 2024).

Throughout this process, portfolio decisions must be made under conditions of extreme technical risk and commercial uncertainty. While project management emphasizes executing individual programs efficiently, portfolio management focuses on selecting the right projects—those that maximize strategic and financial value under constrained resources. This challenge raises critical questions regarding asset entrance, progression, delay, and termination, along with portfolio selection:

- **Asset prioritization and development speed:** Which assets should be actively advanced, and at what development cadence, given capital constraints, probability of technical success, and anticipated value inflection points?
- **Decision readiness and evidentiary standards:** What minimum data packages, decision criteria, and analytical rigor are required to support informed progression, delay, hold, and terminate decisions at each stage of asset maturation?
- **Inter-asset and portfolio interdependencies:** What scientific, operational, or commercial linkages exist across assets—such that outcomes or decisions for one program materially influence the technical and/or commercial viability of others—and how should these dependencies be explicitly modeled and governed?
- **Stage-gate progression discipline:** Should attainment of predefined timelines or milestones automatically trigger advancement to the next development phase, or should progression remain contingent on reassessment of risk-adjusted value, strategic alignment, and opportunity cost?
- **Resource intensity and risk-based funding:** Which programs warrant full resource commitment, which should receive constrained or staged investment, and which should be pursued under *at-risk* funding models to preserve option value while limiting downside exposure?
- **Capital reallocation and opportunity cost management:** At what point should investment in an asset—regardless of prior sunk costs—be discontinued or redeployed toward higher-value internal programs or externally sourced opportunities?

- **Strategic resource allocation across portfolios:** How should financial, human, and operational resources be optimally distributed across projects, programs, and sub-portfolios (e.g., therapeutic area (TA) portfolios) to achieve enterprise-level strategic and financial objectives?
- **Enterprise value maximization in complex organizations:** In multi-TA or multi-business-unit (BU) organizations, how can resources be allocated to maximize aggregate enterprise value while maintaining sufficient autonomy to foster innovation, scientific differentiation, and sustainable growth within individual TAs or BUs?

The purpose of this book is to provide structured guidance and practical answers to these foundational portfolio management questions.

1.1.1 Drivers of Productivity in Biopharma

Over the past three decades, a small handful of biopharma companies—most notably Eli Lilly—have dissected the components of R&D productivity. As early as 2010, Paul et al. published a seminal article aimed at examining ways to improve R&D productivity, in which a *pharmaceutical value equation* was defined as follows:

$$P \propto \frac{WIP \times p(TS) \times V}{CT \times C}$$

where P = productivity, WIP = phase-specific work in process, *p*(TS) = probability of technical success, V = value, CT = cycle time, and C = cost. This expression reflects the productivity of a pipeline in terms of securing a single product from a specific phase of research or development.

We have proposed modest modifications to this value equation in two forms which are referred to as the *non-risk-adjusted productivity index* (PI) and the *expected (or risk-adjusted) productivity index* (ePI) that reflect the productivity of a pipeline across all phases of research and development. Accordingly, PI is defined as:

$$PI \propto \frac{\sum_{i=0}^{n} Mean\ CVGS_i}{\sum_{i=0}^{n} Mean\ Cost_i}$$

where $CVGS_i$ = present value of commercial cash flows (i.e., from the point of product launch) given technical and regulatory success and $Cost_i$ = present value of cost cash (out)flows from a specific phase until launch.

Similarly, ePI can be expressed as:

$$ePI \propto \frac{\sum_{i=0}^{n} Mean\ eCVGS_i}{\sum_{i=0}^{n} Mean\ eCost_i}$$

where $eCVGS_i$ = risk-adjusted, present value of commercial cash flows given technical and regulatory success and $eCost_i$ = present value of risk-adjusted cost cash (out)flows from a specific phase until launch. In this way, cycle time (CT) is accounted for in the calculation of the present value of cost cash (out)flows, whether risk-adjusted or not.

Using some of the data from Paul et al. (2010) and Sertkaya et al. (2024), we have adjusted cost per phase in 2010 and 2018 dollars to 2025 dollars using an annual inflation rate of 2.62% and have

Table 1.1 R&D pipeline throughput attrition model incorporating risk, cost, and cycle time to generate a single new product launch (Modified from Paul et al. 2010 and Sertkaya et al. 2024).

	TARGET-TO-HIT	HIT-TO-LEAD	LEAD OPTIMIZATION	PRECLINICAL	PHASE I	PHASE IIa	PHASE IIb	PHASE III	SUBMISSION TO LAUNCH	LAUNCH
p (Phase Success)	80%	75%	85%	69%	60%	48%	75%	66%	88%	
OUT-OF-POCKET COST IN 2025 $ (M)	1.5	3.7	14.7	7.4	8.5	10.1	15.1	107.0	3.1	171.1
CAPITALIZED OUT-OF-POCKET COST IN 2025 $ (M)	6.1	13.9	48.7	21.4	20.8	21.0	27.1	146.3	3.3	308.5
EXPECTED OUT-OF-POCKET COST IN 2025 $ (M)	33.3	66.7	200.3	85.1	67.8	48.3	34.7	184.2	3.5	723.9
EXPECTED CAPITALIZED OUT-OF-POCKET COST IN 2025 $ (M)	139.1	251.0	661.8	246.9	165.4	100.3	62.2	252.0	3.8	1,882.3
CYCLE TIME (Years)	1.0	1.0	1.5	1.0	2.3	0.8	2.0	3.2	1.4	14.2
PTRS	4.4%	5.5%	7.4%	8.7%	12.5%	20.9%	43.6%	58.1%	88.0%	88.0%
AVERAGE # PROJECTS REQUIRED FOR 1 NEW PRODUCT LAUNCH	22.7	18.1	13.6	11.6	8.0	4.8	2.3	1.7	1.1	1
% TOTAL EXPECTED OUT-OF-POCKET COST PER NEW PRODUCT LAUNCH	4.6%	9.2%	27.7%	11.8%	9.4%	6.7%	4.8%	25.4%	0.5%	

created an R&D pipeline throughput attrition model (see Table 1.1) which allows for an analysis of the cost of generating a commercial product on three metrics as defined by Sertkaya et al. (2024):

1. **Direct out-of-pocket (OOP) cost:** This represents the cash outlay required to bring a single product to approval, excluding any investments associated with failed projects.
2. **Risk-adjusted or expected OOP cost:** This extends the OOP estimate by accounting for the cost of failures due to attrition at any phase of R&D and registration. As such, this metric distributes the cost burden of attrition across a full portfolio of R&D investments.
3. **Expected capitalized OOP cost:** This elaborates the expected OOP estimate by incorporating the time value of money. It factors in phase-specific timelines, overall program duration, and the opportunity cost of capital accumulated throughout the R&D and registration process.

Given the phase-specific risks shown in Table 1.1, approximately 23 entities are required at the *target-to-hit* stage to generate, on average, a new product launch. As entities are terminated during the research phases, nearly 12 survive to enter preclinical, of which almost 8 advance to the first phase of clinical development. On average, from eight NMEs that enter Phase I, one new product is generated. Because this is a discrete, deterministic analysis, we do not know the likelihood that exactly eight preclinical molecules will enter Phase I, nor do we know that one, less than one, or more than one new product will emerge from these 23 target-to-hit or eight Phase I entities. This is solved by conducting a simulation analysis of the NMEs entering and exiting each phase—the launch results of which are shown in Figures 1.2A and 1.2B. The range of possible entrants from target-to-hit to Phase I is 0–19, with a mean and mode of eight (see Figure 1.2A). Similarly, while the mean and modal number of launches is one, the range of possible launches is 0–7, of which one launch has the highest probability of occurrence at 38% (see Figure 1.2B). It is noteworthy that the second-highest likely outcome is zero launches with a probability of 35%, meaning the probability of securing at least one product launch from 23 target-to-hit NMEs is 65%. This probabilistic analysis demonstrates that while calculations of averages are useful, they should be treated with caution as there are distributions of outcomes associated with each calculated mean.

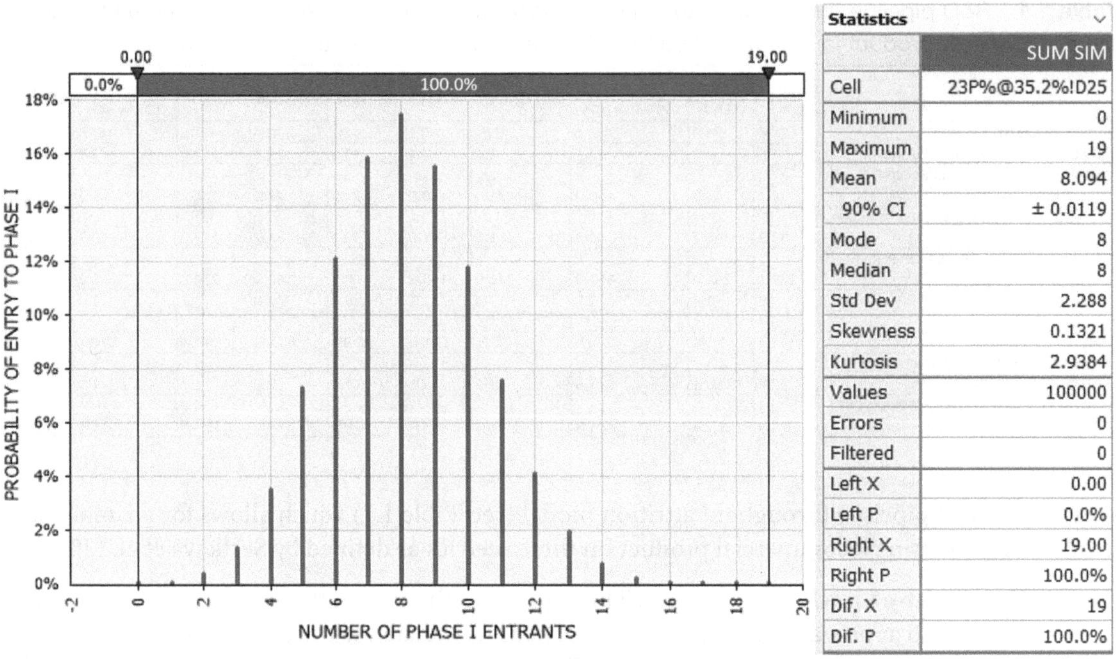

Figure 1.2A Distribution of Phase I entrants from 23 target-to-hit entrants.

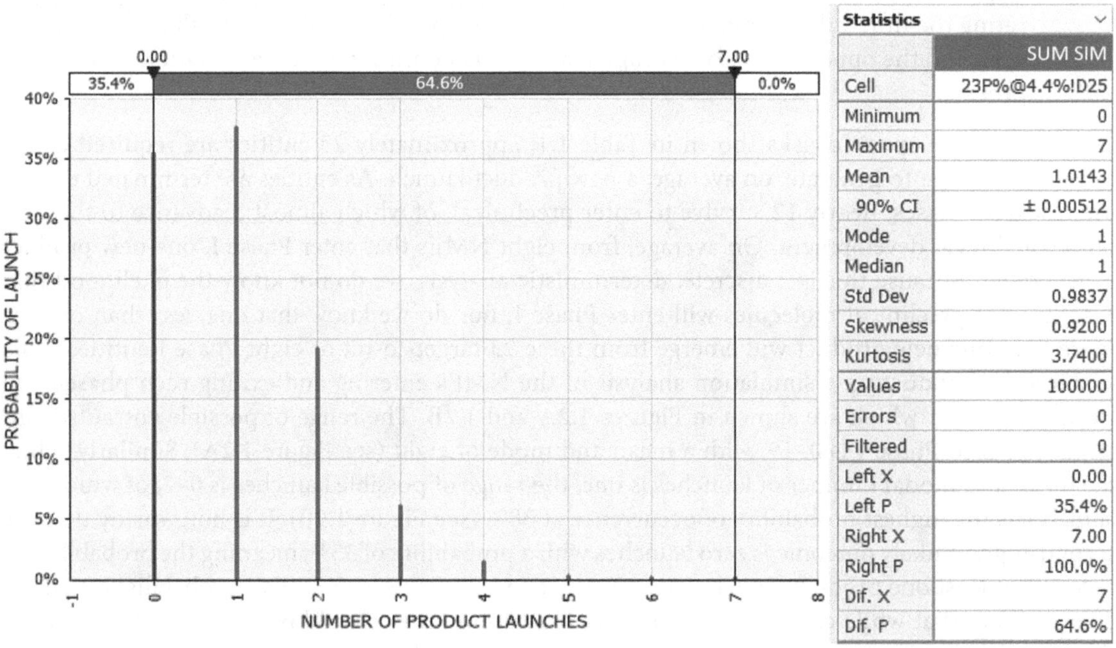

Figure 1.2B Distribution of product launches from 23 target-to-hit entrants.

In building our pipeline throughput model (data shown in Table 1.1), we have done the following:

- **Risk**—Used phase-specific *p*(Success) data for the four stages of research from Paul et al. (2010) and for the four stages of clinical development and registration from Sertkaya et al. (2024). Additionally, while keeping the *p*(Success) of Phase II at 0.36 (Sertkaya et al. 2024), we have disaggregated this phase into Phase IIa and Phase IIb and assigned a lower *p*(Success) for Phase IIa (0.48) than for Phase IIb (0.75).
- **Cost**—Taken phase-specific OOP cost in 2010 dollars for the four stages of research from Paul et al. (2010) and for the four stages of clinical development and registration in 2018 dollars from Sertkaya et al. (2024) and adjusted for inflation to 2025 dollars. The Phase II OOP cost from Sertkaya et al. (2024) has been disaggregated into Phase IIa and Phase IIb with Phase IIb being more expensive ($15.1M) than Phase IIa ($10.1M).
- **Cycle time**—Modified phase-specific cycle time for the four stages of research from Paul et al. (2010) and applied cycle time for the four stages of clinical development and registration from Sertkaya et al. (2024). The Phase II cycle time from Sertkaya et al. (2024) has been disaggregated into Phase IIa and Phase IIb with Phase IIb taking longer (2.0 years) than Phase IIa (0.8 years).
- **Inflation and cost of capital**—Utilized 2.62% annually and 11.0%, respectively.

Beginning with the target-to-hit phase, and without accounting for the cost of failures or opportunity cost of capital, the OOP cost to bring an NME product through the pipeline to launch is $171M. When opportunity cost is considered, this amount rises to $309M (see Table 1.1). When the cost of failures, but not the opportunity cost of capital, is taken into consideration, the expected OOP cost of securing a single NME product launch is $724M. On an expected capitalized basis, i.e., when both the cost of failures and opportunity cost of capital are accounted for, the OOP cost of successfully navigating every phase of research, clinical development, and registration to bring a single NME product to market is $1.882B (see Table 1.1). This estimate is lower than that of Paul et al. ($2.621B in 2025 dollars; based on self-reported data by members of the Pharmaceutical Benchmarking Forum) and higher than the average estimate of Sertkaya et al. ($1.053B in 2025 dollars; based on publicly available data).

While including the post-approval phase (Phase IV), but without accounting for costs and durations associated with target-to-hit, hit-to-lead, and lead optimization phases, Sertkaya et al. (2024), using publicly available databases, have reported phase-specific costs by TA as shown in Table 1.2.

In the study of Sertkaya et al. (2024), aggregate, OOP costs for non-clinical, clinical, regulatory, and Phase IV phases range from a low of $72.5M ($52.7M–$83.7M) for the genitourinary TA to a high of $297.2M ($82.8M–$515.9M) for pain and anesthesia. When accounting for the cost of failures due to the probability of successfully transitioning from one phase to another, the corresponding costs are $244.0M ($152.4M–$345.8M) for genitourinary and $887.9M ($325.3M–$1,476.9M) for pain and anesthesia TAs. Finally, on a capitalized basis, the analogous costs are $394.9M ($228.9M–$626.7M) for genitourinary and $1,756.2M ($648.5M–$3,171.5M) for pain and anesthesia (see Table 1.2). While it is clear that the cost of drug development remains idiosyncratic to a TA and, more important, to a specific disease, in our opinion, it is incumbent on analysts to utilize a transparent, financially defensible methodology based on after-tax cash flows from which to assess the impact of revenues and costs on R&D productivity. Further, as Sertkaya et al. (2024) have shown, the use of average costs to compute the cost of drug discovery and development by any metric can be quite misleading unless adequate attention is paid to the intended TA(s) of origin.

When investigated by TA, Sertkaya et al. (2024) have shown that from non-clinical research, mean aggregate drug development cycle time to launch ranges from 121 months (genitourinary) to

Table 1.2 Phase-specific costs across TAs measured by OOP, expected OOP, and expected capitalized OOP costs (reproduced with permission: Sertkaya et al. 2024).

THERAPEUTIC AREA	PTRS FROM NON-CLINICAL %, (95% CI)	OOP $M	EXPECTED OOP $M	EXPECTED CAPITALIZED OOP $M
ANTI-INFECTIVE	15.5 (10.9-21.7)	109.4 (65.3-186.7)	251.0 (166.7-360.3)	378.7 (244.6-556.1)
CARDIOVASCULAR	7.1 (3.6-15.3)	153.5 (73.7-593.3)	519.3 (273.9-2,168.1)	890.3 (445.0-3,807.9)
CENTRAL NERVOUS SYSTEM	6.7 (4.0-12.9)	113.0 (84.9-132.0)	463.7 (302.5-714.3)	894.6 (526.1-1,550.3)
DERMATOLOGY	8.5 (8.5-8.5)	143.8 (60.0 - 253.0)	419.1 (221.5-688.9)	683.1 (349.5-1,252.2)
ENDOCRINE	11.1 (6.2-17.7)	181.2 (139.2-226.8)	492.0 (359.4-728.4)	780.0 (541.0-1,243.2)
GASTROINTESTINAL	8.2 (6.4-10.3)	216.7 (101.4-363.6)	591.1 (285.2-898.3)	963.5 (451.5-1,564.5)
GENITOURINARY SYSTEM	11.8 (7.8-16.3)	72.5 (52.7-83.7)	244.0 (152.4-345.8)	394.9 (228.9-626.7)
HEMATOLOGY	17.8 (17.8-17.8)	152.9 (82.5-230.7)	381.9 (214.3-578.0)	720.4 (348.2-1,295.1)
ONCOLOGY	4.1 (2.2-8.6)	84.7 (65.5-121.6)	595.5 (344.3-1,041.0)	1,209.2 (624.6-2,388.7)
RESPIRATORY SYSTEM	8.6 (5.3-11.7)	167.8 (94.0-332.2)	409.3 (246.3-685.5)	686.4 (399.5-1,164.1)
OPHTHALMOLOGY	13.9 (11.6-16.2)	256.2 (101.2-381.8)	787.6 (333.4-1,179.7)	1,191.6 (496.3-1,910.8)
PAIN & ANESTHESIA	8.5 (8.5-8.5)	297.2 (82.8-515.9)	887.9 (325.3-1,476.9)	1,756.2 (648.5-3,171.5)
IMMUNOMODULATION	11.9 (7.5-18.0)	119.3 (74.0-180.1)	363.7 (201.7-586.2)	605.8 (306.1-1,041.9)
ALL	8.5 (3.9-16.1)	172.7 (132.5-197.9)	515.8 (327.0-773.2)	879.3 (416.9-1,307.3)

171 months (oncology). If Phase IV is included, between 29.9 and 45.7 months need to be added for these two TAs (see Figure 1.3). While oncology boasts the shortest cycle time necessary for FDA registration (9.6 months), it takes the longest in Phase I (31.9 months) among all TAs, marginally longer than that of the central nervous system in Phase II (40.3 months), and second only to hematology in Phase III (57.7 months). If modeled on the basis of the average of all 13 TAs, eight TAs would have shorter durations, while five TAs would show longer cycle times than the average.

As it is with phase-specific costs, it is tempting to draw conclusions with respect to which TAs show shorter or longer cycle times in drug development. However, it is more important to recognize that over the past three decades, the lines of distinction between clinical development phases have become intentionally blurred. What was previously deemed a Phase I study to investigate safety and tolerability in humans has, in many cases, morphed into a longer, riskier study to investigate early efficacy using either a single ascending drug dose or multiple ascending doses. Understandably, this type of trial lends itself to a longer cycle time with higher cost and higher risk. The same is true of Phase II, which is a combination of Phase IIa to demonstrate that the drug conveys physiologic benefit in accordance with its hypothesized target(s) and Phase IIb, which is geared toward finding the optimal dose (and regimen) necessary to conduct registrational trials in Phase III. The authors therefore remain cautious about drawing inferences and conclusions from data sets, no matter how large they may be.

While phase-specific cost and cycle time affect the denominator of the R&D productivity equation, phase-specific risk remains a keenly debated topic as a critical numerator. Paul et al. (2010) have shown, quite elegantly, the key areas for improving R&D productivity in which clinical risk dominates all other parameters. We have used our model to conduct a similar sensitivity analysis, which shows that risk—defined as 1-*p*(Success)—and, in particular, clinical risk, remains the most important determinant of R&D productivity.

The mean OOP expenditure required to advance an NME from discovery to market authorization is approximately $171M, with late-stage clinical development representing the primary cost driver. Phase III clinical development alone accounts for roughly $107M of this total. Sensitivity

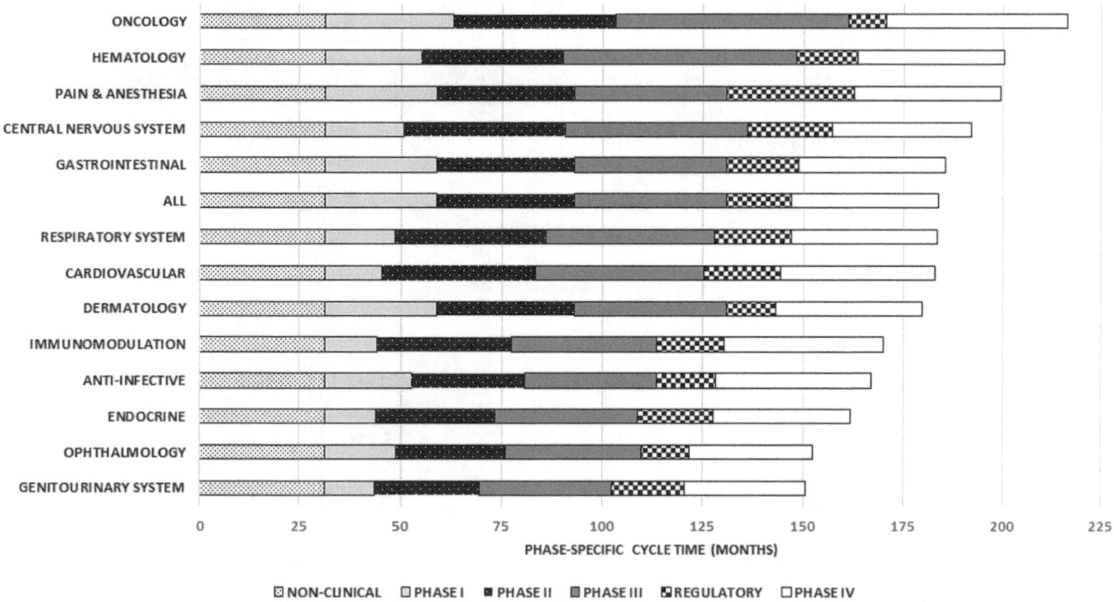

Figure 1.3 Phase-specific cycle times across TAs (Sertkaya et al. 2024).

analysis demonstrates that a ±25% variation in Phase III base cost produces a corresponding range in total OOP expenditure from approximately $144M to $198M. In contrast, equivalent proportional changes applied to OOP costs in other development phases exert only modest effects on aggregate mean OOP cost, underscoring the disproportionate influence of Phase III spending (see Figure 1.4).

When costs are evaluated on an expected-value basis that incorporates phase-specific probabilities of technical and regulatory success, the average OOP investment required to achieve a single marketed NME increases substantially to approximately $724M. Under this framework, the expected OOP cost is highly sensitive to changes in success probabilities across most development stages, with the largest contributions arising from Phases IIa and III. For example, adjusting the probability of success in Phase IIa from its base value of 48% by ±15 percentage points results in an expected OOP cost range spanning approximately $605M to $952M (see WAV Figure 1.5). A comparable perturbation applied to the Phase III success probability (base value 66%) yields a range from roughly $591M to $936M. Among phase-specific OOP expenditures, lead optimization—despite its relatively modest absolute cost ($14.7M)—emerges as the most influential cost variable on expected OOP outcomes, producing a swing of nearly $100M when varied by ±25%.

Capitalized OOP cost, which accounts for development duration and the time value of money, exhibits a distinct sensitivity profile. Within this framework, Phase III direct costs again dominate all other phase-specific cost inputs. A ±25% change in Phase III OOP expenditure results in a capitalized cost range of approximately $272M to $345M per successfully launched NME. Among cycle-time assumptions, Phase III duration (base value 3.2 years) exerts the greatest leverage on capitalized cost; varying this parameter by ±25% drives capitalized OOP cost between approximately $290M and $329M. As expected, the assumed cost of capital is a critical determinant of capitalized investment. A one percentage point decrease in the cost of capital reduces capitalized OOP cost to approximately $292M, while a one percentage point increase raises it to roughly $327M.

When all key inputs—including phase-specific probabilities of success, OOP costs, development cycle times, and cost of capital—are simultaneously subjected to sensitivity analysis, the resulting

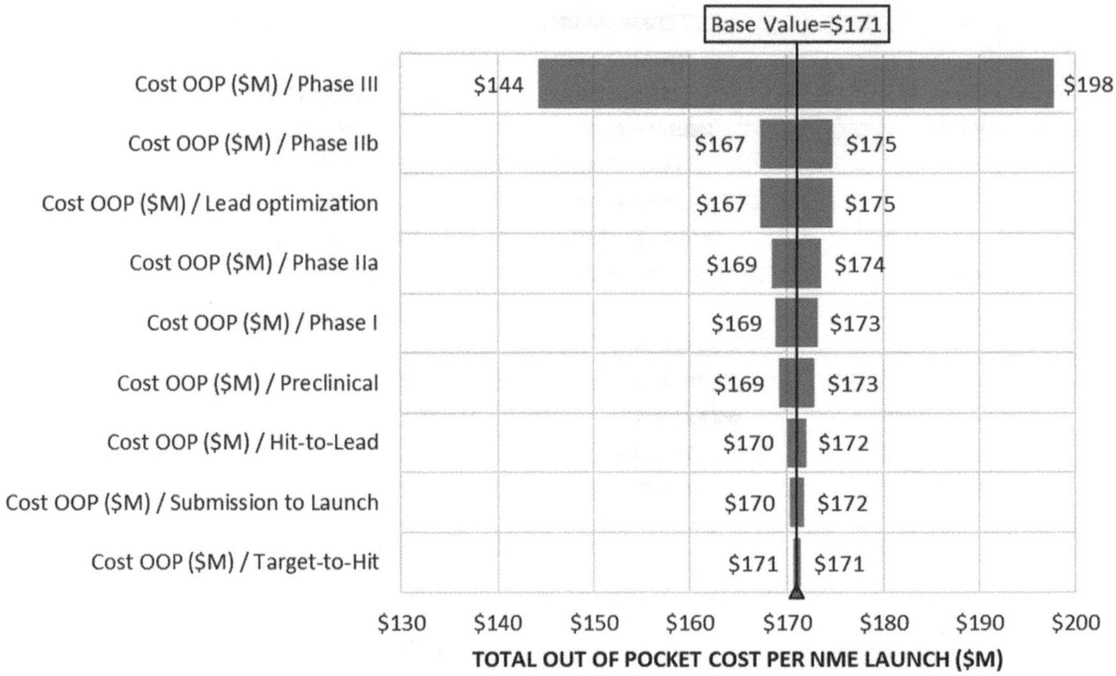

Figure 1.4 Impact of phase-specific cost variables to OOP cost per NME launch.

variability in expected capitalized OOP cost is substantial (see Figure 1.6). The probability of success in Phase IIa emerges as the single most influential parameter. Increasing this probability from its base value of 48% to 63% lowers expected capitalized OOP cost to approximately $1.51 billion, whereas reducing it to 33% increases expected cost to roughly $2.59 billion. The probability of success in Phase III represents the second most impactful variable, shifting expected capitalized OOP cost from approximately $1.53 billion to $2.44 billion when varied by ±15 percentage points from its base value of 66%.

Notably, the cost of capital exerts greater influence on expected capitalized OOP cost than any individual phase-specific cost or cycle-time variable. A reduction in the cost of capital from 11% to 10% lowers the expected capitalized OOP cost to approximately $1.72B, while an increase to 12% raises it to approximately $2.06B. Among direct cost inputs, lead optimization remains the most sensitive driver, whereas Phase III duration dominates among cycle-time parameters in shaping the overall capitalized investment required to bring an NME to market.

Across all TAs, Sertkaya et al. (2024) report no differentiation in attrition rates in the non-clinical phase of research. In Phase I, the probability of success ranges from a low of 60% (dermatology and pain and anesthesia) to a high of 86% (ophthalmology) (see Table 1.3).

In Phase II, where attrition rates are at their highest, a probability of success of greater than 50% is observed in only two TAs (hematology and ophthalmology), with a low of 27% reported for oncology and diseases of the respiratory system. Approximately one-third of all projects in five TAs—cardiovascular, central nervous system, dermatology, gastrointestinal, and pain and anesthesia—succeed in Phase II. This phase of clinical development is often referred to as the *valley of clinical death* and will be discussed in detail in this chapter.

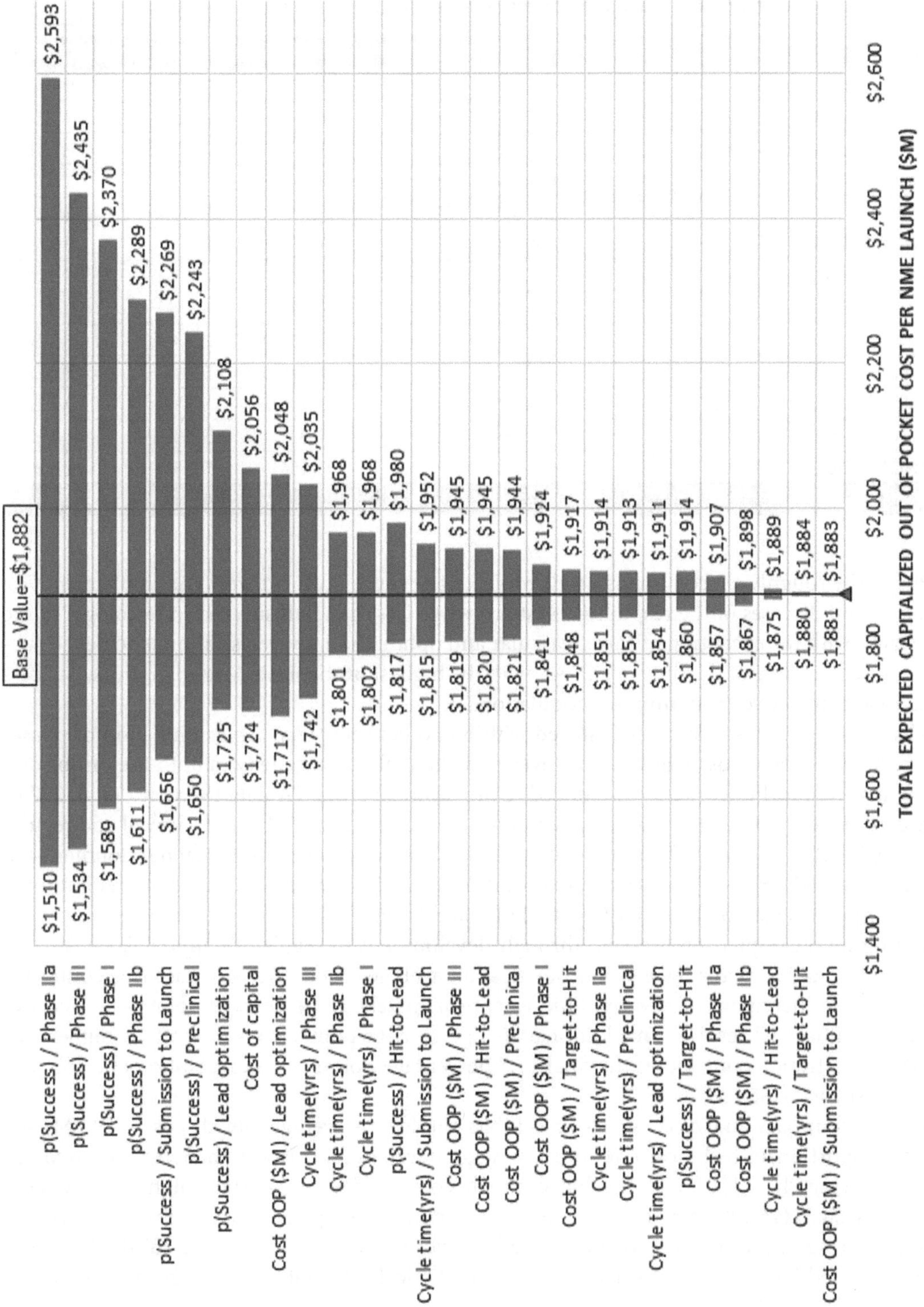

Figure 1.6 Impact of phase-specific risk, cost, and cycle time, and cost of capital variables to expected capitalized OOP cost per NME launch.

Table 1.3 Phase-specific probabilities of success across TAs (Sertkaya et al. 2024).

THERAPEUTIC AREA	NON-CLINICAL	PHASE I	PHASE II	PHASE III	REGULATORY
ANTI-INFECTIVE	68.0%	65.9%	49.6%	74.1%	94.4%
CARDIOVASCULAR	68.0%	65.0%	37.1%	57.6%	75.5%
CENTRAL NERVOUS SYSTEM	68.0%	61.5%	33.1%	55.9%	87.0%
DERMATOLOGY	68.0%	60.2%	35.9%	65.5%	88.3%
ENDOCRINE	68.0%	68.0%	46.3%	63.9%	81.3%
GASTROINTESTINAL	68.0%	71.6%	35.3%	55.3%	86.2%
GENITOURINARY SYSTEM	68.0%	62.9%	44.9%	71.4%	85.7%
HEMATOLOGY	68.0%	73.3%	56.6%	75.0%	84.0%
ONCOLOGY	68.0%	61.5%	26.8%	42.7%	85.5%
RESPIRATORY SYSTEM	68.0%	68.5%	27.4%	75.6%	89.5%
OPHTHALMOLOGY	68.0%	86.0%	52.7%	58.3%	77.5%
PAIN & ANESTHESIA	68.0%	60.2%	35.9%	65.5%	88.3%
IMMUNOMODULATION	68.0%	69.9%	40.1%	65.4%	95.3%
ALL	68.0%	60.2%	35.9%	65.5%	88.3%

In Phase III, roughly three-fourths of all projects are successful in three TAs (anti-infectives, hematology, and diseases of the respiratory system) while less than half succeed in oncology. In the United States, with the exception of cardiovascular and ophthalmology, registration is achieved successfully more than 80% of the time in all TAs, with the highest probabilities of registrational success reported in anti-infectives and immunomodulation.

At what cost is phase-specific risk resolved within and between TAs? An examination of phase-specific risk resolution across selected TAs reveals substantial heterogeneity in both the timing and cost at which technical uncertainty is reduced. Using average cross-TA data (Sertkaya et al. 2024), projects entering non-clinical development exhibit a cumulative probability of technical and regulatory success (PTRS) to launch of approximately 8.5%. Completion of non-clinical development—at an average cost of $11.8M—increases cumulative PTRS to 12.5%, corresponding to the resolution of roughly four percentage points of risk (see Table 1.4 and Figure 1.7).

Assuming uninterrupted progression through development to regulatory approval, Phase I resolves an additional 8.3 percentage points of risk at an average cost of $7.1M. Phase II accounts for the largest single increment of risk resolution—approximately 37 percentage points—at a cost of $21.0M, while Phase III resolves an additional 30.5 percentage points at a substantially higher cost of $89.3M. Registration contributes the final 11.7 percentage points of risk resolution at a comparatively modest cost of $2.6M.

Across the four TAs analyzed, oncology exhibits the lowest average launch cost ($69.9M) but retains disproportionate technical risk into late-stage development. Fewer than 10% of oncology programs entering Phase II ultimately achieve launch, compared with materially higher rates in central nervous system (16.1%), anti-infective (34.7%), and hematology (35.7%). As a result, oncology programs entering Phase III succeed at a markedly lower rate (36.5%), reflecting substantial unresolved risk carried forward from earlier phases. Parenthetically, hematology has the highest launch cost burden at $121.8M.

Table 1.4 The cost of phase-specific risk resolution (I) (Sertkaya et al. 2024).

AVERAGE	*p*(SUCCESS)	COST ($M)	AVERAGE CUM *p*(SUCCESS)	AVERAGE CUM COST ($M)
NON-CLINICAL	68.0%	11.8	8.5%	11.8
PHASE I	60.2%	7.1	12.5%	18.9
PHASE II	35.9%	21.0	20.8%	39.9
PHASE III	65.5%	89.3	57.8%	129.2
REGISTRATION	88.3%	2.6	88.3%	131.8
ONCOLOGY	*p*(SUCCESS)	COST ($M)	ONCOLOGY CUM *p*(SUCCESS)	ONCOLOGY CUM COST ($M)
NON-CLINICAL	68.0%	7.0	4.1%	7.0
PHASE I	61.5%	8.1	6.0%	15.1
PHASE II	26.8%	14.5	9.8%	29.6
PHASE III	42.7%	37.7	36.5%	67.3
REGISTRATION	85.5%	2.6	85.5%	69.9
CENTRAL NERVOUS SYSTEM	*p*(SUCCESS)	COST ($M)	CENTRAL NERVOUS SYSTEM CUM *p*(SUCCESS)	CENTRAL NERVOUS SYSTEM CUM COST ($M)
NON-CLINICAL	68.0%	8.7	6.7%	8.7
PHASE I	61.5%	6.8	9.9%	15.5
PHASE II	33.1%	16.1	16.1%	31.6
PHASE III	55.9%	59.4	48.6%	91.0
REGISTRATION	87.0%	2.6	87.0%	93.6
ANTI-INFECTIVE	*p*(SUCCESS)	COST ($M)	ANTI-INFECTIVE CUM *p*(SUCCESS)	ANTI-INFECTIVE CUM COST ($M)
NON-CLINICAL	68.0%	9.4	15.5%	9.4
PHASE I	65.9%	2.8	22.9%	12.2
PHASE II	49.6%	22.3	34.7%	34.5
PHASE III	74.1%	41.6	70.0%	76.1
REGISTRATION	94.4%	2.6	94.4%	78.7
HEMATOLOGY	*p*(SUCCESS)	COST ($M)	HEMATOLOGY CUM *p*(SUCCESS)	HEMATOLOGY CUM COST ($M)
NON-CLINICAL	68.0%	17.9	17.8%	17.9
PHASE I	73.3%	17.3	26.1%	35.2
PHASE II	56.6%	22.1	35.7%	57.3
PHASE III	75.0%	61.9	63.0%	119.2
REGISTRATION	84.0%	2.6	84.0%	121.8

In contrast, anti-infective and central nervous system programs achieve their largest reductions in risk between Phase II and Phase III—resolving approximately 35.3 and 32.5 percentage points, respectively—at costs of $22.3M and $16.1M. Hematology and central nervous system are the most capital-intensive TAs for late-stage risk resolution—requiring approximately $61.9M and $59.4M, respectively—during Phase III to resolve 21.0 and 38.4 percentage points of remaining risk.

Beginning from the non-clinical phase of research, the aggregate probability of technical and regulatory success to a successful registrational drug is highest for hematology (17.8%), anti-infective (15.6%), and ophthalmology (13.9%) and is lowest across oncology (4.1%), central nervous system (6.7%), and cardiovascular (7.1%) (see Figure 1.8).

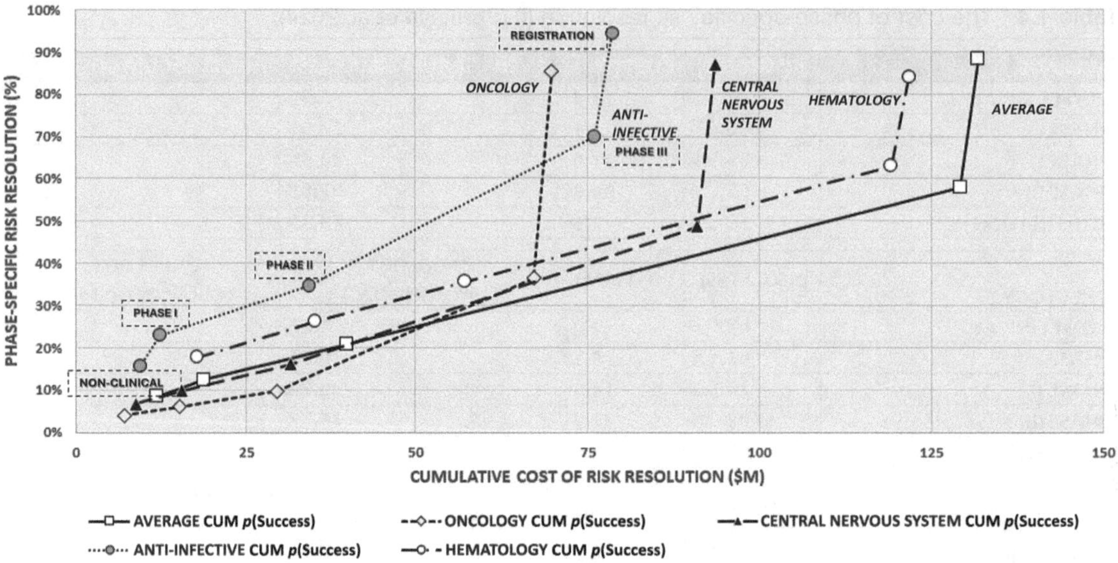

Figure 1.7 The cost of phase-specific risk resolution (II) (Sertkaya et al. 2024).

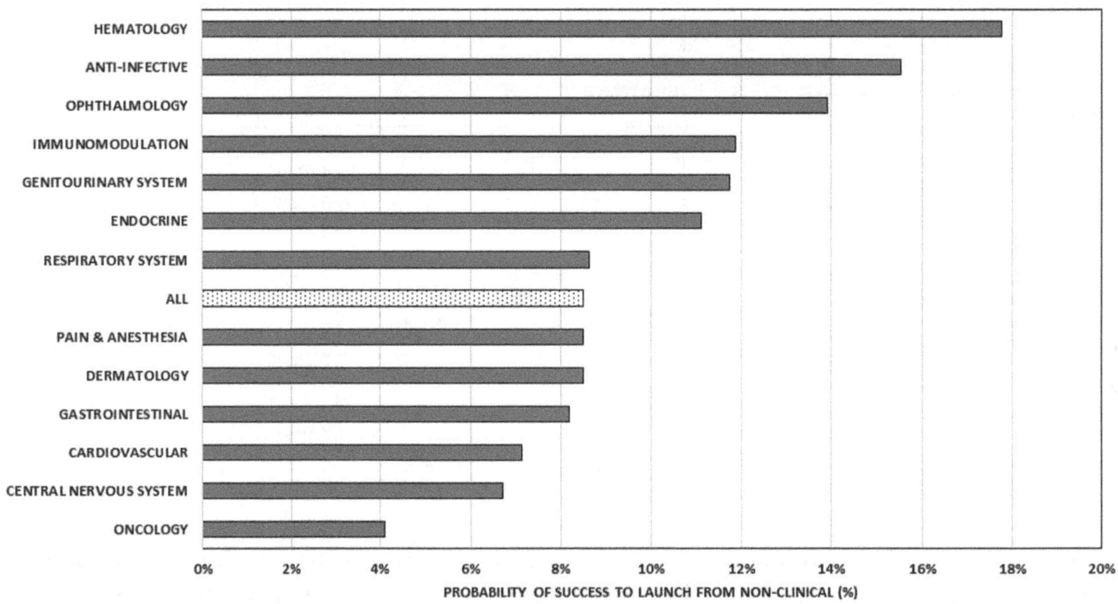

Figure 1.8 Probability of success to launch from non-clinical across all TAs (Sertkaya et al. 2024).

1.2 THE VALLEY OF CLINICAL DEATH

As shown in Table 1.3, clinical probability of success accounts for the highest levels of attrition in R&D productivity. In particular, Phase II, often described as the *valley of clinical death*, continues to be a graveyard for the majority of NMEs. On average, across all TAs, the likelihood of emerging successfully from Phase II is 35.9% with a range of 26.8% (oncology) and 56.6% (hematology) after

$39.9M in OOP cost (range = $29.6M in oncology and $57.3M in hematology) has been spent from the beginning of non-clinical (see Table 1.4) (Sertkaya et al. 2024). It is clear that reducing attrition in Phase II remains the strongest lever for improving R&D productivity and, consequently, to better manage the cost of generating new drug products from novel molecular entities.

From a study of a decade's worth of drug attrition across several TAs (1991–2000), Kola and Landis (2004) have shown that while drug failure due to PK and bioavailability had been reduced dramatically over this period, efficacy, safety, and toxicology remained the major drivers of drug failure. This study points to the broad distinction between largely controllable risks (e.g., bioavailability due to alternative formulations and drug delivery systems) and largely uncontrollable risks (e.g., efficacy due to poor target validation and druggability). In an effort to reduce Phase II (and Phase III) attrition, Eli Lilly has pointed to two key approaches: (a) better target selection—specifically target validation and druggability—and (b) achievement of earlier clinical PoC via the use of biomarkers and surrogate endpoints (Paul et al. 2010). While target selection is a function of earlier target identification, it represents a critical step in the protracted commitment of resources to downstream drug development. The same can be stated of clinical PoC, which sets the stage for the final and most expensive phase of drug development—Phase III. It is here that the biopharma industry continues to struggle with decision making based on imperfect clinical data and information as it attempts to minimize both Type I (false positive) and Type II (false negative) errors.

In response to rising development costs and persistent late-stage attrition, Eli Lilly introduced an alternative clinical development paradigm in the early 2000s aimed at accelerating early human data generation to support earlier investment decisions. Launched in 2002, this model—internally known as *Chorus*—was designed to prioritize early demonstration of human PoC, ideally during Phase I, rather than following a conventional, sequential escalation of clinical investment. By front-loading critical data acquisition, Chorus sought to accelerate go/no-go decisions and limit capital exposure to high-risk assets (Paul et al. 2010; Owens et al. 2015).

The Chorus framework embodied a *quick-win or fast-failure* philosophy, explicitly seeking to maximize changes—positive or negative—in a program's probability of technical success over the shortest feasible timeframe and at minimal cost. This approach later became known as *lean-to-proof-of-concept* (L2POC). Assets selected for L2POC typically had below-average baseline probabilities of success due to uncertainties in target biology, translational relevance, or clinical feasibility, but were considered suitable for early risk reduction through well-defined biomarkers or clinically interpretable endpoints (Owens et al. 2015).

Under this model, the primary objective prior to Phase II was not demonstration of efficacy, but resolution of key mechanistic and translational uncertainties. Chorus programs focused on establishing whether adequate target engagement could be achieved in humans within acceptable safety margins, whether a clinically viable dosing regimen could be defined, and whether PK and PD variability could be sufficiently controlled. Collectively, these criteria defined achievement of PoM, which served as the threshold for committing additional resources to later-stage development (Owens et al. 2015).

Retrospective evaluation of the Chorus model demonstrated a reduction in the number of candidates advancing into Phase II, but a marked improvement in the quality of those that did. Phase II success rates increased from approximately 34% under traditional development to nearly 50% under Chorus. Importantly, early PoC was achieved at substantially lower cost—approximately $6M (2010 dollars) versus ~$22M using conventional approaches (Paul et al. 2010).

Across a ten-year comparison (2002–2012), Eli Lilly reported that cumulative probabilities of technical and regulatory success remained constant at roughly 6% under both paradigms, while total development costs were consistently lower for Chorus programs. Inflation-adjusted OOP, capitalized, and risk-adjusted capitalized costs were all meaningfully reduced, demonstrating that early, disciplined risk resolution can materially improve capital efficiency without compromising overall success rates (Owens et al. 2015).

Notably, concerns that lean development might significantly extend timelines were not supported by the data. Aggregate development cycle time under Chorus was only approximately six months longer than under traditional development, indicating that improved early-stage decision quality and capital efficiency can be achieved without materially delaying market entry.

It is noteworthy that Owens et al. (2015) have assumed that 50% of Phase II risk is resolved at PoC in the Chorus model. This is a debatable assumption, as we believe there is a far greater risk in achieving a successful PoC (equivalent to Phase IIa) than in conducting a successful dose-finding study (equivalent to Phase IIb). Unfortunately, to our knowledge, there are no published reports of differences between Phase IIa and Phase IIb probability-of-success estimates. To this extent, we have conducted sensitivity analyses of the impact of phase-specific, non-clinical (target-to-hit to preclinical) and clinical (Phase I to Phase III) risks on the expected capitalized cost of bringing one NME successfully to market. A synopsis of the results is as follows:

- When varied by ±15 percentage points, the *p*(Success) of preclinical is the most dominant risk variable in research to total expected capitalized OOP cost. At its base value of 69%, expected capitalized OOP cost is $1.882B; when the base value increases by 2.5%, 5.0%, 7.5%, 10.0%, 12.5%, and 15.0%, the corresponding expected capitalized OOP costs are $1.837B, $1.795B, $1.755B, $1.718B, $1.683B, and $1.650B. Conversely, when the base value is lowered by 2.5%, 5.0%, 7.5%, 10.0%, 12.5%, and 15.0%, the corresponding expected capitalized OOP costs are $1.931B, $1.984B, $2.041B, $2.102B, $2.170B, and $2.243B (see WAV Figure 1.9A).
- Sensitivity analysis indicates that *p*(Success) of Phase IIa is the single most influential driver of total expected capitalized OOP cost in clinical development. Within a ±15 percentage point variation range, changes in this parameter produce the largest swing in capitalized spend relative to all other clinical risk inputs (see WAV Figure 1.9B). At the reference assumption of a 48% Phase IIa *p*(Success), the modeled expected capitalized OOP cost is $1.882B. Incremental improvements in Phase IIa success probability materially alter expected cost. Specifically, increasing *p*(Success) by 2.5, 5.0, 7.5, 10.0, 12.5, and 15.0 percentage points reduces expected capitalized OOP cost to $1.805B, $1.735B, $1.671B, $1.613B, $1.559B, and $1.510B, respectively. Conversely, deterioration in Phase IIa performance assumptions results in rapid cost escalation. Reductions of 2.5, 5.0, 7.5, 10.0, 12.5, and 15.0 percentage points from the base case increase expected capitalized OOP cost to $1.968B, $2.064B, $2.172B, $2.294B, $2.433B, and $2.593B, respectively.

Collectively, these results underscore Phase IIa success probability as the dominant economic risk lever in clinical development, with relatively modest changes in early PoC performance translating into several hundred million dollars of downstream capital efficiency impact.

In 2014, Pfizer conducted a review of 44 NMEs that had reached a decision point in Phase II over the 2005–2009 period and found that only one-third of these programs had achieved a positive clinical PoC (Morgan et al. 2012). While the primary reason for such high attrition was insufficient (or a lack of) clinical efficacy, a more granular analysis led to the conclusion that three fundamental elements—referred to by Pfizer as the *three pillars of survival*—needed to be demonstrated in order to achieve Phase II success (see Figure 1.10):

1. **Pillar 1—Target-site exposure:** Achieving and maintaining adequate drug concentrations at the site of pharmacologic action for a sufficient duration to enable the intended mechanism of action. This includes ensuring appropriate systemic and tissue exposure profiles,

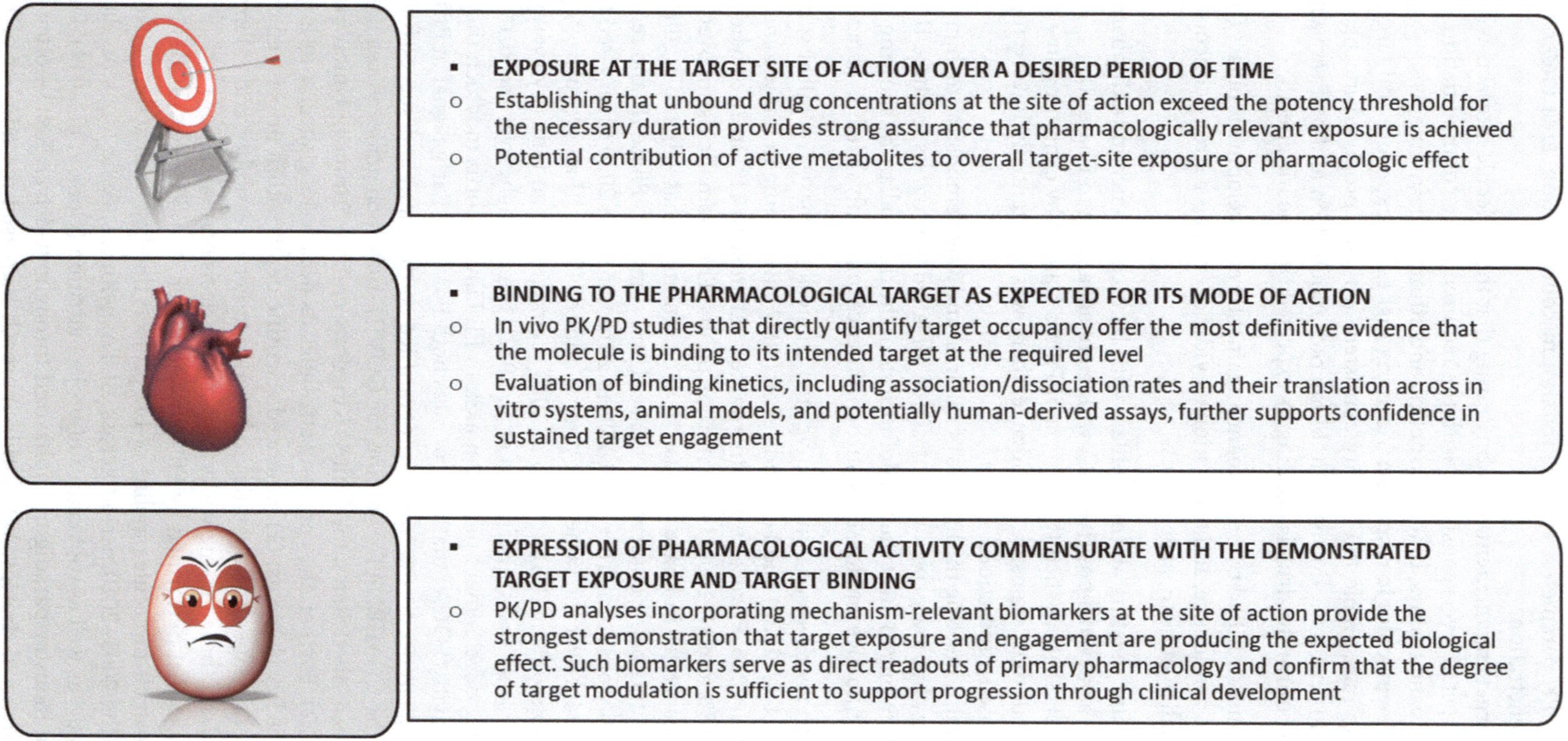

Figure 1.10 Pfizer's three pillars of survival for improving Phase II success (Morgan et al. 2012).

informed by PK parameters such as maximum concentration, area under the curve, and target-site penetration.

2. **Pillar 2—Target engagement:** Demonstrating that the molecule interacts with its intended biological target at levels consistent with the required degree of modulation. Evidence may include occupancy assays, binding kinetics, and affinity measurements.
3. **Pillar 3—Pharmacologic response:** Verifying that target exposure and target engagement translate into measurable PD activity consistent with the proposed mechanism. This encompasses, for example, downstream biomarker modulation and pathway activation/inhibition that confirm the drug is exerting its expected therapeutic effect.

Evaluation of programs stratified by performance against three development pillars (pharmacology confidence, exposure confidence, and translational evidence) reveals a strong concordance between early evidence of quality and late-stage success:

1. Of 14 programs meeting all three pillars with high confidence in both pharmacology and exposure, 12 achieved a positive PoC, with eight advancing to Phase III.
2. From 12 assets that satisfied the criteria of Pillars 1 and 2 by exhibiting low pharmacology confidence but high exposure confidence, there was substantially lower progression, with only two Phase III initiations.
3. Of six programs meeting the hurdles of Pillars 2 and 3 by demonstrating high confidence in pharmacology but low exposure confidence, all failed to progress to Phase III.
4. All 12 assets failing to meet the full definitions of the three pillars by showing both low pharmacology confidence and low exposure confidence uniformly failed to advance to Phase III.

Further evidence of Pfizer's success in positive step changes in Phase II clinical success has been documented by Wu et al. (2020) and Fernando et al. (2022). Beginning in Phase I and from a low of 4% PTRS to launch in 2016, Pfizer's cumulative success rate rose to 21% in 2020 when compared to industry averages of 8% and 11% over the same time points. This dramatic rise in overall success rate is due to higher success rates in Phase I, Phase II, and combined Phase III and regulatory, the most pronounced contributors being observed in Phase II and combined Phase III and regulatory where (a) Phase II probability of success rose from 15% in 2016 to 52% in 2020 (compared to an industry average of 29% and 34% over the same time points) and (b) Phase III and regulatory probability of success increased from 67% in 2016 to 85% in 2020 (compared to an industry average of approximately 70% over the same years) (Fernando et al. 2022). Interestingly, over the 2010–2020 period, Pfizer adopted a drug development paradigm akin to Eli Lilly's Chorus in which they have labeled Signs of Clinical Activity (SOCA) that incorporates both PoM and Early Signal of Efficacy (ESOE). It is believed that adherence to the SOCA program led to the rapid and successful development with BioNTech of the COMIRNATY vaccine for COVID-19, which achieved regulatory approval for adult usage in the United States, the United Kingdom, and the European Union in eight months (Fernando et al. 2022). Further, the company attributes its decade-long success in improving R&D productivity to three key elements: (1) enhanced scientific depth within prioritized therapeutic areas, (2) expansion of therapeutic modalities to broaden target accessibility, and (3) implementation of quantitative, evidence-based frameworks for portfolio decision making.

In a vein similar to that performed by Eli Lilly and Pfizer, AstraZeneca (in 2011) began its own investigations of its small-molecule pipeline of 142 projects that had been active between preclinical and Phase II over the 2005–2010 period across all TAs with a specific focus on data and subsequent decisions made, as well as root causes of project attrition (Cook et al. 2014). During this period, the prevailing industry paradigm emphasized throughput-driven R&D—often characterized as a *multiple-shots-on-goal* strategy. While this approach increased the number of assets entering development, it had the predictable effect of diverting resources away from rigorous interrogation of disease pathophysiology and therapeutic rationale. Over time, this bias toward volume eroded both the scientific integrity and long-term durability of R&D portfolios. In an environment optimized for advancement velocity, Ringel et al. (2013) posited that progress-oriented decision making tended to displace hypothesis-driven, truth-seeking behaviors that could have yielded deeper biological insight and more robust target validation.

Internal performance metrics at AstraZeneca illustrate the downstream consequences of this strategy. Preclinical transition success rates were broadly aligned with external benchmarks, at slightly above 60%. Phase I outcomes were comparatively strong, with a 59% success rate compared to the industry median of 48%. However, this apparent early-stage strength did not translate into mid-stage development performance. The median probability of success in Phase II was approximately half the industry benchmark of 29%, indicating a pronounced attrition inflection. These data suggest that increased Phase I throughput was achieved at the expense of asset quality, leaving programs insufficiently differentiated or mechanistically substantiated to demonstrate efficacy in Phase II.

This quality deficit manifested in poor late-stage productivity. Between 2005 and 2010, AstraZeneca's NME launch rate was reported at approximately 2%, materially below the industry median of 6% (Cook et al. 2014). Failure mode analysis further reinforced the conclusion that the core issue was not early safety, but inadequate biological and translational validation. Safety-related liabilities dominated terminations in preclinical development (82% of failures, largely due to compound-related organ toxicities) and in Phase I (62% of failures). In contrast, lack of efficacy was the principal driver of attrition in Phase IIa (57%) and Phase IIb (88%), underscoring weaknesses in target selection, mechanism validation, and clinical translation.

Comprehensive post hoc reviews of portfolio outcomes led AstraZeneca to redefine its progression framework around three foundational scientific criteria: (1) a robust understanding of disease biology, (2) a clearly established causal link between the target and the disease, and (3) a well-characterized and testable mechanism of action. Building on this reframed philosophy, and as detailed by Cook et al. (2014) and Morgan et al. (2018), the organization identified five critical technical determinants of project success termed the *5R framework* (see Figure 1.11), which collectively shifted decision making from volume-driven advancement to evidence-based value creation:

1. The strength and quality of target validation—the right target
2. Demonstration of target engagement—the right tissue
3. Adequate safety margins—the right safety
4. Patient stratification plans—the right patient
5. Medical value proposition—the right commercial potential

Cook et al. (2014) emphasize that the 5R framework was never intended to require unequivocal strength across all five dimensions in order for a program to advance. Rather, its primary value lies in systematically surfacing the most consequential sources of technical risk—those most likely to precipitate failure during preclinical or clinical development—so they can be explicitly tested, mitigated, or used as a basis for early termination decisions.

The impact of this approach on AstraZeneca's R&D productivity, particularly in navigating the so-called *valley of clinical death*, has been substantial. Analyses spanning two development eras, as reported by Morgan et al. (2018), demonstrate a marked inflection in end-to-end performance. Aggregate success from candidate nomination through Phase III completion increased from approximately 4% during the 2005–2010 period to 19% during the 2012–2016 period. This improvement coincided with a deliberate shift toward more stringent target rationale at the research stage and a fundamental rebalancing of portfolio scale and composition.

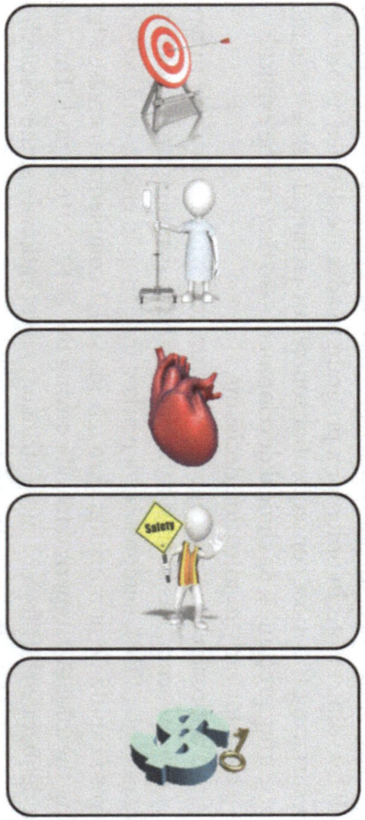

- **RIGHT TARGET**
  - Demonstration of strong mechanistic association between biological target and disease process
  - Clear potential for clinically meaningful differentiation in therapeutic activity relative to existing or emerging therapies
  - Availability or feasibility of developing predictive biomarkers to support target engagement

- **RIGHT PATIENT**
  - Evidence that the candidate achieves sufficient systemic and relevant tissue exposure
  - Establishment of validated PD biomarkers to quantify biological response
  - Comprehensive characterization of preclinical and clinical PK/PD relationships to inform dose and regimen selection
  - Assessment of metabolic pathways and interactions to predict and mitigate drug–drug interaction risks

- **RIGHT TISSUE**
  - Characterization of exposure margins that support a differentiated and favorable safety profile
  - Evaluation of secondary pharmacology liabilities, including off-target interactions
  - Assessment of reactive metabolite formation, genotoxicity potential, and interaction risks that could impact safety or efficacy
  - Understanding of target-related safety liabilities, including on-target toxicity mechanisms

- **RIGHT SAFETY**
  - Identification of patient subgroups most likely to exhibit an optimal therapeutic response based on mechanistic, biomarker, or genetic insights
  - Quantitative assessment of benefit–risk profiles across the intended indication and target population(s)

- **RIGHT COMMERCIAL POTENTIAL**
  - Clear articulation of a differentiated clinical and economic value proposition relative to anticipated future standards of care
  - Integration of market access considerations, including payer expectations and provider adoption barriers
  - Development of a precision-medicine strategy where appropriate, incorporating companion diagnostics, predictive biomarkers, and patient-selection tools

Figure 1.11 AstraZeneca's 5R framework for improving clinical success (Cook et al. 2014; Morgan et al. 2018).

Prior to adoption of the 5R framework, AstraZeneca advanced 287 small-molecule projects between 2005 and 2010. In contrast, only 76 such projects entered development from 2012 to 2016. Importantly, this contraction reflected not only a reduction in activity but a strategic elevation of quality thresholds. Backup programs were aggressively pruned, declining from roughly 28% of the portfolio to less than 7%. Simultaneously, the company reshaped its target class mix, increasing kinase-focused programs from 21% to 36% while reducing reliance on G protein-coupled receptor targets from 25% to 5%, and placing greater emphasis on assets with differentiated and novel mechanisms of action (Morgan et al. 2018).

These strategic shifts translated into measurable improvements in development efficiency. Before implementation of the framework, only 23% of discovery programs yielded leads of sufficient quality to progress into lead optimization, corresponding to an approximately 4:1 ratio of lead generation efforts to viable lead optimization candidates. Following adoption, this transition probability rose to 48%, effectively halving the number of projects required to produce a development-ready lead. In parallel, cycle times were substantially compressed: time from project initiation to lead optimization decreased from 26 to 19 months, Phase II duration was reduced from 31 to 14 months, and cost to PoC fell from approximately $100M to $69M—well below the industry benchmark of $119M.

The nature of project attrition also evolved. Across the two periods, terminations driven by strategic considerations (35%) or failure to identify viable leads (32%) declined sharply, while failures increasingly reflected explicit target validation outcomes, which accounted for 77% of closures in the later period. This shift underscores the centrality of target validation within the 5R construct and highlights its role as the dominant determinant of downstream success.

Comparative analyses against industry benchmarks further illustrate the magnitude of change. During 2005–2010, AstraZeneca underperformed the industry in preclinical (66% vs. 71%) and Phase II (15% vs. 22%) success rates, exceeded industry norms in Phase I (59% vs. 51%), and matched them in Phase III (60%). By contrast, in the 2012–2016 period, the company outperformed industry comparators across all development stages, with the most dramatic differential observed in Phase II (43% vs. 18%). As a result, AstraZeneca's cumulative probability of technical and regulatory success from preclinical development to launch increased from 4% to 19%, while the industry average remained flat at approximately 4% across both time periods.

Failure mode distributions also shifted in a manner consistent with earlier, more decisive risk resolution. In the earlier period, safety considerations dominated preclinical and Phase I failures (82% and 63%, respectively), while lack of efficacy was the primary driver in Phase II (64%). In the later period, safety remained the leading cause of preclinical attrition (50%), but efficacy emerged as an equally important factor in Phase I (38.5%) and became overwhelmingly dominant in Phase II (84%), reflecting more effective early elimination of safety liabilities and a sharper focus on biological and translational validity.

Both Cook et al. (2014) and Morgan et al. (2018) note that the effectiveness of the 5R framework was contingent on a broader cultural transformation. The expanded use of interim analyses further reinforced this discipline, enabling earlier and more confident determinations of futility or acceleration, and materially improving the quality of development decisions across the portfolio.

It is noteworthy that AstraZeneca has added a sixth dimension to the 5R framework—the *right* digital health technology solution in clinical trials aimed at improving patient and clinical trial site experience (Duran et al. 2023). Across three TAs, the positive impact of digital technology in clinical trials was demonstrated in cardiovascular (DAPA-MI study in myocardial infarction), inflammation (CRESCENDO study in COPD), and oncology (TROPION-PanTumor01 study in stomatitis that accompanies anticancer therapy).

Although the experiences of Eli Lilly, Pfizer, and AstraZeneca clearly illustrate how rigorous science can move the evidentiary burden for critical go/no-go decisions upstream in the R&D continuum, scientific excellence alone has not been sufficient to drive sustained productivity gains. Equally consequential has been the implementation of disciplined, fit-for-purpose decision frameworks that enable timely, consistent, and high-quality portfolio choices.

The true cost of suboptimal decision making in biopharma R&D is inherently difficult to quantify and is rarely disclosed. Nevertheless, a compelling case can be made that a substantial proportion of pipeline attrition is attributable not to unavoidable scientific uncertainty, but to opaque, inconsistent, or poorly governed decision processes. In the absence of structured mechanisms that enforce transparency, explicitly surface assumptions, and align evidence thresholds with development risk, organizations are prone to advance programs based on incomplete data, implicit biases, or organizational momentum. As a result, assets may progress further than warranted, accumulating avoidable cost and risk before ultimately failing.

1.3 THE STAGE-GATE REVIEW PROCESS

The stage-gate research (SGR) framework (Cooper et al. 2001) provides a structured governance mechanism for determining whether to commit capital and organizational resources to a molecular entity as it advances through the R&D continuum. At the initial decision point (Gate G0), leadership determines whether a scientific concept warrants formal investment and entry into discovery research (see Figure 1.12).

For a typical small-molecule NME, early development follows a largely sequential pathway encompassing target identification and validation, assay development, compound screening, hit-to-lead activities, and iterative lead optimization. Successful completion of these activities enables nomination of a development candidate eligible for advancement through Gate G1 and formal entry into preclinical development.

Upon clearance of Gate G1, the asset undergoes a defined package of nonclinical studies, including PK, PD, and toxicology assessments in relevant animal and cellular models. These data are evaluated at Gate G2 to determine readiness for FIH studies and transition into Phase I clinical development.

Phase I trials represent the initial component of early clinical development and are primarily designed to characterize safety, tolerability, and human PK, typically in healthy volunteers or, where appropriate, patient populations. Secondary objectives often include exploratory biomarkers and early PD signals. Assets that meet prespecified criteria may then advance into Phase IIa, which is commonly focused on PoC. As established across therapeutic areas, PoC represents the highest-risk inflection point in the clinical life cycle, given its emphasis on demonstrating clinically meaningful efficacy.

Although not always depicted formally in high-level process diagrams, a governance decision point almost invariably exists between Phase I and Phase IIa across most therapeutic areas. Successful demonstration of PoC enables consideration at Gate G3 for entry into Phase IIb, the initial stage of late clinical development. Phase IIb studies are typically dose-ranging in design, with the objective of defining the optimal therapeutic dose by balancing efficacy against safety and tolerability constraints. The selected dose regimen is subsequently carried forward into Phase III registrational trials, the second component of late clinical development.

If Phase IIb and Phase III studies collectively demonstrate statistically robust efficacy and an acceptable benefit–risk profile, the program may advance through Gate G4 to pursue regulatory

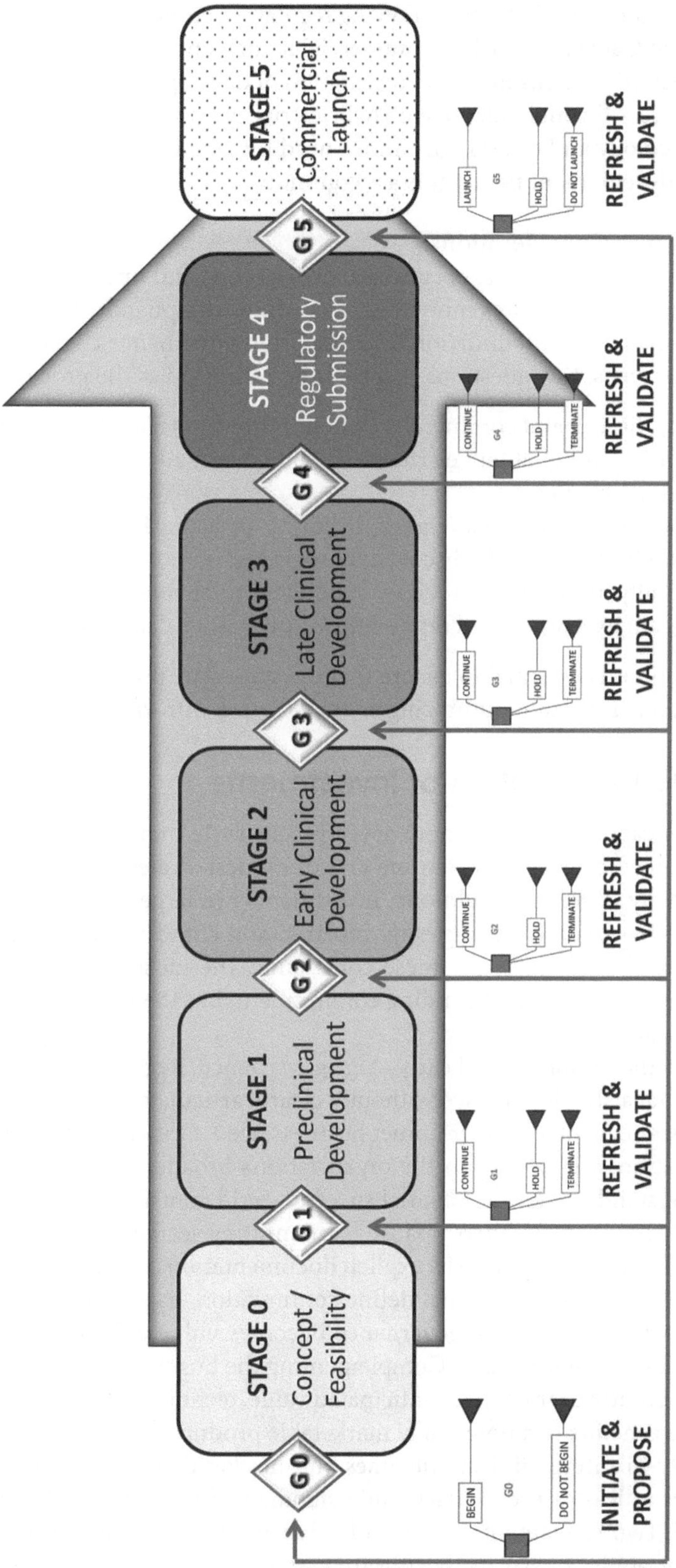

Figure 1.12 Stage-gate review process for drug research and development.

approval across one or more jurisdictions. Following regulatory authorization and label definition, a final decision is made at Gate G5 regarding commercial launch and life-cycle management strategy.

It should be emphasized that this linear representation of development is an abstraction. In practice, many NMEs progress through adaptive or hybrid trial designs (e.g., Phase I/IIa or Phase IIb/III), particularly in oncology and other high-unmet-need therapeutic areas, where regulatory flexibility and accelerated development paradigms are common.

Strategic and Governance Considerations

Prima facie, the stage-gate model appears conceptually rigorous and operationally tractable. However, effective implementation requires more than procedural compliance. The SGR framework, in isolation, does not resolve several foundational strategic and governance challenges that materially influence portfolio outcomes. Key questions that must be addressed as the process matures include:

- What scientific, clinical, and commercial evidence thresholds justify initial asset investment?
- Which quantitative and qualitative criteria should govern advancement, pause, or termination decisions at each gate?
- Should externally sourced assets (e.g., in-licensing or acquisitions) be evaluated and governed differently from internally discovered programs?
- How systematically are post hoc reviews conducted to enable organizational learning, improve project team execution, and refine future governance decisions?

Addressing these questions is essential to ensure that the stage-gate process functions not merely as a control mechanism, but as a value-optimizing system aligned with enterprise-level R&D strategy.

1.3.1 Rationale for Initiation of Investments

In industries with low capital intensity, short development cycles, and limited downside risk, initiating investment in new ideas is often routine and of modest strategic consequence. Biopharma operates under markedly different conditions: investments are large, timelines are long, and the probability of failure—particularly during early research and clinical translation—far exceeds the likelihood of commercial success. Under these constraints, the most consequential decision in a drug's life cycle is often not a late-stage funding commitment, but the initial allocation of resources to explore a scientific concept.

This reality elevates the importance of early-stage governance. Regardless of scientific novelty or internal advocacy, no idea should advance without a clearly articulated business case that links the concept to a defined strategic objective or unmet business need. Critically, the opportunity must be evaluated, not in isolation, but relative to the organization's broader portfolio of actionable alternatives. Only by understanding how a potential success would contribute to portfolio strategy and long-term financial objectives can leadership justify committing scarce resources. Accordingly, each investment proposal should be supported by explicit documentation of its business rationale, including prescriptive and measurable criteria that define continuation, pause, or termination thresholds. Absent such predefined decision rules, governance becomes vulnerable to subjectivity, organizational politics, and hierarchical influence. Complementing the business case, a credible project plan is required that transparently characterizes anticipated value, technical and regulatory risk, resource requirements, and the timeline to a potentially marketable product.

Despite uncertainty in future clinical outcomes and market dynamics, constructing a credible business case requires at least a preliminary understanding of the commercial environment over the product life cycle. Two fundamental drivers of value are the magnitude of unmet patient need and the willingness or ability of payers, healthcare systems, or patients to fund access. Perceived product uniqueness is therefore closely linked to the degree to which a therapy alters disease burden. Disease-modifying interventions are typically valued more highly than purely symptomatic treatments, although exceptions exist—for example, symptom-focused therapies are valued in chronic conditions such as osteoarthritis, where disease-modifying options are limited.

In competitive markets, differentiation is a primary determinant of value. Experience in lipid-lowering demonstrates that being best in class, as with atorvastatin (Lipitor), can be more commercially advantageous than being first in class, as exemplified by lovastatin (Mevacor). Consequently, new therapeutic investments must align tightly with corporate strategy while supporting financial objectives. Organizations that allow individual assets to dictate strategy often struggle to achieve sustained success due to gaps in expertise, execution, or competitive positioning.

It is sometimes argued that assigning financial value to early research-stage assets is inappropriate, given limited data. While early estimates are inherently uncertain, the absence of a valuation framework leaves decision makers unable to compare opportunities with differing risk profiles, resource demands, and time horizons. We contend that all investment opportunities should be associated with a bounded range of commercial value estimates, spanning plausible scenarios from narrow indications to broader, multi-indication applications. As evidence accumulates, this range should be progressively refined and narrowed.

Although not all drivers of value can be precisely quantified, a defensible business case remains essential to sound investment governance in high-uncertainty, asymmetric-risk environments. Where conventional financial valuation is impractical, utility theory—integrating quantitative and qualitative attributes into a common unit of measure—offers a pragmatic and robust decision-support alternative.

1.3.2 Decision Criteria for Progression, Hold, and Termination

A broad range of financial evaluation methodologies—applied in nominal or risk-adjusted form—are used to support gated investment decisions across the drug development life cycle. Despite their widespread adoption, no single metric is universally accepted or without conceptual limitations. Net present value (NPV) and expected NPV (eNPV) remain the most commonly applied once assets enter clinical development, as they consolidate discounted future cash flows. However, both suffer from a key limitation: they lack an intrinsic benchmark for absolute attractiveness, as there is no consensus on what constitutes an acceptable NPV or eNPV across organizations, TAs, or development stages.

The same discounted cash-flow models can be used to derive return-based metrics, namely internal rate of return (IRR) and expected IRR (eIRR). Unlike value-based measures, rates of return can be evaluated directly against an organization's weighted average cost of capital, providing a more intuitive assessment of whether an investment is expected to create or destroy shareholder value after accounting for time and risk. For this reason, eIRR is particularly useful as a primary financial decision criterion for assets in clinical development, especially when comparing programs with differing capital requirements, timelines, and risk profiles.

At the portfolio level, we also strongly advocate the use of the PI and its risk-adjusted form, ePI, particularly under conditions of constrained resources. These metrics capture the efficiency with which capital is converted into expected value and are especially informative for evaluating marginal investments near portfolio capacity. While no industry standard defines a *healthy* PI or ePI by development phase, their practical value lies in revealing the characteristic concave efficiency curve observed in Pareto analyses. This curve should be interpreted as an efficiency frontier for individual investments ranked by reward-to-cost ratio, rather than as a classical efficient frontier of optimized asset combinations. The application of PI and ePI to incremental portfolio optimization is discussed further in Chapter 6.

Because most biopharma organizations do not operate with explicit corporate targets for NPV, eNPV, PI, or ePI, alternative metrics are often needed to support gated decision making. Given that long-range planning is frequently anchored in sales growth objectives, we propose the non-risk-adjusted present value of cumulative sales (PV Sales) and the risk-adjusted present value of cumulative sales (PV eSales) as complementary portfolio-relevant measures. These metrics directly link individual asset investments to enterprise-level growth expectations. Once a program has met predefined clinical thresholds—such as acceptable safety margins, clinically meaningful efficacy, and viable dosing—its projected contribution to future portfolio sales growth should be a key determinant of strategic priority and continued funding.

While financial metrics can be applied with reasonable consistency during clinical development, their utility diminishes in earlier research stages, where commercial assumptions are highly speculative. Nevertheless, the absence of financial valuation does not obviate the need for disciplined progression frameworks. Instead, early-stage decisions should rely on predefined scientific criteria, including target exposure and occupancy, target engagement, downstream biological activity, and demonstrable physiological effect. When established in advance, these criteria provide a defensible basis for progression, pause, or termination.

In practice, the primary weakness in research decision making is not the lack of metrics but the failure to define explicit continuation and, critically, discontinuation thresholds. Without clear termination rules, organizations tend to perpetuate investments beyond the point where evidence supports further development, often driven by sunk-cost bias. This results in a systematic bias toward continuation, an excess of Type I errors (advancing programs that should be terminated), and heightened fear of Type II errors (terminating programs that might have succeeded). Over time, this imbalance erodes portfolio efficiency and diverts resources from higher-value opportunities aligned with the organization's strategic, scientific, and financial objectives.

1.3.3 The Attractiveness of External Assets

When comparing the strategic and economic merit of internally originated programs with externally sourced assets, organizations are often subject to asymmetric decision biases that can materially distort portfolio outcomes. Although these biases operate in opposing directions, both undermine objective capital allocation and risk-adjusted value creation.

Internal assets are typically evaluated with a high degree of scrutiny due to deep institutional knowledge of their discovery history, preclinical liabilities, translational uncertainties, and prior development challenges. While this transparency should be advantageous, it can foster overly conservative or pessimistic assessments of technical and regulatory viability. At the same time, internal programs may be driven by optimism bias that is fueled by organizational attachment, scientific advocacy, and the belief that shortcomings can be addressed in later stages. The result is a paradox in which internal assets are both rigorously critiqued at the technical level and implicitly assumed to succeed eventually, even when near- or mid-term evidence does not fully justify continued investment.

Externally sourced assets exhibit a complementary but equally problematic bias. Limited visibility into their full development history, failure modes, and unresolved risks can cause these programs to appear cleaner, more differentiated, or more advanced than internal comparators. Information asymmetry, selective disclosure, and reliance on sponsor narratives often produce an inflated perception of their risk/benefit profile, particularly during early diligence. This effect is frequently reinforced by incentive structures within business development organizations, where successful acquisitions or partnerships are more visibly rewarded than decisions to out-license, divest, or deprioritize internally generated programs that are scientifically weak, strategically misaligned, or financially unattractive.

Collectively, these dynamics create a structural imbalance in how internal and external opportunities are framed, debated, and funded. To mitigate this, we advocate a unified, origin-agnostic decision framework in which all assets are evaluated against the same predefined thresholds, value drivers, and risk-adjusted metrics. These criteria should encompass scientific robustness, probability of technical and regulatory success, capital intensity, time-to-value inflection, strategic fit, and opportunity cost relative to competing portfolio alternatives.

Applied consistently and transparently, such a framework aligns gated investment decisions with evolving enterprise strategy, financial objectives, and resource constraints. It reinforces the principle that capital deployed toward external acquisitions is economically indistinguishable from capital invested in internal programs—both represent irreversible commitments of scarce resources. By enforcing equivalence in evaluation standards and decision rigor, leadership can increase confidence that portfolio resources are allocated to the highest-quality opportunities, thereby maximizing the likelihood of sustainable value creation across the development pipeline.

1.3.4 Learning from Footprints in the Sand

Leading biopharma organizations—including Eli Lilly, Pfizer, and AstraZeneca—have shown that rigorous retrospective analysis can be a powerful strategic tool when applied systematically. The purpose of these reviews is not simply to catalog past successes or failures, but to identify what has demonstrably improved over time, what has remained static or underperformed, and which decision behaviors, operating assumptions, or execution patterns should be deliberately replicated or avoided. When used effectively, retrospective analysis functions as an institutional learning mechanism that informs forward-looking capital allocation and risk governance.

At the project level, formal *lessons-learned* reviews are a standard component of professional project management and are routinely conducted at major milestones or program closure. These reviews examine timelines, costs, technical outcomes, and execution efficiency in detail. By contrast, comparable analytical rigor is far less commonly applied to portfolio- or enterprise-level decision making, particularly with respect to gated investment decisions that determine whether assets are initiated, advanced, paused, or terminated.

One explanation for this imbalance is discomfort with separating decision quality from eventual outcomes. In drug development, well-reasoned, data-driven decisions may still result in clinical or regulatory failure due to factors beyond managerial control, whereas poorly substantiated decisions may occasionally lead to success driven by chance or unforeseen biology. This outcome ambiguity can obscure causal relationships and discourage critical examination of the decision logic itself, independent of the result.

In the absence of clear executive accountability for both decision processes and downstream consequences, governance mechanisms—especially stage-gate reviews—can become diffuse. Programs may persist through multiple investment cycles without explicit reaffirmation of value, advancing by default rather than deliberate recommitment. Termination decisions are often deferred until negative data become unequivocal, by which point disproportionate capital, talent, and organizational focus have already been consumed, eroding portfolio efficiency.

Our experience suggests that executive committees are not inherently reluctant to make advancement or termination decisions; rather, the limiting factor is the lack of structured ownership, post hoc evaluation, and feedback loops that link decision rationale to observed outcomes over time. Without these mechanisms, organizations miss the opportunity to refine decision frameworks, calibrate risk tolerance, and improve the consistency of portfolio-level judgments.

From a governance standpoint, embedding disciplined retrospective review of gated decisions—analogous to project-level lessons-learned processes—offers a significant opportunity to strengthen enterprise decision quality. By explicitly examining what decisions were made, why they were made, and on what assumptions and evidence, biopharma leaders can institutionalize learning, reduce systemic bias, and enhance the long-term productivity of their R&D portfolios.

2

SCENARIO ANALYSIS AND STRATEGIC PLANNING

2.1 SCENARIO ANALYSIS AS A STRATEGIC INSTRUMENT UNDER UNCERTAINTY

Scenario analysis is a structured approach for examining alternative future states that may emerge from the combined influence of favorable and unfavorable uncertainties. Rather than representing forecasts, scenarios delineate bounded, internally consistent depictions of how the future might plausibly unfold. These future states are shaped by the interaction of known trends and unresolved uncertainties, whose aggregate effects define the opportunity–risk landscape facing an organization.

Although some futures may be characterized as so-called *black swan* events—that is, outcomes with a very low likelihood but disproportionately high impact (Taleb 2007)—the practical value of scenario analysis is greatest when attention is focused on credible, decision-relevant uncertainties. In such cases, the range of outcomes is constrained to events that, while uncertain, are sufficiently plausible to merit strategic consideration. Depending on the nature and magnitude of uncertainty, future environments may fall into one of four broad categories:

1. **Relatively predictable futures**—where causal relationships are largely understood
2. **Discrete alternative outcomes**—where outcomes diverge along a limited number of well-defined paths
3. **Continuous ranges of futures**—where variability exists within identifiable bounds
4. **Truly ambiguous outcomes**—where outcomes and their drivers remain fundamentally unclear

Scenario analysis does not require assigning explicit probabilities to events or uncertainties. Instead, its value lies in articulating the logic that connects trends, uncertainties, and outcomes. By doing so, scenarios sharpen managerial judgment, direct attention to the most consequential sources of uncertainty, and support disciplined strategic reasoning in environments where probabilistic forecasting may be infeasible.

One established methodology for constructing scenarios under deep uncertainty is the *extreme-world framework* (see Figure 2.1), as articulated by Goodwin and Wright (2010). This approach deliberately explores the outer bounds of plausible futures in order to stress-test strategic assumptions and expose vulnerabilities. The method proceeds through the following eight stages:

1. **Define the focal issue and planning horizon:** Clearly articulate the strategic question of interest—such as market access, pipeline sustainability, or regulatory evolution—and specify the temporal window over which scenarios will be developed.

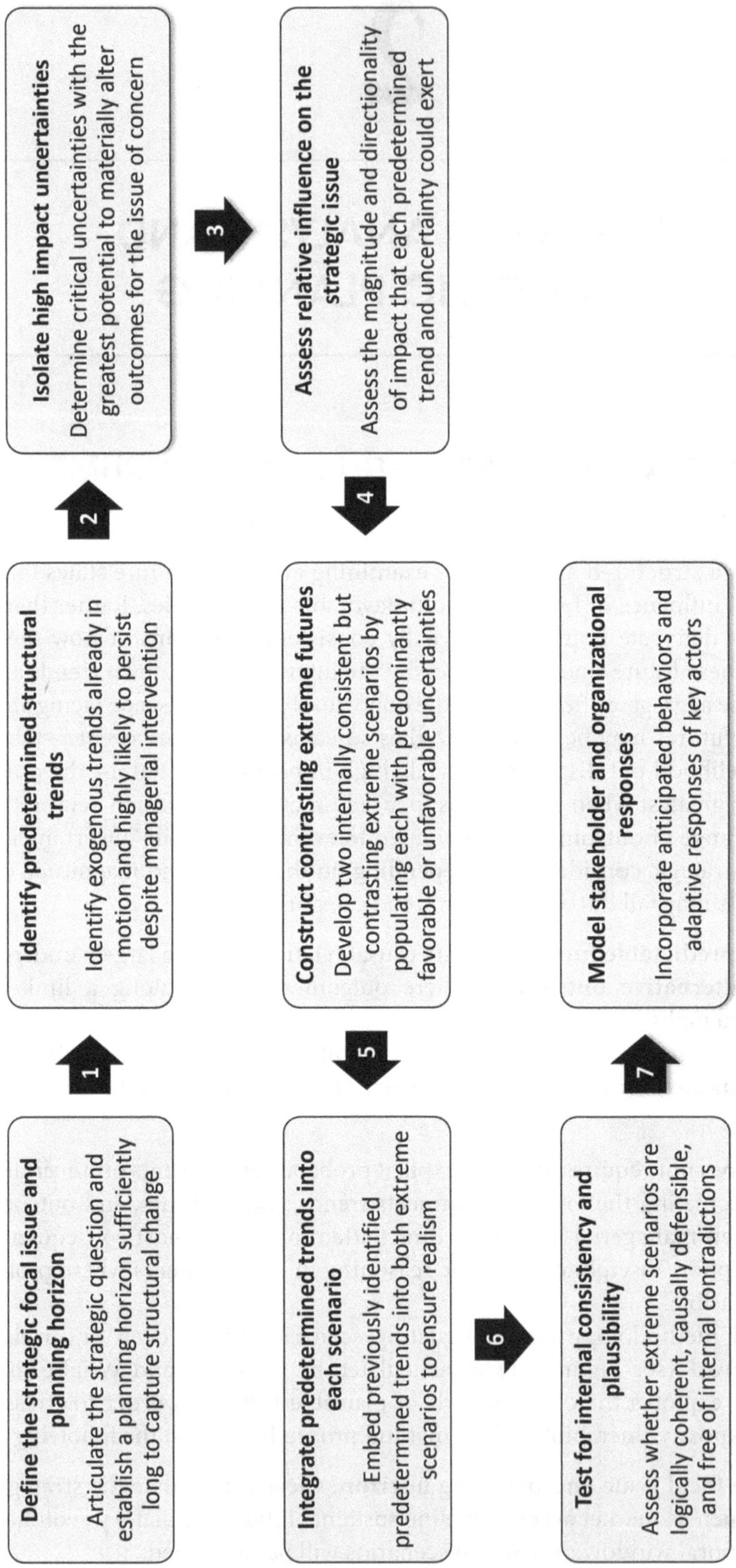

Figure 2.1 The extreme-world scenario analysis framework (Goodwin and Wright 2010).

2. **Identify predetermined trends:** Enumerate trends that are already in motion and largely insensitive to managerial influence (e.g., demographic shifts, long-cycle scientific advances, or regulatory baselines).
3. **Isolate critical uncertainties:** Determine the uncertainties with the greatest potential to influence outcomes, whether positively or negatively, such as reimbursement regimes, technological discontinuities, or competitive behaviors.
4. **Evaluate relative impact:** Assess how strongly each trend and uncertainty could affect the focal issue, recognizing that not all drivers are equally consequential.
5. **Construct opposing extreme worlds:** Develop two contrasting scenarios by aggregating uncertainties with predominantly favorable effects into one extreme and predominantly adverse effects into another. Each scenario is extended to a plausible end state through a coherent narrative that links trends and uncertainties via causal or reinforcing relationships.
6. **Integrate predetermined trends:** Embed the previously identified trends into both scenarios to ensure realism and continuity with known structural forces.
7. **Test for internal consistency:** Examine whether the assumed combinations of trends and uncertainties can plausibly coexist within a single future state.
8. **Incorporate stakeholder responses:** Model how key actors—firms, regulators, payers, clinicians, and patients—might respond to each scenario, recognizing that adaptive behavior can amplify or dampen outcomes.

Between these two extremes lies a spectrum of intermediate futures, including a baseline or status quo scenario in which the future largely represents a logical extension of current conditions.

Goodwin and Wright (2010) also propose a more analytically nuanced methodology, the *driving forces framework*, which accommodates varying degrees of predictability and uncertainty and does not necessarily privilege extreme outcomes. In this approach, factors relevant to the focal issue are positioned within a matrix based on their perceived impact and predictability.

The analytical process begins by examining low-probability, high-impact elements—often associated with black swan-type risks—and identifying clusters of interrelated factors and their underlying driving forces, particularly structural uncertainties. The extreme manifestations of these driving forces are then explored to understand the boundaries of potential impact. This process is repeated across the remaining quadrants of the matrix, systematically evaluating permutations of driving force extremes.

For each configuration, analysts develop a narrative that emphasizes interdependencies, causal pathways, and reinforcing mechanisms. The distinguishing feature of the driving forces framework is its explicit linkage to strategy formulation. Rather than stopping at the articulation of future states, the methodology culminates in the generation of strategic options tailored to each scenario. In doing so, it moves beyond descriptive foresight toward prescriptive decision support, enabling organizations to translate uncertainty into actionable strategic choices.

An alternative and widely adopted methodology is the *Oxford Scenario Planning approach* (Ramirez and Wilkinson 2018), which emphasizes collaborative sensemaking and organizational learning. This approach examines uncertainty across two interconnected levels of interaction.

First, it considers the relationship between the organization and its transactional environment, encompassing industry competitors, suppliers, customers, potential entrants, substitutes, regulators, investors, employees, and advocacy groups—many of which align with Porter's Five Forces framework. Second, it expands the lens to the interaction between the transactional environment and the broader contextual environment, including macroeconomic conditions, geopolitical dynamics, public policy, technological evolution, and societal values.

Oxford Scenario Planning is inherently participatory, engaging stakeholders at all hierarchical levels and across functional disciplines. Through structured dialogue and iterative exploration of periods characterized by turbulence, uncertainty, novelty, and ambiguity (TUNA conditions), the process generates shared insights into how alternative futures could affect organizational viability. Equally important, it often challenges prevailing assumptions about the present, prompting organizations to reassess their current strategies and mental models.

In practice, no organization—particularly in biopharma—can credibly prepare for every conceivable future. Consequently, strategic planning typically centers on a primary future state that serves as an anchor for investment and resource allocation. This focal plan, however, should not be rigid. Rather, it should be complemented by a portfolio of strategic options designed to mitigate downside risk or capture upside potential associated with alternative scenarios.

This approach preserves managerial flexibility. As leading indicators signal the emergence of a nonbaseline scenario, the organization can pivot rapidly, deploying pre-positioned options rather than reacting ad hoc. Organizations that excel in this regard tend to exhibit strong learning capabilities, continuously monitoring their competitive and regulatory environments and anticipating the likely responses of rivals and stakeholders.

Preparing for multiple plausible futures requires the deliberate construction of *strategic alternatives* that align with the opportunities and threats revealed by scenario analysis. These alternatives should not be abstract or detached from operational reality. Instead, robust strategies are characterized by three defining attributes—the *CAD principle*:

- **Compelling in nature:** Each alternative addresses a clearly articulated strategic challenge or opportunity
- **Actionable in design:** The strategy can be translated into concrete decisions and investments
- **Differentiable in variety:** Alternatives are meaningfully distinct from one another, enabling genuine choice

Strategic alternatives should be clearly separated from policy decisions (which define where and how the organization competes regardless of future conditions) and operational decisions (which concern downstream execution). One effective tool for generating alternatives is the strategy table (described in detail in Chapters 6 and 7), which systematically combines strategic choices across multiple dimensions. However, combinatorial explosion is a practical risk: for example, four strategic dimensions with three options each yield $3^4 = 81$ theoretically distinct, mutually exclusive, and collectively exhaustive strategies, excluding hybrid combinations.

Accordingly, analysts should aggressively narrow the field to a manageable subset—typically three to five CAD-compliant strategies—that perform robustly across multiple scenarios before committing to detailed quantitative evaluation.

Once a short list of strategic alternatives has been identified, decision makers must determine how best to evaluate them. Options include expected value or expected utility analysis using decision trees (discussed in Chapter 6), as well as multicriteria prioritization methods such as Multiple Objective Decision Analysis (MODA) or the Analytic Hierarchy Process (AHP) (explored in Chapter 8). While ranking methods can be useful, there is a strong case for fully quantifying leading alternatives to support transparent, evidence-based decision making.

Scenario analysis is frequently complemented by *SWOT analysis*, which serves as a bridge between environmental scanning and strategy formulation. SWOT distinguishes internal attributes (strengths and weaknesses) from external conditions (opportunities and threats), enabling a holistic assessment of strategic positioning (see Figure 2.2). Guiding questions include:

INTERNAL	**STRENGTHS** • What are the organization's distinctive core competencies? • What does it do better than its main competitors and why? • Does it possess any unique capabilities and how imitable are they? • Does it have any clear cost advantages? • In what areas does it innovate and lead? • What does the competition see as the organization's strengths?	**WEAKNESSES** • In what areas can the organization improve? • What does it do worse than its main competitors and why? • In what areas does the organization purposely allocate fewer resources than its main competitors? • What needs to change in order to succeed against the current and future leading competition? • What does the competition see as the organization's weaknesses? • How easily can the competition imitate the organization?
EXTERNAL	**OPPORTUNITIES** • What opportunities are open to the organization? • What market trends can it take advantage of? • What products and services will customers demand in the future? • How can the organization turn its strengths into new products and services? • What will affect the loyalty of customers?	**THREATS** • What threats could undermine the organization's main business(es)? • Who are the newest competitors and why are they threats? • What disruptive technologies are on the horizon and how do they impact the organization? • What are the major obstacles to the organization's growth? • What threats arise from the organization's internal weaknesses?

Figure 2.2 SWOT analysis of an organization.

- **Strengths:** What differentiates the organization today, and how defensible are these advantages over time?
- **Weaknesses:** Where does the organization underperform, and what structural changes are required to close these gaps?
- **Opportunities:** Which emerging trends can be leveraged now to meet future customer and market expectations?
- **Threats:** What barriers constrain growth, and how might these impediments intensify or evolve?

When integrated with scenario analysis, SWOT provides a disciplined pathway from uncertainty to strategy, supporting resilient decision making in complex, high-stakes environments such as biopharma development and commercialization.

2.2 CORE CONCEPTS OF STRATEGIC PLANNING UNDER UNCERTAINTY

Strategic planning can be rigorously defined as a disciplined process through which an organization designs and commits to courses of action intended to achieve a clearly articulated future state within a specified time horizon. This future state—rather than near-term actions—serves as the anchor for multiperiod strategy formulation and the iterative development of strategic plans. Effective strategic planning, therefore, begins not with annual targets, but with a long-range perspective on where the organization intends to compete and how it expects its competitive environment to evolve.

In practice, this requires leadership to adopt an explicit forward-looking orientation. Organizations must assess not only the markets in which they currently participate, but also those they anticipate entering or exiting over time. Scenario analysis plays a critical enabling role in this process by helping organizations develop informed perspectives on how industry structures, competitive dynamics, and external forces may plausibly evolve. Only by first developing a reasoned view of the future through scenario analysis can an organization determine how best to position itself for durable success with the aid of strategic planning.

Assuming the future represents a smooth extrapolation of historical trends, the fallacy of linear continuity is, at best, naïve and, at worst, strategically negligent. Equally misguided is the belief that disruptive or transformative strategies are accessible only to incumbents with scale, capital, or market dominance. Organizations that fail to continuously adapt—particularly in anticipation of foreseeable competitive, technological, or macroeconomic shifts—expose themselves to significant downside risk.

When adverse conditions materialize, such as economic contractions or reimbursement shocks, many organizations default to short-term defensive actions, most notably workforce reductions framed as *cost discipline*. While such measures may temporarily preserve selected financial ratios, they often erode organizational capability by shedding critical talent. The result is frequently a prolonged recovery period and diminished competitive resilience, rather than a rapid return to successful performance.

The distinction between scenario analysis and strategic planning is especially salient in complex industries such as healthcare and biopharma. Scenario analysis is concerned with exploring future states over which the organization has limited or no control, whereas strategic planning focuses on decisions and actions within the organization's sphere of influence.

For example, scenario analysis in healthcare may examine how the increasing adoption of personalized medicine could reshape (a) the development and deployment of targeted therapies and (b) the business models of firms involved in diagnostics, therapeutics, and delivery technologies.

While industry participants such as lobbyists can exert influence over regulatory pathways, pricing mechanisms, and innovation incentives, they have far less control over how patient expectations (e.g., better targeted therapeutic intervention) evolve regarding individualized care. Scenario analysis explicitly acknowledges this asymmetry of control, whereas strategic planning is concerned with how the organization responds to these external realities.

Most organizations engage in some form of scenario thinking, but the depth, rigor, and strategic follow-through associated with such thinking vary dramatically and help distinguish leaders from laggards. Lagging organizations tend to rely on incremental logic, implicitly asking: *Given where we have been, what does the past suggest our future business model should resemble?*

In contrast, leading organizations deliberately challenge historical inertia by asking a fundamentally different question: *Which critical uncertainties could fundamentally disrupt our current business model, and how should we position ourselves to capture upside while minimizing negative consequences?*

This distinction—between extrapolation and challenge—largely determines whether scenario analysis becomes a perfunctory exercise or a genuine source of strategic advantage. Once an organization has identified a limited number of plausible future states on which to focus, the discipline of strategic planning begins in earnest. A coherent strategic plan is explicitly grounded in the organization's mission, vision, and values (see Figure 2.3), which collectively define its *raison d'être*—its enduring purpose and aspirational direction:

- **Mission**—Why does the organization exist, and whom does it serve?
- **Vision**—What future position does the organization seek to attain, and by when?
- **Values**—What principles govern decision making and organizational behavior?

These elements provide the normative framework within which strategic choices are evaluated and prioritized.

Consider a hypothetical biopharma company whose stated vision is to rank among the top five global industry players by revenue. Its strategic plan may be structured around three enterprise-level goals (see Figure 2.4):

1. Delivering integrated healthcare solutions across diagnostics, therapeutics, and delivery platforms
2. Demonstrating exemplary corporate citizenship and societal responsibility
3. Establishing a reputation as an employer of choice

Each strategic goal is operationalized through a set of measurable objectives that define the pathways by which the organization intends to achieve its vision. Collectively, these objectives translate abstract ambition into actionable intent. In this example, the healthcare company has articulated an

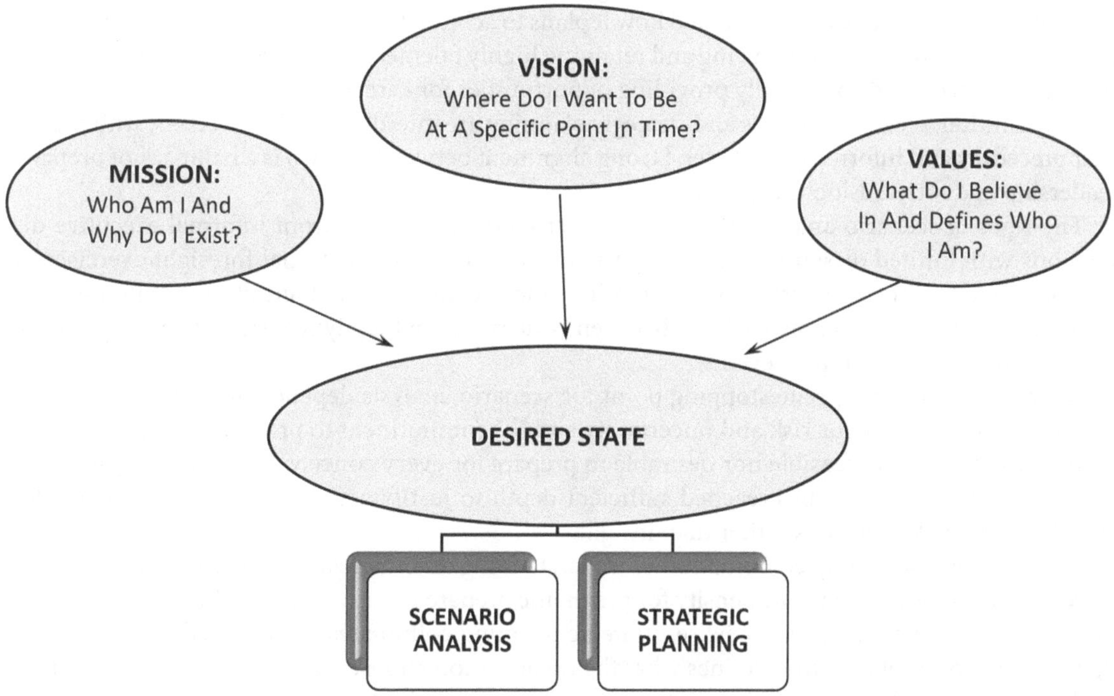

Figure 2.3 Organizational mission, vision, and values.

Figure 2.4 Example of strategic planning in a biopharma company.

expressed intent to become a premier provider of healthcare through its integrated suite of diagnostic, drug delivery, and therapeutic products. In pursuit of its goal to be seen as a highly responsible corporate citizen, the organization has outlined objectives that revolve around listening to its customers, volunteering its services, and adopting the highest standards of environmental friendliness. In addition, the company has articulated how it plans to achieve the goal of creating a safe and fulfilling work environment through hiring and retaining highly talented individuals, providing enriching work experiences, and continually providing opportunities for career growth.

By definition, scenario analysis and strategic planning are interdependent processes, with the former preceding and informing the latter. Strong alignment between the two is a hallmark of prepared leadership and forward-looking governance.

The rigor of scenario analysis varies widely across organizations, from informal executive discussions with limited dissemination to highly structured, cross-functional foresight exercises incorporating internal and external expertise. While most analyses begin logically, they often struggle to reach closure. Debates may oscillate between concerns about analytical exhaustion and fears of overlooking rare, high-impact events.

In practice, the appropriate stopping point for scenario analysis depends on two organizational attributes: (a) tolerance for risk and uncertainty and (b) commitment to pragmatic preparedness.

Because it is neither feasible nor desirable to prepare for every conceivable future, organizations must judge when analysis has reached sufficient depth to justify action. Beyond that point, additional analysis risks paralysis rather than insight.

In reality, only a healthy combination of understanding an organization's orientation to risk and uncertainty, coupled with its propensity for pragmatic preparedness, will determine when a scenario analysis should be concluded. Likewise, strategic planning culminates in the articulation of a strategy—explicit or implicit—that defines what the organization chooses to do and not do. The quality of this strategy is influenced by three interdependent forces (Abraham 2006):

1. **Strategic thinking:** The ability to understand how markets, competitors, and technologies may evolve, and to generate differentiated strategic alternatives that deliver renewed customer value
2. **Business model:** A clear description of how the organization creates and captures value, including how objectives will be achieved and performance measured over time
3. **Value proposition:** The totality of experiences delivered to customers, where value is realized when perceived net utility exceeds total cost

Many organizations invest heavily in operational and contingency planning while underinvesting in genuine strategic thinking. Incentive systems often favor near-term, measurable outcomes, which, by themselves, are antithetical to preparing for uncertainty and uncontrollability, making operational planning more attractive than preparing for uncertainty over longer horizons.

Effective planning requires sustained collaboration among strategy, portfolio management, and finance. When these functions operate in silos, portfolio management is often reduced to retrospective analysis rather than proactive decision support. A robust planning cadence demands early and continuous alignment on strategic intent, financial targets, and resource availability.

Finally, while deliberate strategy is essential, realized strategy often reflects a combination of planned intent and emergent adaptation. Organizations that balance control with learning are best positioned to abandon failing strategies, capitalize on unforeseen opportunities, and avoid cognitive traps such as groupthink or peripheral blindness (Mintzberg et al. 1998).

In this sense, strategic planning is not merely a tool for setting direction, but a continuous organizational capability for navigating uncertainty, learning from experience, and sustaining competitive relevance over time.

A further, and entirely predictable, dynamic arises from the pressure to deliver short-term performance targets. In pursuit of annual growth commitments, organizations frequently gravitate toward metrics that are easily measured and readily communicated—such as achieving a 10% compound annual revenue growth rate. While disciplined attention to quantitative objectives is a hallmark of performance-oriented enterprises, such metrics can become dangerously dominant when treated as ends in themselves rather than indicators of strategic health.

When numerical targets assume primacy, the implicit logic becomes that any strategic pathway capable of delivering the stated outcome is acceptable. In effect, the organization risks adopting a *results-at-all-costs* mindset, where the quality, sustainability, and strategic coherence of decisions are subordinated to near-term attainment of the desired endpoint. This condition mirrors the cautionary lesson often attributed to *Alice in Wonderland*: if the destination is all that matters, then the choice of route becomes irrelevant.

Over time, this narrow focus on annual objectives can erode the very capabilities that underpin long-term competitive advantage. Core competencies may deteriorate in a largely foreseeable manner as management attention and capital are disproportionately allocated to low-risk, incremental initiatives—such as successive product line extensions designed to bolster near-term revenues—at the expense of sustained investment in foundational or platform technologies. In research-intensive industries like biopharma, this imbalance can gradually hollow out scientific and technical capabilities, undermining the organization's ability to innovate meaningfully in the future.

Compounding this risk, even organizations that formally adhere to sound strategic planning practices may neutralize their effectiveness by relegating strategy formulation to staff functions that lack decision authority. When strategic planning is decoupled from executive power, it becomes advisory rather than directive, increasing the likelihood that well-intentioned analyses translate into fragmented actions and unintended downstream consequences. As observed by Mintzberg and colleagues (1998) (see Figure 2.5), such structural separation between strategy development and

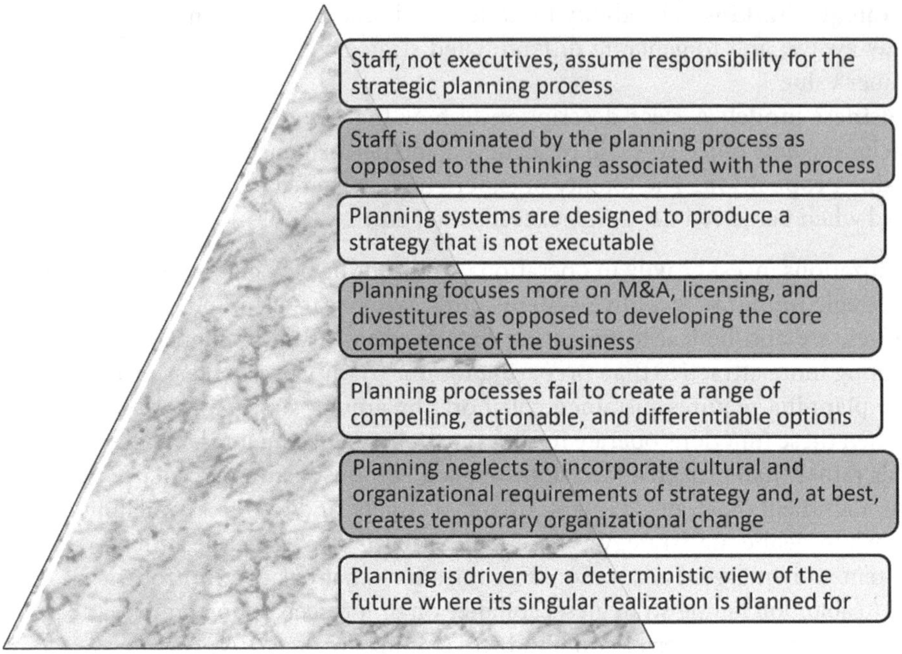

Figure 2.5 The seven deadly sins of strategic planning (Mintzberg et al. 1998).

strategic decision making often leads to misalignment, inertia, and outcomes that diverge sharply from original intent.

When strategic planning lacks sustained sponsorship from a visible and empowered senior leader, it rapidly risks devolving into a procedural exercise rather than a vehicle for rigorous strategic insight. In such circumstances, the emphasis shifts from cultivating high-quality strategic thinking to maintaining the mechanics of the planning process itself—calendars, templates, and governance rituals—often at the expense of substantive debate about competitive positioning and long-term value creation.

A common manifestation of this degradation is an overemphasis on externally driven portfolio expansion strategies, such as mergers, acquisitions, or partnerships, as substitutes for the more difficult task of strengthening internal capabilities. While inorganic growth mechanisms can play an important role in value creation, their dominance in strategic plans frequently reflects an avoidance of hard choices about capability development, scientific investment, or business model evolution. As a result, strategic recommendations may appear ambitious in scope but prove largely inactionable because they are disconnected from the organization's core competencies and execution capacity.

At the same time, in an effort to avoid being perceived as detached or overly theoretical, strategic planning functions may converge on a single, highly deterministic view of the future. This tendency constrains the exploration of uncertainty and limits the generation of differentiated strategic options. The outcome is a narrow set of strategies that offer little resilience to alternative future states and provide minimal guidance for adaptive decision making.

Finally, when insufficient attention is paid to the organization's cultural orientation toward risk, ambiguity, and change, even well-designed strategic plans may yield only a transient impact. Initial momentum can dissipate as entrenched behaviors, legacy incentives, and comfort with established practices reassert themselves. In these cases, the familiarity of the present and recent past ultimately overwhelms the discomfort associated with preparing for an uncertain future, leaving the organization structurally unchanged despite the appearance of strategic activity.

A defining characteristic of a disciplined, mature, and forward-looking organization is the explicit differentiation of purpose across its principal planning processes—namely, strategic, portfolio, and business planning (see Figure 2.6). Each of these planning layers serves a distinct role and operates over a different time horizon; clarity regarding their respective *raison d'être* is essential for coherent decision making.

In the absence of such clarity, even well-intentioned stakeholders may work at cross-purposes, inadvertently undermining one another's efforts as they attempt to satisfy the objectives of competing plans. This misalignment is often exacerbated when individual functions—such as portfolio management, finance, and resource management—maintain independent versions of portfolio data, assumptions, and metrics. The resulting fragmentation can produce inconsistent analyses, divergent strategic interpretations, and conflicting conclusions presented to senior leadership, eroding confidence in the planning process and slowing decision making.

Effective strategic planning is inherently long-term in orientation. Its purpose is not to optimize near-term performance, but to establish a durable direction for the enterprise over multiple planning cycles. In research-intensive sectors such as biopharma, where product development timelines are long and resource commitments are substantial, it is common—and appropriate—for strategic plans to extend over a 10- to 15-year horizon. This extended time frame enables leadership to address fundamental questions of competitive positioning, capability development, and portfolio evolution that cannot be meaningfully resolved within annual or short-term planning cycles.

At its core, a strategic plan serves two fundamental purposes. First, it establishes the arenas in which the organization intends to compete by defining the scope of markets, technologies, and value

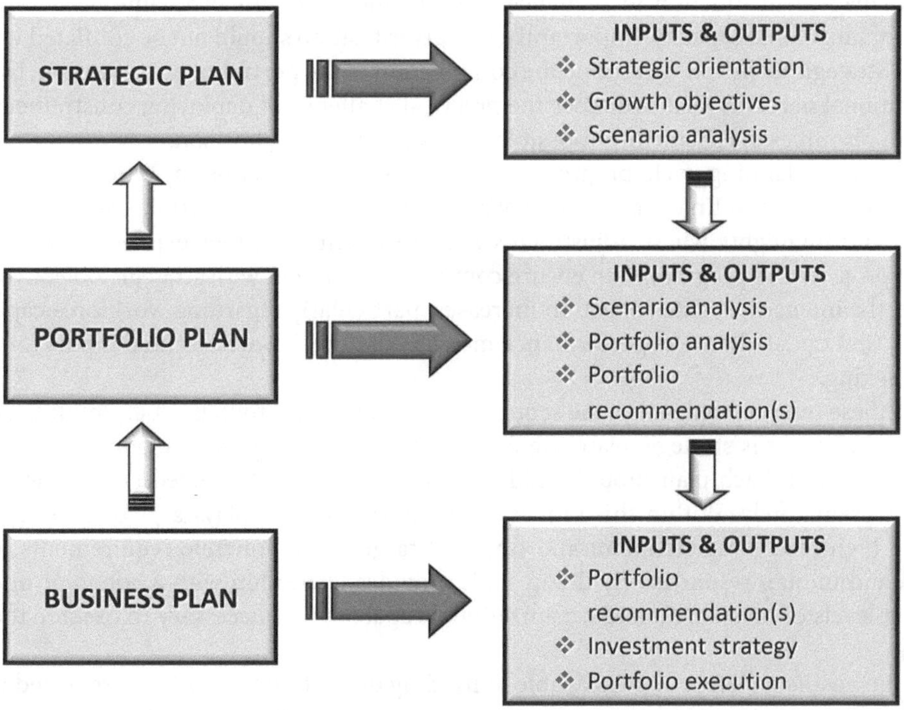

Figure 2.6 Temporal dependence between strategic, portfolio, and business plans.

chains it will actively pursue or deliberately avoid. Second, it sets the parameters for assembling and managing a portfolio of initiatives that collectively satisfy the organization's strategic ambitions and financial commitments across multiple time horizons—near-term, intermediate, and long-term.

As the organization transitions from strategy formulation to execution, the existing portfolio of projects and programs undergoes a comprehensive reassessment. This review updates both the strategic relevance and the financial contribution of each initiative in light of the newly articulated strategic direction. The objective is not merely to validate ongoing activities, but to ensure that each element of the portfolio remains aligned with evolving priorities and resource realities.

The resulting strategic plan then becomes the primary input to portfolio planning. The portfolio plan is explicitly tasked with operationalizing the strategy by enabling achievement of its objectives over time. This requires more than incremental portfolio adjustments; it demands a deliberate and analytically rigorous selection of portfolio components, ideally through formal portfolio optimization techniques. Such optimization balances expected value, risk, and strategic fit while explicitly accounting for constraints on resources, talent, and organizational capacity. Through this process, the portfolio becomes a dynamic mechanism for translating strategic intent into executable investment choices.

It is important to distinguish this activity from strategic decision making. Portfolio *optimization* determines *what* the organization should invest in to advance its strategic objectives, whereas *prioritization* addresses *when* those investments can be executed given real-world constraints. As such, prioritization is inherently operational, reflecting the practical necessity of allocating finite financial and human resources to maximize throughput, minimize bottlenecks, and preserve delivery discipline over the planning horizon.

Following the definition of an optimized portfolio, individual initiatives are evaluated through a disciplined prioritization process that supports effective execution planning. This step enables the resource management function to sequence projects and programs over time in alignment with available organizational capacity. Importantly, this prioritization should not be conflated with strategic choice. Strategic decisions determine the composition of the portfolio; prioritization, by contrast, is an operational necessity that addresses the practical challenge of deploying constrained financial and human resources as efficiently as possible across the planning horizon.

As the annual planning cycle progresses, the business plan assumes a critical role in calibrating portfolio ambition to financial reality. Specifically, it informs the portfolio plan of affordability thresholds and highlights where adjustments may be required—either expanding or contracting the mix of projects and programs—to ensure continued alignment with strategic objectives. During this phase, the intensity of data collection increases, particularly regarding workforce capacity, skill availability, and updated project performance metrics, all of which are essential inputs for informed decision making.

Despite these interdependencies, the sequencing of planning activities is paramount. The appropriate temporal order is strategic planning first, portfolio planning second, and business planning last (see Figure 2.6). Each plan should build upon the outputs of its predecessor, rather than constraining it prematurely. Within this sequence, the portfolio plan plays a pivotal integrative role, translating high-level strategic intent and financial targets into concrete requirements for capital allocation and human resources. By doing so, it provides leadership with a coherent understanding of what levels of investment and organizational capability are necessary to execute the strategy effectively.

When the business plan proves incapable of funding or staffing the portfolio required to execute the strategy, the organization must confront this misalignment directly. At that point, leadership is left with only two viable corrective actions: either recalibrate the ambition embedded in the strategic plan or deliberately relax the resource constraints imposed by the business plan. Allowing the inconsistency to persist without adjustment undermines both credibility and performance.

Despite this logic, many otherwise high-performing organizations reverse the intended planning sequence by allowing the business plan—typically driven by near-term budgetary considerations—to precede and constrain portfolio planning. In such cases, the portfolio plan is forced to operate within resource limits that were defined without regard to strategic intent. The predictable consequence is a portfolio that is structurally incapable of delivering the level of value assumed by the strategic plan. Over time, this disconnect manifests as repeated failure to achieve strategic and financial targets, often culminating in forced consolidation through acquisition or, conversely, the need to seek acquisition as a means of survival.

A more coherent construct is to view planning through a hierarchy of dependencies (see Figure 2.7). Corporate goals and long-term strategic objectives establish the overarching expectations for value creation and growth. These expectations then shape the portfolio strategy—defining the scale, composition, and risk profile of investments required to meet those goals. Only after these requirements are clearly articulated should the business plan determine the specific funding levels, operating budgets, and resource allocations necessary to support the portfolio. This dependency structure ensures that financial planning serves strategy, rather than inadvertently dictating it.

When the business plan—effectively the organization's research and development (R&D) and commercial investment blueprint—is allowed to override either the portfolio strategy or the corporate goals established through the strategic plan, sustained performance over multiple planning cycles becomes highly improbable. In such cases, near-term financial constraints take precedence over long-term value creation, constraining the organization's ability to execute against its stated ambitions.

In practice, however, it is not uncommon for the annual budgeting process embedded within the business plan to dictate the scope, scale, and composition of the portfolio. As a result, portfolio

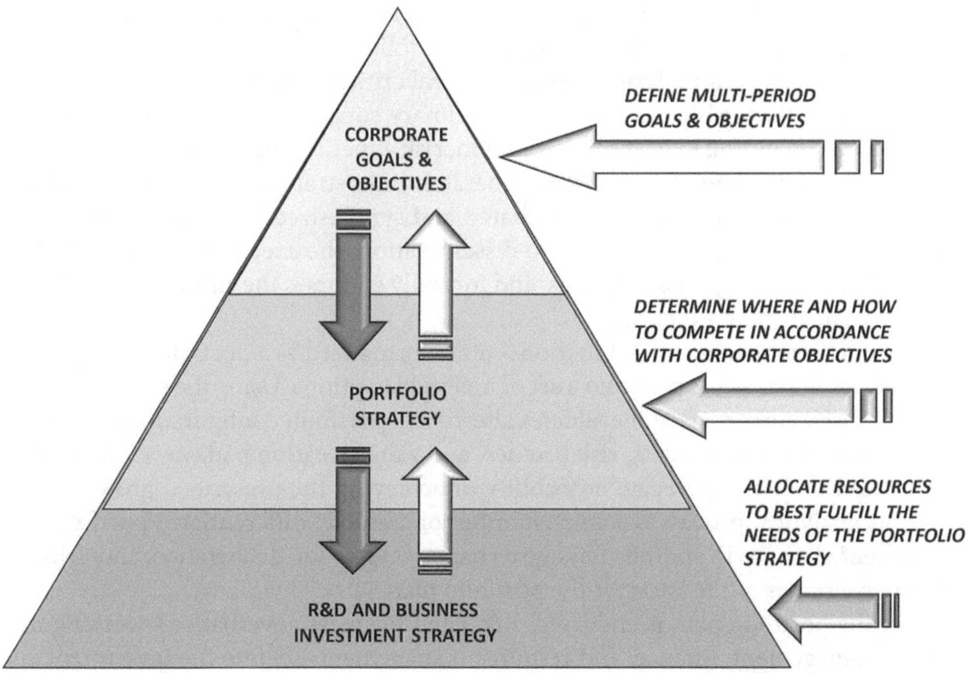

Figure 2.7 Hierarchy of dependencies created by corporate goals and objectives.

planning becomes reactive to budget ceilings rather than strategically driven, while the strategic plan itself remains largely static and progressively detached from operational reality. Over time, this dynamic renders the strategic plan obsolete—not because it was poorly conceived, but because it is no longer meaningfully connected to the investment decisions that determine what the organization actually does.

To reduce the chronic risk of underperforming against both strategic and financial objectives—often driven by misalignment between strategy formulation and budget execution—it is critical that meaningful dialogue between finance and portfolio management occur *before* annual budget decisions are locked in. From a portfolio perspective, few outcomes are more valuable than explicit alignment between these functions on the level and composition of resources required to realistically deliver the ambitions set forth in the strategic plan.

Despite the apparent soundness of this approach, organizational realities frequently undermine its execution. Functional silos and entrenched fiefdoms often result in fragmented decision-making processes, particularly with respect to how strategic growth targets are defined and how financial constraints—most notably budget ceilings—are established. Rather than being developed collaboratively, these parameters are typically determined sequentially and in isolation.

As a consequence, portfolio management is frequently excluded from upstream strategic and financial deliberations and instead positioned as a downstream analytical function. Its role becomes one of assessing feasibility after strategic objectives and budgetary limits have already been finalized, rather than shaping those decisions in the first place. This dynamic significantly reduces the organization's ability to reconcile ambition with execution capacity.

For those who question whether this pattern is pervasive, a simple diagnostic question is instructive: *When was the last time your chief financial officer actively sought input from the head of portfolio management to inform budget-setting decisions?* If such engagement is rare or nonexistent, it raises a fundamental question about whether the organization is truly integrating strategy, portfolio design, and financial planning—or merely coordinating them after the fact.

A structured analysis that examines **s**uppliers, **i**nputs, **p**rocesses, **o**utputs, and **c**ustomers (SIPOC) clarifies the roles and handoffs across the strategic, portfolio, and business planning continuum (as illustrated in Figure 2.8). This analysis highlights several critical dynamics.

First, the executive leadership team acts as the primary source of strategic direction. It provides foundational inputs, including long-term orientation, risk appetite, and annual growth expectations. Through an integrated planning process that deliberately links strategic, portfolio, and business considerations, multiple future scenarios are evaluated and synthesized into a coherent strategic plan aligned with these growth objectives. Prior to dissemination, the executive committee reviews the underlying analyses, challenges assumptions, and formally endorses the strategy before it is released to the portfolio management function.

Second, the portfolio management function—often formalized as a portfolio management office (PMO)—translates the strategic plan into a set of investable options. Using the strategy as its guiding framework, the PMO constructs and evaluates alternative portfolio configurations, each representing a different combination of assets, risk profiles, and value-creation pathways. These alternatives are assessed based on their aggregate probability of achieving the prescribed growth targets. The outcome of this analytical process is a small number of credible, differentiated portfolio strategies, which are presented to R&D and business governance bodies for deliberation and selection. The chosen alternative serves as the basis for the portfolio plan.

Third, once the portfolio plan is endorsed, R&D and business governance—working in concert with portfolio management, finance, and resource management—define the investment and capability requirements necessary to execute it. This includes articulating an R&D investment plan that

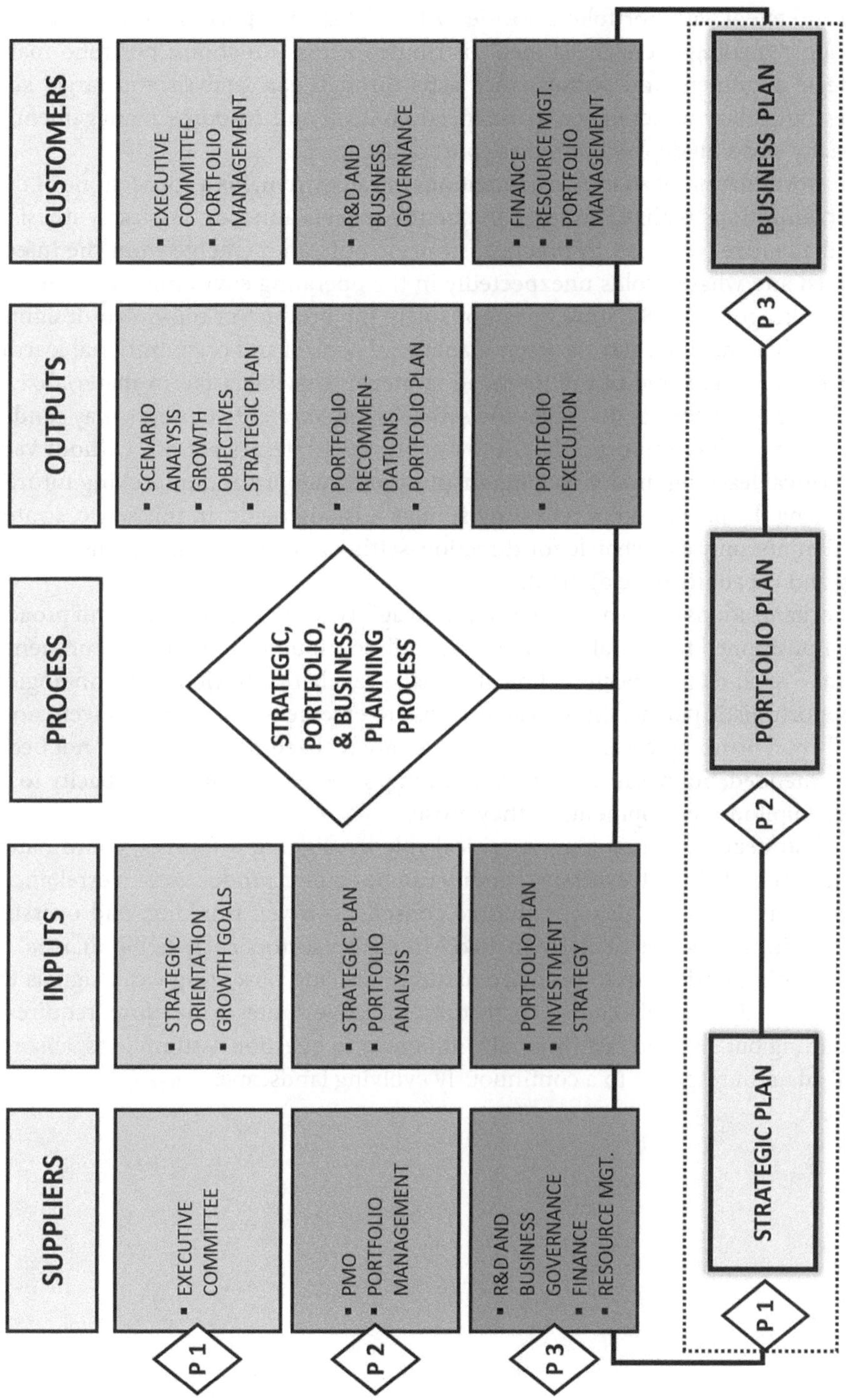

Figure 2.8 Temporal dependence between strategic, portfolio and business plans—SIPOC.

aligns funding and talent with portfolio priorities. At this stage, the portfolio transitions from planning to execution. Ongoing oversight is then distributed across functions: portfolio management monitors strategic alignment and performance gaps through gap analysis and target simulation; finance ensures budgetary adherence and financial control; and resource management manages workforce capacity and skill deployment to support delivery.

Although the formulation of an explicit, intentional strategy remains a cornerstone of disciplined planning, an organization's realized strategy frequently reflects a more complex synthesis of deliberate intent and adaptive response. In practice, strategic outcomes emerge from the interaction of what was planned and what unfolds unexpectedly in the operating environment. As articulated by Mintzberg and colleagues (1998), strategy is not solely the product of top-down design; rather, it evolves through a dynamic interplay between managerial control and organizational learning.

Within this framework, some of the intended strategies inevitably fail to materialize. External disruptions, adverse regulatory shifts, scientific setbacks, or competitive moves may render certain plans obsolete or infeasible. Importantly, these unrealized strategies are not without value. They often serve as critical learning inputs, shaping institutional memory and improving future decision making by reducing the likelihood of repeating similar misjudgments. In this sense, strategic planning will function not only as a vehicle for direction-setting but also as a mechanism for organizational learning and capability development.

Conversely, organizations characterized by cultural agility, scientific curiosity, and proactive leadership are often positioned to capitalize on favorable discontinuities in their environment. Positive inflection points—such as unexpected clinical signals, regulatory flexibility, technological breakthroughs, or opportunistic partnerships—can give rise to emergent strategies that were not contemplated during formal planning cycles. These strategies are realized retrospectively, not because they were originally intended, but because the organization possessed the adaptive capacity to recognize and exploit serendipitous developments as they arose.

While formal strategic planning plays a critical role in aligning resources, coordinating effort, and articulating corporate intent, excessive rigidity can be counterproductive. Overreliance on fixed plans may constrain peripheral vision, reinforce consensus-driven thinking, and oversimplify an inherently uncertain future—particularly in R&D-intensive sectors such as biopharma. Such distortions risk anchoring leadership to outdated assumptions and obscuring weak signals that could otherwise inform timely strategic course correction. Effective strategy, therefore, requires not only disciplined planning but also the institutional willingness to question assumptions, absorb new information, and adapt in response to a continuously evolving landscape.

3

DECISION ANALYSIS AND PORTFOLIO MANAGEMENT

3.1 INTRODUCTION TO DECISION ANALYSIS AND ITS UTILITY IN DECISION MAKING

Decision analysis is a formal, quantitative discipline for supporting complex decision making under uncertainty. It provides a structured framework for defining decision alternatives, characterizing uncertainty, quantifying outcomes, and evaluating trade-offs, thereby enabling consistent, risk-adjusted comparison of competing strategies. The objective is not merely to support *good* decisions, but to identify alternatives that dominate others in expected or risk-adjusted value.

This approach is particularly well-suited to biopharma research, development, and commercialization decisions, which commonly involve multiple plausible strategic alternatives, long development timelines, staged capital commitments, and substantial technical, regulatory, and commercial uncertainty. Decisions are often made sequentially—across discovery, preclinical, clinical, regulatory, and launch phases—requiring early commitments before risks are fully resolved, while new information is generated over time.

Historically, many portfolio and strategic decisions in biopharma have relied on expert judgment, experience, heuristics, and advocacy-driven discussion, with quantitative analysis playing variable roles. A frequent pitfall arises when qualitative considerations, such as strategic alignment and competitive positioning, are blended indiscriminately with quantitative metrics such as expected net present value (eNPV) or internal rate of return (IRR). While this can create the appearance of balance, it often obscures economic trade-offs and can result in the endorsement of *acceptable* decisions that fall short of maximizing value.

Human intuition, though valuable, is difficult to evaluate for internal consistency or completeness (Skinner 1999). While intuitive judgment can lead to value creation, it does not reliably ensure value maximization. Decision analysis explicitly seeks to identify the alternative that maximizes expected value, subject to uncertainty and risk preferences, while recognizing that its principal contribution lies in improving insight, structure, and creativity rather than mechanically producing a single *optimal* answer (Keeney 1982).

Importantly, recommendations emerging from decision analysis are conditionally prescriptive. They depend on explicit assumptions, probability judgments, value assessments, and risk tolerances (Goodwin and Wright 2010). When analytical results diverge from leadership intuition, the framework provides a transparent basis for constructive dialogue, enabling decision makers to examine whether disagreements stem from data, beliefs, risk attitudes, or strategic priorities. This process frequently strengthens both the analysis and the final decision.

Decision analysis focuses on decision quality rather than outcome certainty. While it increases the likelihood of favorable outcomes on average, it cannot eliminate the stochastic nature of real-world performance. Superior decisions may still yield unfavorable outcomes, while inferior decisions may occasionally succeed by chance (see Figure 3.1). Throughout repeated decisions and portfolios, however, disciplined decision-making practices systematically outperform intuition alone.

A simplified investment example illustrates the importance of explicitly modeling uncertainty. Two development alternatives each require a $10M investment. Under point estimates, Alternative A appears superior, with an expected first-year revenue of $100M compared to $95M for Alternative B. However, when uncertainty is considered ($80M–$105M for A and $90M–$120M for B), probabilistic estimation reverses the ranking, yielding expected revenues of approximately $97.5M for A and $98.3M for B. Apparent dominance under deterministic assumptions can therefore be overturned once uncertainty is incorporated.

Beyond evaluation, decision analysis often derives much of its value from expanding the decision space. In biopharma, teams may become anchored to a narrow set of development or commercialization pathways that appear logical but are artificially constrained. By forcing explicit consideration of nonobvious or contingent strategies, decision analysis reduces the risk of unintentionally foregone value.

Drug development decisions are particularly amenable to decision analysis because they naturally decompose into staged commitments. The methodology clarifies which decisions must be made immediately and which can be deferred until additional data are available, thereby aligning capital deployment with uncertainty resolution and supporting resource efficiency.

Over time, several structured decision analysis frameworks have been proposed. In practice, hybrid approaches—drawing on methods such as the dialogue decision process (Strategic Decisions Group 2005) and the decision analysis cycle (Skinner 1999)—have been widely applied in biopharma R&D and portfolio management (see Figure 3.2). These approaches emphasize transparency, structured dialogue, and iterative refinement:

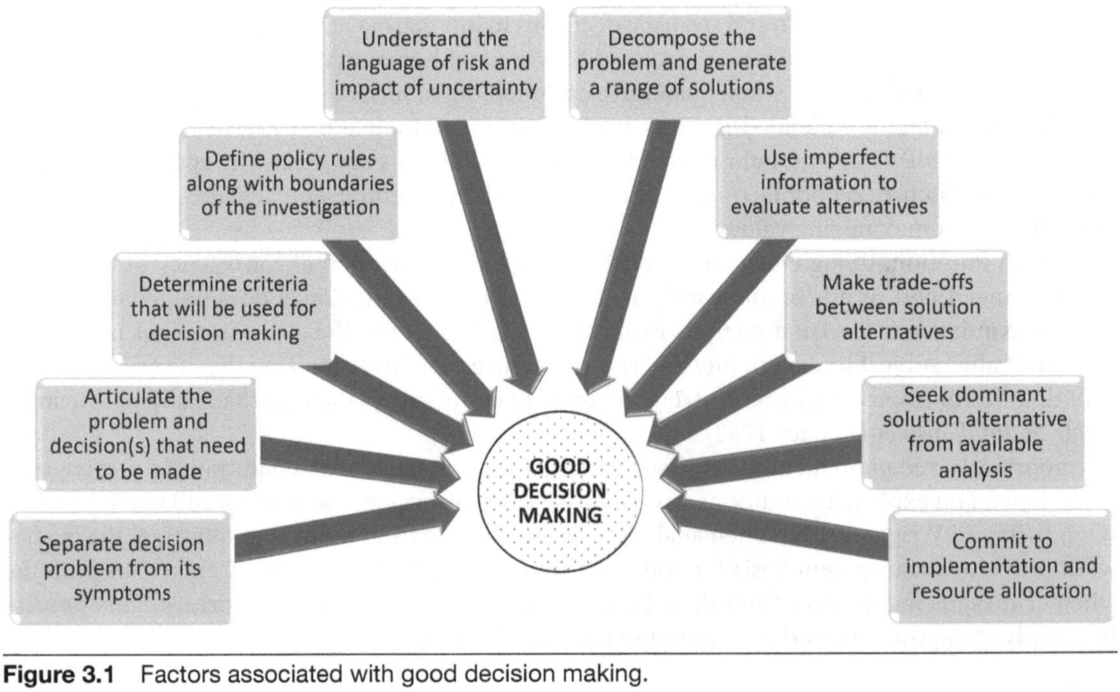

Figure 3.1 Factors associated with good decision making.

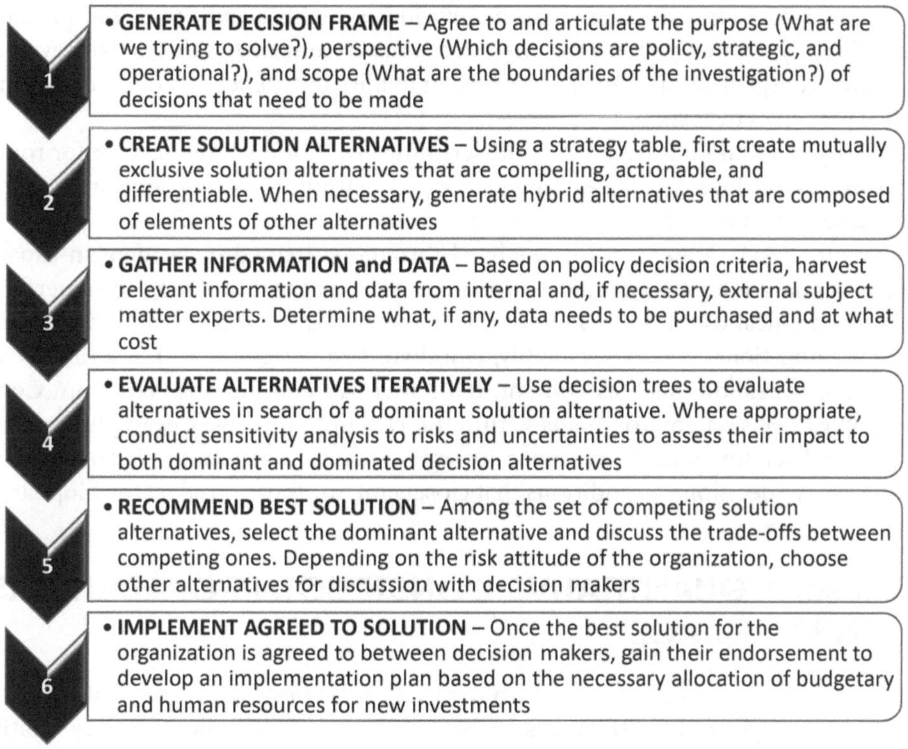

Figure 3.2 Six-step Decision Analysis Process (Strategic Decisions Group 2005; Skinner 1999).

- **Step 1: Establishing the decision frame:** The process begins by defining the decision problem, scope, boundaries, and authority. Clear alignment at this stage prevents misdirected analysis, inappropriate mixing of policy constraints with strategic choices, and premature narrowing of the solution space.
- **Step 2: Generating strategic alternatives:** A structured approach, often using strategy tables, is used to generate a manageable set of mutually exclusive, compelling, actionable, and differentiable alternatives that span the relevant strategic space.
- **Step 3: Information and data development:** Teams identify the data required to evaluate alternatives and apply value-of-information principles to ensure that the cost of additional data collection is justified by its expected benefit.
- **Step 4: Risk-adjusted evaluation and sensitivity analysis:** Alternatives are evaluated using formal decision models that explicitly incorporate uncertainty to identify those that dominate others in terms of expected financial value (expected value, EV) or, where appropriate, expected utility (EU). Sensitivity analyses identify key value drivers and determine whether further analysis is likely to change the decision.
- **Step 5: Recommendation and risk preference alignment:** While the highest EV or EU alternative may be recommended under risk neutrality, presenting leading options that illustrate value-risk trade-offs enables alignment with organizational risk preferences.
- **Step 6: Decision endorsement and organizational communication:** Transparent documentation supports executive endorsement and enables effective communication to stakeholders, reinforcing alignment and execution readiness.

Beyond individual decisions, decision analysis provides a durable audit trail that preserves institutional learning. By explicitly recording how decisions were framed, alternatives evaluated, and assumptions made, organizations can assess decision quality independently of outcomes and reproduce effective processes over time.

Decision analysis should not be applied indiscriminately. It offers limited value for routine, low-risk, or small-scale decisions. Its greatest impact arises in settings involving large, irreversible investments, long time horizons, and high uncertainty—conditions characteristic of biopharma research and development (R&D). When institutionalized through sustained investment in capability and culture, decision analysis enhances organizational decision maturity and long-term performance.

The practical application of decision analysis has been documented across a wide range of industries and organizational contexts. Notably, Goodwin and Wright (2010) describe its successful deployment in organizations such as DuPont, ICI Americas, and Phillips Petroleum Company, as well as in complex decision environments within the U.S. military. These examples underscore the versatility of the discipline and its relevance to organizations confronting high-stakes, uncertain, and capital-intensive decisions—conditions that closely mirror those faced by the biopharma sector.

3.2 RISK AS A QUANTITATIVE CONSTRUCT: PROBABILITY AS THE COMMON LANGUAGE

Statements such as "Project A is most likely to succeed" or "Project B is highly unlikely to succeed" are common in strategic and portfolio discussions. While such assertions convey relative beliefs about risk, they do not imply deterministic outcomes. Rather, they reflect judgments about long-run frequencies—suggesting that if similar projects were undertaken repeatedly, one would be expected to succeed more often than the other. In biopharma R&D, however, decisions are typically made once, not across repeated trials, creating the challenge of translating these beliefs into actionable guidance for irreversible decisions.

Decision analysis addresses this challenge by representing uncertainty explicitly through probability. Probability of success provides a numerical expression of the likelihood that a future event will occur and serves as a foundational input to quantitative decision models. Probabilities may be defined objectively or subjectively. Objective probabilities are derived from historical data as the ratio of observed outcomes to opportunities (X/Y) and are applicable when stable, repeatable processes and robust datasets exist. In contrast, many drug discovery and development decisions involve novel mechanisms, indications, or regulatory pathways, requiring the use of subjective probabilities. Subjective probabilities reflect informed judgment based on available data, scientific understanding, and experience, and are expressed on a scale from 0 to 1.0 (Skinner 1999).

By forcing uncertainty into a numerical framework, probability provides a common language for discussing risk. Quantification replaces ambiguous descriptors such as *likely* or *high risk* with explicit assumptions, reducing misinterpretation and enabling transparent dialogue. For example, *most likely to succeed* may imply a probability of 0.9 for one individual and slightly above 0.5 for another; numerical expression makes such differences explicit.

Resistance to assigning numerical probabilities arises for several reasons, including discomfort with linguistic precision, concern about overstating certainty, reputational risk if predictions prove incorrect, perceived insufficiency of information, and philosophical objections to probabilities for single events. Because subjective probabilities cannot be empirically validated for one-time outcomes, they cannot be *correct* in an absolute sense; they can only be judged on coherence, transparency, and defensibility given the information available at the time of decision making.

To address these challenges, decision analysis employs structured elicitation techniques to support disciplined and minimally biased probability assessment. Methods such as the five-step elicitation process described by Kirkwood (1996) improve consistency and credibility and ensure adherence to the axioms of probability theory, making estimates suitable for formal decision models.

CORE CONCEPTS IN PROBABILITY THEORY FOR DECISION ANALYSIS

Probability theory provides the formal language for representing uncertainty and ensuring internal consistency in decision models. Two structural concepts are essential when defining uncertain outcomes (Goodwin and Wright 2010):

- **Mutual exclusivity**—Events cannot occur simultaneously
- **Collective exhaustiveness**—All possible outcomes are represented, ensuring total probability sums to one

These concepts are grounded in Kolmogorov's axioms, which require probabilities to be nonnegative and normalized between 0 and 1.

The Addition Rule

For mutually exclusive events, the probability that one or the other occurs is the sum of their probabilities:

$$p(A \text{ or } B) = p(A) + p(B)$$

This rule is commonly applied in development timeline modeling. When events are not mutually exclusive, the joint probability must be subtracted to avoid double-counting:

$$p(A \text{ or } B) = p(A) + p(B) - p(A \text{ and } B)$$

This principle is illustrated using hypothetical epidemiological data on retired professional boxers (see Table 3.1). Accounting for overlap between brain trauma and Alzheimer's disease (AD) yields a combined probability of 0.60 for experiencing at least one of these outcomes:

$$\begin{gathered} p(\text{High level of insults or AD}) = p(\text{High level of insults}) + \\ p(\text{AD}) - p(\text{High level of insults and AD}) = \\ (550/1000) + (400/1000) - (350/1000) = 0.60 \end{gathered}$$

Table 3.1 Frequency of AD in 1,000 retired professional boxers.

	Low Level of Insults	High Level of Insults	Total
Alzheimer's Disease	50	350	400
No Alzheimer's Disease	400	200	600
Total	450	550	1,000

Complementary Events

For any event A, the probability of its complement is:

$$p(\bar{A}) = 1 - p(A)$$

This relationship ensures completeness in outcome modeling, such as defining failure probabilities from estimated probabilities of technical and regulatory success.

Marginal and Conditional Probabilities

Marginal probabilities describe the likelihood of an event irrespective of other conditions, while conditional probabilities reflect the likelihood given specific information. Using hypothetical patient data (see Table 3.2), the marginal probability of AD is 0.055 (55/1,000), while the conditional probability given a family history increases to 0.091 (40/440). Patients without a family history have a correspondingly lower conditional probability of 0.027 (15/560), demonstrating how conditioning factors influence risk.

Table 3.2 Results of a diagnosis of hypothetical patients for AD.

	Patients Diagnosed with Disease	Patients not Diagnosed with Disease	Total Number of Patients
Family history of disease	40	400	440
No family history of disease	15	545	560
Total number of patients	55	945	1,000

These distinctions underpin applications such as clinical success estimation, launch forecasting, and portfolio risk aggregation.

Bayesian Updating and Diagnostic Testing

In decision analysis, probabilities must be revised as new information becomes available. Consider a population in which 10% of individuals aged 65 or older have mild cognitive impairment (MCI). A PET imaging test has sensitivity (true positive rate) of 0.90 and specificity (true negative rate) of 0.95. The probability of a positive test result—regardless of whether the individual truly has MCI—is calculated by summing the probabilities of the two mutually exclusive pathways that lead to a positive result:

$$p(\text{Positive PET}) = p(\text{Positive PET} \mid \text{MCI}) \cdot p(\text{MCI}) + p(\text{Positive PET} \mid \text{No MCI}).$$

$p(\text{No MCI})$ i.e.,

$$p(\text{Positive PET}) = (0.90 \times 0.10) + (0.05 \times 0.90) = 0.135$$

This relationship is represented in an uncertainty tree (see Figure 3.3A).

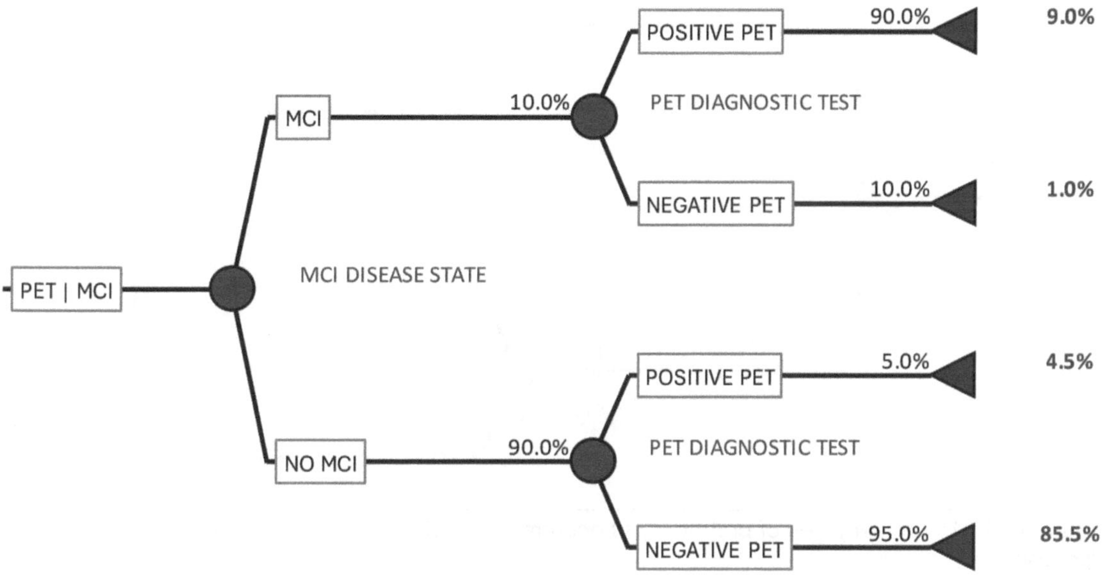

Figure 3.3A Uncertainty tree of probability of occurrence of MCI and predictability of PET diagnoses.

The posterior probability that an individual has MCI given a positive PET result (see Figure 3.3B) is calculated using conditional probability:

$$p(\text{MCI} \mid \text{Positive PET}) = \frac{p(\text{Positive PET} \mid \text{MCI}) \cdot p(\text{MCI})}{p(\text{Positive PET})}$$

$$= \frac{0.90 \times 0.10}{0.135} = 0.667$$

or Bayes' theorem:

$$p(\text{MCI} \mid \text{Positive PET})$$

$$= \frac{p(\text{Positive PET} \mid \text{MCI}) \cdot p(\text{MCI})}{[p(\text{Positive PET} \mid \text{MCI}) \cdot p(\text{MCI})] + [p(\text{Positive PET} \mid \text{No MCI}) \cdot p(\text{No MCI})]}$$

$$= \frac{0.90 \times 0.10}{(0.90 \times 0.10) + (0.05 \times 0.90)} = 0.667$$

Despite high test accuracy, the result highlights the critical influence of baseline prevalence on interpretation.

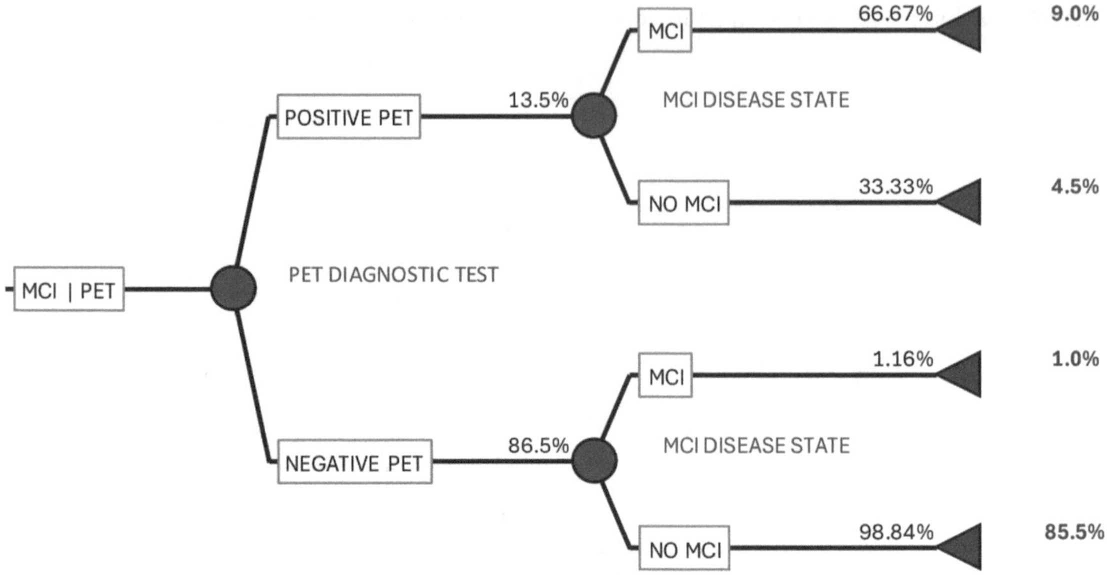

Figure 3.3B Uncertainty tree of probability of occurrence of MCI given the predictability of PET diagnoses.

Independent and Dependent Events

Events are independent if the occurrence of one does not affect the probability of the other:

$$p(A \mid B) = p(A) \text{ and } p(B \mid A) = p(B)$$

In such cases, joint probability is the product of individual probabilities:

$$p(A \text{ and } B) = p(A) \times p(B)$$

When events are dependent, joint probability must account for conditionality:

$$p(A \text{ and } B) = p(A) \times p(B \mid A)$$

This distinction is essential in modeling linked clinical outcomes, such as progression from MCI to AD.

Applied rigorously, these probability principles enable the systematic elicitation of risks associated with unique, nonrepeatable events, such as clinical trial transitions. While subjective, adherence to probability theory ensures internal consistency and analytical defensibility, providing a robust quantitative foundation for high-stakes decision making across the drug development life cycle.

3.3 UNCERTAINTY—THE LANGUAGE OF OUTCOMES

Once a strategic decision is executed, its consequences manifest as outcomes that are only partially influenceable and largely beyond managerial control. Outcomes rarely resolve to a single deterministic result; instead, they span a distribution of possible states. For example, post-launch commercial performance may range from strong to weak penetration, reflecting uncertainty rather than a binary success or failure. Organizations can influence outcome probabilities through decisions such as investment in medical affairs, promotion, or market access. However, they cannot control exogenous drivers, including prescriber behavior, payer decisions, competitive responses, or pricing dynamics. Consequently, control over decisions does not equate to control over outcomes, making uncertainty an inherent feature of realized performance.

Traditional project risk management has focused primarily on mitigating threats to execution. More recently, uncertainty management has expanded this view to include balancing downside risk with upside value creation (Goodwin and Wright 2010). Decision analysis is central to this perspective, as it explicitly models how alternative strategies reshape the distribution of possible outcomes and their associated risk-adjusted value. This is particularly relevant in biopharma, where aggressive strategies may increase both expected value and downside exposure, while conservative approaches may reduce variance at the cost of foregone upside.

Despite widespread acknowledgment of uncertainty, many organizations continue to rely on single *base case* forecasts expressed as point estimates. Such forecasts implicitly suppress uncertainty and raise critical questions: Does the base case represent the mean, median, or mode of the outcome distribution, or merely an informed guess? In practice, a point estimate is deterministic and functions as a prediction rather than an estimate, conveying a level of precision unsupported by evidence.

Interrogating a Point Estimate: A Commercial Example

Consider the statement: *Peak U.S. sales for drug ABC in 2035 are estimated at $1B.* This assertion prompts essential questions for decision making:

- Where does $1B lie within the distribution of possible outcomes—downside, most likely, or upside?
- What assumptions regarding pricing, access, promotion, and adoption underpin the estimate, and how sensitive is the result to these inputs?
- What competitive landscape is assumed, and what market share is required to achieve $1B?
- If uncertainty were explicitly modeled, how would the probability-weighted expected value compare to the stated figure?

The objective of addressing these questions is not analytical perfection, but determining how much uncertainty must be represented to support a decision aligned with organizational risk tolerance—particularly when decisions involve large, staged investments over long time horizons.

A common objection to uncertainty analysis is that it introduces additional assumptions and subjectivity. However, all estimates are inherently subjective. The relevant issue is not subjectivity per se, but whether assumptions are transparent, internally consistent, traceable, and defensible. Decision analysis does not eliminate subjectivity; it structures and exposes it to scrutiny, thereby improving decision quality even when outcomes remain uncertain.

EV Versus Risk Exposure

Consider two investment opportunities, each requiring $1,000 up front and each with an average payoff of $10,000. When evaluated solely on EV, a risk-neutral decision maker would be indifferent between them (see Figure 3.4A).

When uncertainty is introduced, however, the equivalence dissolves. As shown in Figure 3.4B, both investments retain the same EV, but their outcome distributions differ substantially.

Investment A exhibits no downside risk, while Investment B includes a 50% probability of zero payoff. Although EV alone cannot distinguish between them, their risk profiles are fundamentally different. This highlights a key limitation of EV as a sole decision criterion, particularly for large, infrequent, or irreversible investments.

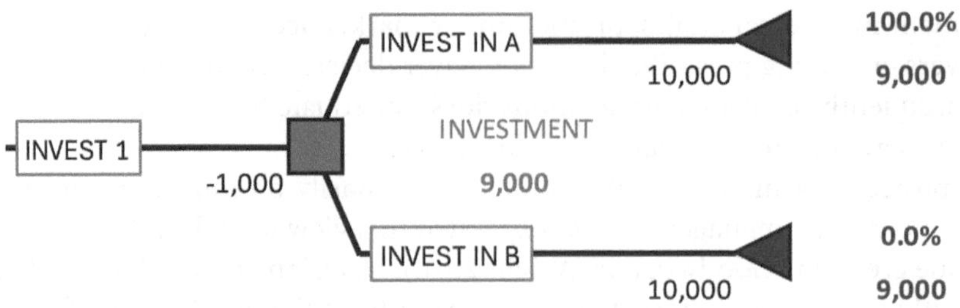

Figure 3.4A Investment opportunities with the same EV, version I.

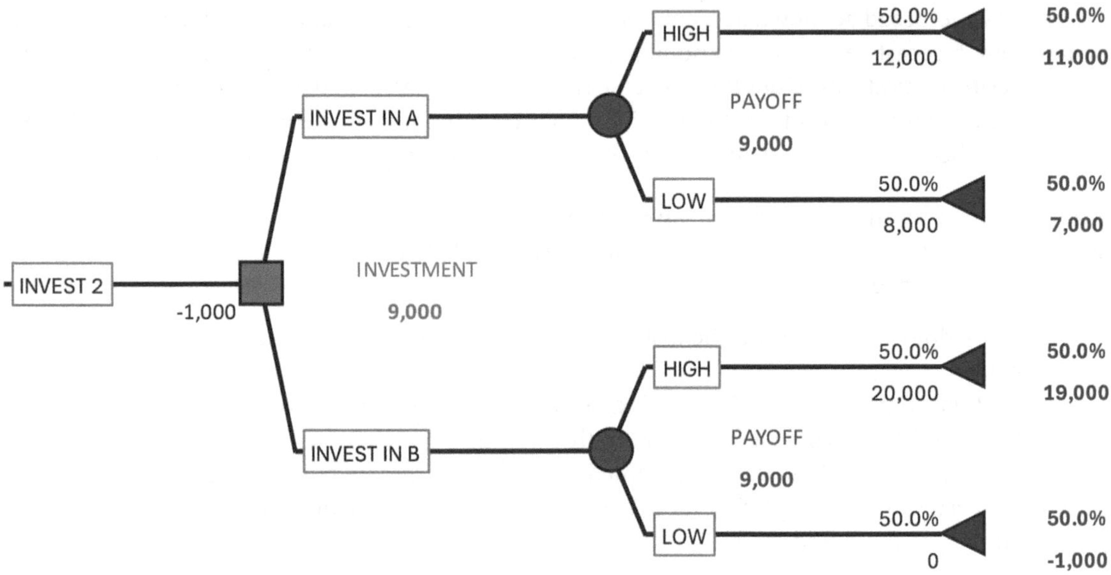

Figure 3.4B Investment opportunities with the same EV, version II.

Paradoxically, early-stage R&D projects—where uncertainty is highest—are often evaluated using deterministic point estimates of value, risk, cost, and time. These simplified business cases are frequently justified on the grounds that uncertainty will be addressed later. The authors argue the opposite: limited information should be represented as wider uncertainty ranges, not as unjustifiably precise estimates.

At a minimum, early business cases should specify ranges associated with defined confidence intervals (e.g., an 80% confidence range). Persistent patterns of underperformance—lower value realization, higher failure rates, cost overruns, and schedule delays—are not merely unfortunate; they are portfolio-level value destructive. Moreover, early-stage optimism combined with institutional resistance to termination contributes to the escalation of commitment and inefficient capital allocation.

A business case that explicitly represents uncertainty across value, risk, cost, and time does not reduce uncertainty, but it materially improves decisions about when, where, and how much to invest. As evidence accumulates, uncertainty ranges should narrow at a consistent confidence level, reflecting learning rather than increased analytical complexity. This approach reinforces the essential characteristics of robust business cases: transparency, traceability, consistency, and defensibility.

The role of uncertainty in shaping decision outcomes is explored further in the following section through *tornado sensitivity analysis*, which identifies the uncertainties that most strongly influence EV. By highlighting key value drivers, this tool guides targeted data collection and analysis prior to committing to high-risk, high-cost decisions.

3.4 DECISION MAKING UNDER CONDITIONS OF RISK AND UNCERTAINTY

Investment decisions in biopharma R&D are characterized by large, staged expenditures, long development timelines, and substantial technical, regulatory, and commercial uncertainty. Under such conditions, intuitive or deterministic decision approaches are often inadequate, motivating the use of formal decision analysis.

To illustrate, consider Drugs 'R' Us, which is evaluating two competing investment strategies:

- **Strategy A:** Invest in a lead compound for AD at a total cost of $350M
- **Strategy B:** Invest concurrently in the AD lead compound and a backup compound for generalized anxiety disorder (GAD) at a combined cost of $625M

Both strategies face uncertainty related to technical and regulatory success, as well as commercial performance driven by competitive dynamics, pricing, payer access, and adoption. Available information includes probabilities of technical success, probabilities of high or low commercial success conditional on technical success, and associated commercial values—summarized in Table 3.3.

Using these inputs, the strategies are represented as a decision tree (see Figure 3.5), beginning with a decision node followed by chance nodes capturing technical/regulatory outcomes and conditional commercial success.

Table 3.3 Risks and uncertainties associated with competing strategies in drug development and commercialization.

STRATEGY A			STRATEGY B		
Technical & Regulatory Outcome	High Commercial Success	Low Commercial Success	Technical & Regulatory Outcome	High Commercial Success	Low Commercial Success
p(Lead succeeds in AD) = 0.60	*p*(Success) = 0.75 Payoff = $1,000M	*p*(Success) = 0.25 Payoff = $800M	*p*(Lead succeeds in AD; Back-up fails in GAD) = 0.60 * 0.80 = 0.48	*p*(Success) = 0.50 Payoff = $1,000M	*p*(Success) = 0.50 Payoff = $800M
p(Lead fails in AD) = 0.40			*p*(Lead fails in AD; Back-up succeeds in GAD) = 0.40 * 0.20 = 0.08	*p*(Success) = 0.50 Payoff = $1,750M	*p*(Success) = 0.50 Payoff = $1,250M
			p(Lead succeeds in AD; Back-up succeeds in GAD) = 0.60 * 0.20 = 0.12	*p*(Success) = 0.50 Payoff = $2,750M	*p*(Success) = 0.50 Payoff = $2,050M
			p(Lead fails in AD; Back-up fails in GAD) = 0.40 * 0.80 = 0.32		

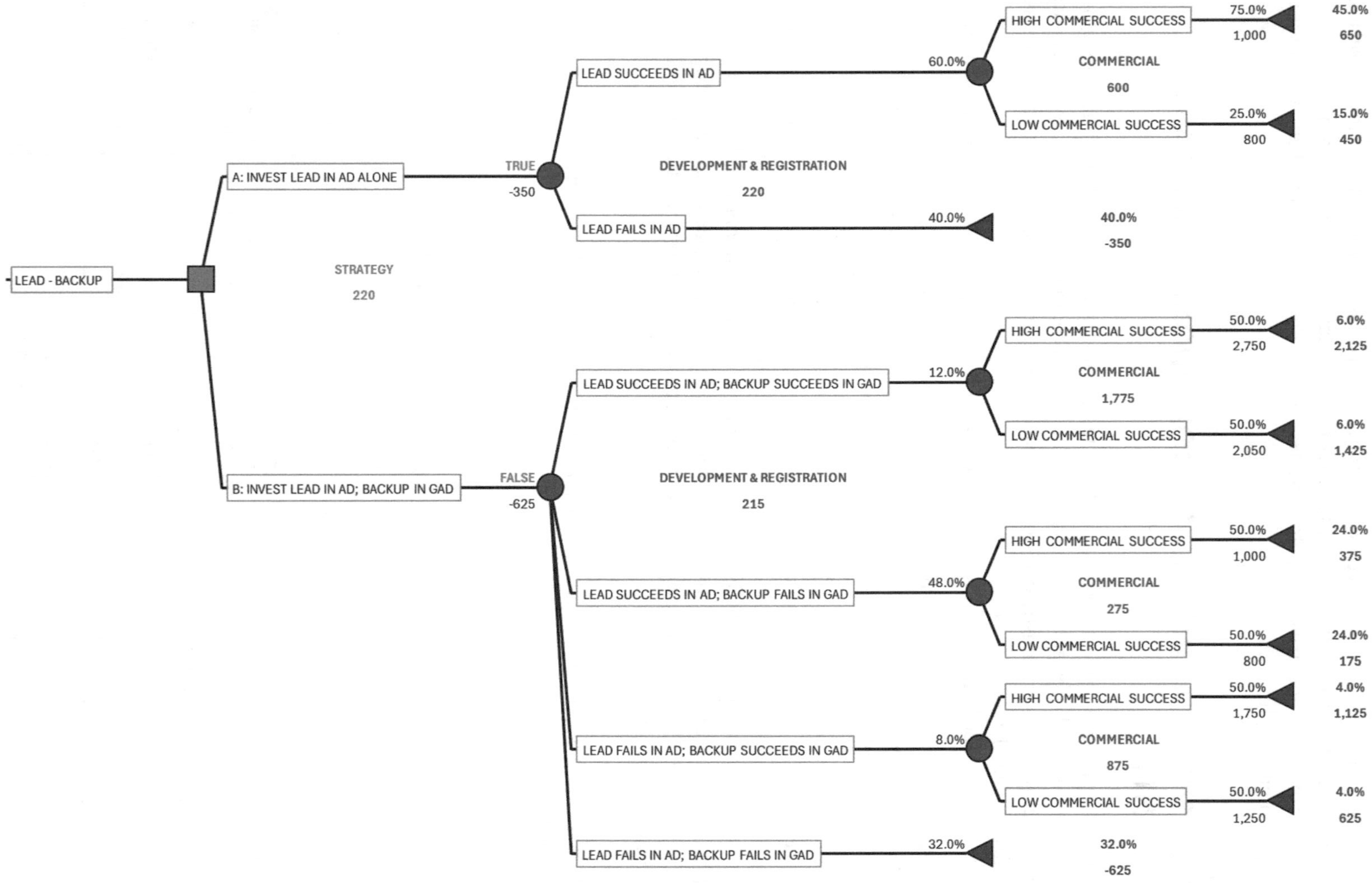

Figure 3.5 Decision tree of investment strategies related to (a) lead compound in AD alone or (b) lead and backup compounds in AD and GAD, respectively.

This structure ensures logical completeness, transparency of assumptions, and probabilistic representation of both risk and uncertainty. EVs are calculated by aggregating probability-weighted payoffs across terminal nodes. Rollback analysis using PrecisionTree® indicates that Strategy A is marginally dominant, with an EV of approximately $220M, compared with $215M for Strategy B (see Figure 3.6A). However, the small difference in EV provides limited confidence given the scale of investment and uncertainty involved.

A one-way sensitivity analysis of the probability of high commercial success for AD under Strategy A (base case = 0.75) shows that EV parity occurs at approximately 0.71, where both strategies converge near $215M (see Figure 3.6B). This result highlights the fragility of Strategy A's dominance and the importance of commercial execution risk in AD.

To explore interactions among uncertainties, a two-way sensitivity analysis, using Excel's data table function, varied both:

- The probability of high commercial success for AD alone, and
- The probability of high commercial success for the combined AD–GAD program.

Results summarized in Table 3.4 show regions of dominance, indifference, and reversal between strategies, demonstrating that modest shifts in either probability can alter the preferred decision.

A parallel two-way analysis on commercial values (see Table 3.5) indicates robust dominance regimes: Strategy B dominates when the AD-alone value is ≤ ~$950M, while Strategy A dominates when the AD-alone value exceeds ~$1.1B, with a conditional region between these thresholds.

Incorporating development risk, a final two-way analysis (see Table 3.6) links technical success probabilities for AD with commercial value assumptions for AD–GAD. High technical confidence (≥ ~0.70) favors Strategy A, while low confidence (≤ ~0.50) favors Strategy B, independent of commercial upside.

Collectively, these analyses demonstrate that strategic dominance is conditional rather than absolute, emerging from the interaction of success probabilities, commercial value, and capital at risk.

To move beyond deterministic sensitivity testing, the decision tree was extended to a stochastic framework using @Risk. Commercial values for AD, GAD, and AD–GAD were modeled using asymmetric triangular distributions (see Figure 3.7), enabling Monte Carlo or Latin Hypercube simulation of EV distributions.

Simulation results show that Strategy A yields EVs ranging from approximately $141M to $271M, with a mean of $210M, and a 36% probability of exceeding its deterministic EV (see Figure 3.8A).

For Strategy B, simulated EVs range from $113M to $257M, with a mean of $192M, and only a 12% probability of exceeding its deterministic EV (see Figure 3.8B).

Tornado analyses further identify the primary value drivers. For Strategy A, high commercial success in AD alone dominates EV variability (see Figure 3.9A), while for Strategy B, AD commercial performance under partial program failure is the largest contributor (see Figure 3.9B).

Overall, this integrated deterministic and stochastic decision analysis demonstrates that biopharma investment decisions should be evaluated probabilistically, with explicit attention to robustness, downside exposure, and the assumptions that most influence strategic dominance. For senior leadership, the principal value lies not in identifying a single *best* strategy, but in understanding the boundaries within which that strategy remains defensible under uncertainty.

A

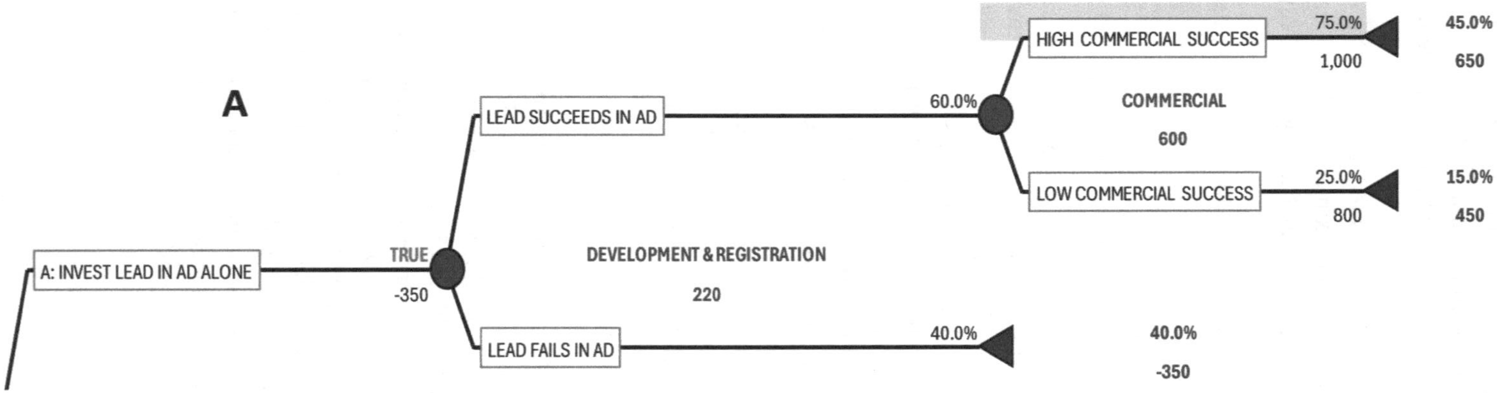

B

	PROBABILITY OF HIGH COMMERCIAL SUCCESS										
	0.65	0.66	0.67	0.68	0.69	0.70	0.71	0.72	0.73	0.74	0.75
220	208	209.2	210.4	211.6	212.8	214	215.2	216.4	217.6	218.8	220

Figure 3.6A/B One-way sensitivity analysis to the probability of high commercial success if the lead compound succeeds in AD.

Table 3.4 Two-way sensitivity analysis to the probability of high commercial success if the lead compound succeeds in AD alone and the lead and back-up compounds succeed in AD and GAD, respectively.

		PROBABILITY OF HIGH COMMERCIAL SUCCESS IN AD ALONE										
	AD ALONE	0.0	0.1	0.2	0.3	0.4	0.5	0.6	0.7	0.8	0.9	1.0
PROBABILITY OF HIGH COMMERCIAL SUCCESS IN AD & GAD	0.0	AD & GAD	AD & GAD	AD & GAD	AD & GAD	AD ALONE	AD ALONE	AD ALONE	AD ALONE	AD ALONE	AD ALONE	AD ALONE
	0.1	AD & GAD	AD & GAD	AD & GAD	AD & GAD	AD & GAD	AD ALONE	AD ALONE	AD ALONE	AD ALONE	AD ALONE	AD ALONE
	0.2	AD & GAD	AD & GAD	AD & GAD	AD & GAD	AD & GAD	AD ALONE	AD ALONE	AD ALONE	AD ALONE	AD ALONE	AD ALONE
	0.3	AD & GAD	AD & GAD	AD & GAD	AD & GAD	AD & GAD	AD & GAD	AD ALONE	AD ALONE	AD ALONE	AD ALONE	AD ALONE
	0.4	AD & GAD	AD & GAD	AD & GAD	AD & GAD	AD & GAD	AD & GAD	AD & GAD	AD ALONE	AD ALONE	AD ALONE	AD ALONE
	0.5	AD & GAD	AD & GAD	AD & GAD	AD & GAD	AD & GAD	AD & GAD	AD & GAD	AD & GAD	AD ALONE	AD ALONE	AD ALONE
	0.6	AD & GAD	AD & GAD	AD & GAD	AD & GAD	AD & GAD	AD & GAD	AD & GAD	AD & GAD	AD ALONE	AD ALONE	AD ALONE
	0.7	AD & GAD	AD & GAD	AD & GAD	AD & GAD	AD & GAD	AD & GAD	AD & GAD	AD & GAD	AD & GAD	AD ALONE	AD ALONE
	0.8	AD & GAD	AD & GAD	AD & GAD	AD & GAD	AD & GAD	AD & GAD	AD & GAD	AD & GAD	AD & GAD	AD & GAD	AD ALONE
	0.9	AD & GAD	AD & GAD	AD & GAD	AD & GAD	AD & GAD	AD & GAD	AD & GAD	AD & GAD	AD & GAD	AD & GAD	AD ALONE
	1.0	AD & GAD	AD & GAD	AD & GAD	AD & GAD	AD & GAD	AD & GAD	AD & GAD	AD & GAD	AD & GAD	AD & GAD	AD & GAD

Table 3.5 Two-way sensitivity analysis to the value of high commercial success if the lead compound succeeds in AD alone, and the lead and back-up compounds succeed in AD and GAD, respectively.

		HIGH COMMERCIAL VALUE IN AD ALONE										
	AD ALONE	850	900	950	1,000	1,050	1,100	1,150	1,200	1,250	1,300	1,350
HIGH COMMERCIAL VALUE IN AD & GAD	2,500	AD & GAD	AD & GAD	AD & GAD	AD ALONE	AD ALONE	AD ALONE	AD ALONE	AD ALONE	AD ALONE	AD ALONE	AD ALONE
	2,600	AD & GAD	AD & GAD	AD & GAD	AD ALONE	AD ALONE	AD ALONE	AD ALONE	AD ALONE	AD ALONE	AD ALONE	AD ALONE
	2,700	AD & GAD	AD & GAD	AD & GAD	AD ALONE	AD ALONE	AD ALONE	AD ALONE	AD ALONE	AD ALONE	AD ALONE	AD ALONE
	2,800	AD & GAD	AD & GAD	AD & GAD	AD ALONE	AD ALONE	AD ALONE	AD ALONE	AD ALONE	AD ALONE	AD ALONE	AD ALONE
	2,900	AD & GAD	AD & GAD	AD & GAD	AD & GAD	AD ALONE	AD ALONE	AD ALONE	AD ALONE	AD ALONE	AD ALONE	AD ALONE
	3,000	AD & GAD	AD & GAD	AD & GAD	AD & GAD	AD ALONE	AD ALONE	AD ALONE	AD ALONE	AD ALONE	AD ALONE	AD ALONE
	3,100	AD & GAD	AD & GAD	AD & GAD	AD & GAD	AD ALONE	AD ALONE	AD ALONE	AD ALONE	AD ALONE	AD ALONE	AD ALONE
	3,200	AD & GAD	AD & GAD	AD & GAD	AD & GAD	AD ALONE	AD ALONE	AD ALONE	AD ALONE	AD ALONE	AD ALONE	AD ALONE
	3,300	AD & GAD	AD & GAD	AD & GAD	AD & GAD	AD & GAD	AD ALONE	AD ALONE	AD ALONE	AD ALONE	AD ALONE	AD ALONE
	3,400	AD & GAD	AD & GAD	AD & GAD	AD & GAD	AD & GAD	AD ALONE	AD ALONE	AD ALONE	AD ALONE	AD ALONE	AD ALONE
	3,500	AD & GAD	AD & GAD	AD & GAD	AD & GAD	AD & GAD	AD ALONE	AD ALONE	AD ALONE	AD ALONE	AD ALONE	AD ALONE

Table 3.6 Two-way sensitivity analysis to the probability of high commercial success if the lead compound succeeds in AD alone, and the value of high commercial success if the lead and back-up compounds succeed in AD and GAD, respectively.

	PROBABILITY OF SUCCESS IN AD ALONE											
	AD ALONE	0.0	0.1	0.2	0.3	0.4	0.5	0.6	0.7	0.8	0.9	1.0
HIGH COMMERCIAL VALUE IN AD & GAD	2,500	AD & GAD	AD & GAD	AD & GAD	AD & GAD	AD & GAD	AD & GAD	AD ALONE	AD ALONE	AD ALONE	AD ALONE	AD ALONE
	2,600	AD & GAD	AD & GAD	AD & GAD	AD & GAD	AD & GAD	AD & GAD	AD ALONE	AD ALONE	AD ALONE	AD ALONE	AD ALONE
	2,700	AD & GAD	AD & GAD	AD & GAD	AD & GAD	AD & GAD	AD & GAD	AD ALONE	AD ALONE	AD ALONE	AD ALONE	AD ALONE
	2,800	AD & GAD	AD & GAD	AD & GAD	AD & GAD	AD & GAD	AD & GAD	AD ALONE	AD ALONE	AD ALONE	AD ALONE	AD ALONE
	2,900	AD & GAD	AD & GAD	AD & GAD	AD & GAD	AD & GAD	AD & GAD	AD & GAD	AD ALONE	AD ALONE	AD ALONE	AD ALONE
	3,000	AD & GAD	AD & GAD	AD & GAD	AD & GAD	AD & GAD	AD & GAD	AD & GAD	AD ALONE	AD ALONE	AD ALONE	AD ALONE
	3,100	AD & GAD	AD & GAD	AD & GAD	AD & GAD	AD & GAD	AD & GAD	AD & GAD	AD ALONE	AD ALONE	AD ALONE	AD ALONE
	3,200	AD & GAD	AD & GAD	AD & GAD	AD & GAD	AD & GAD	AD & GAD	AD & GAD	AD ALONE	AD ALONE	AD ALONE	AD ALONE
	3,300	AD & GAD	AD & GAD	AD & GAD	AD & GAD	AD & GAD	AD & GAD	AD & GAD	AD ALONE	AD ALONE	AD ALONE	AD ALONE
	3,400	AD & GAD	AD & GAD	AD & GAD	AD & GAD	AD & GAD	AD & GAD	AD & GAD	AD ALONE	AD ALONE	AD ALONE	AD ALONE
	3,500	AD & GAD	AD & GAD	AD & GAD	AD & GAD	AD & GAD	AD & GAD	AD & GAD	AD ALONE	AD ALONE	AD ALONE	AD ALONE

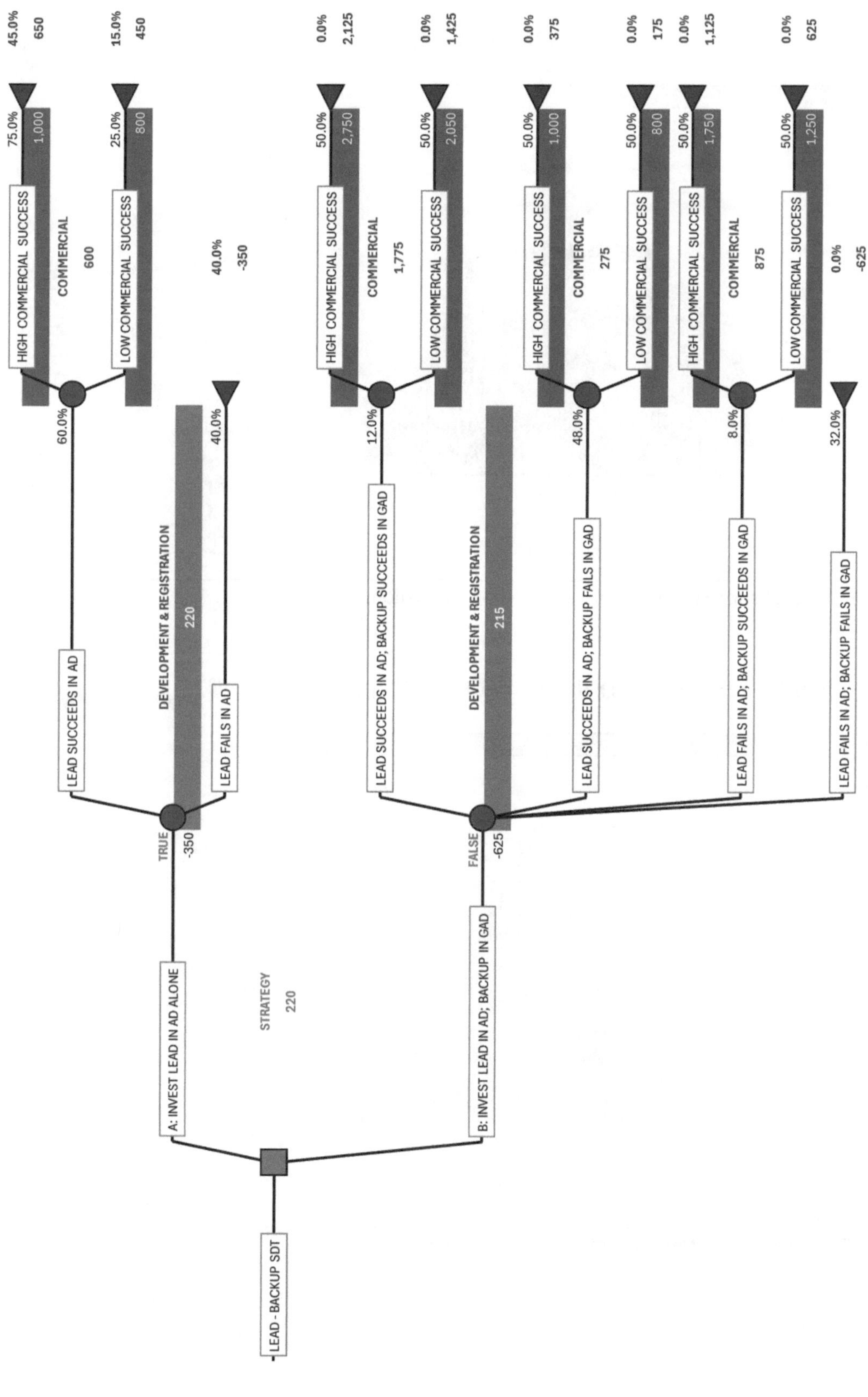

Figure 3.7 Conversion of deterministic decision tree to a stochastic decision tree (SDT), using range estimates to represent uncertain commercial values.

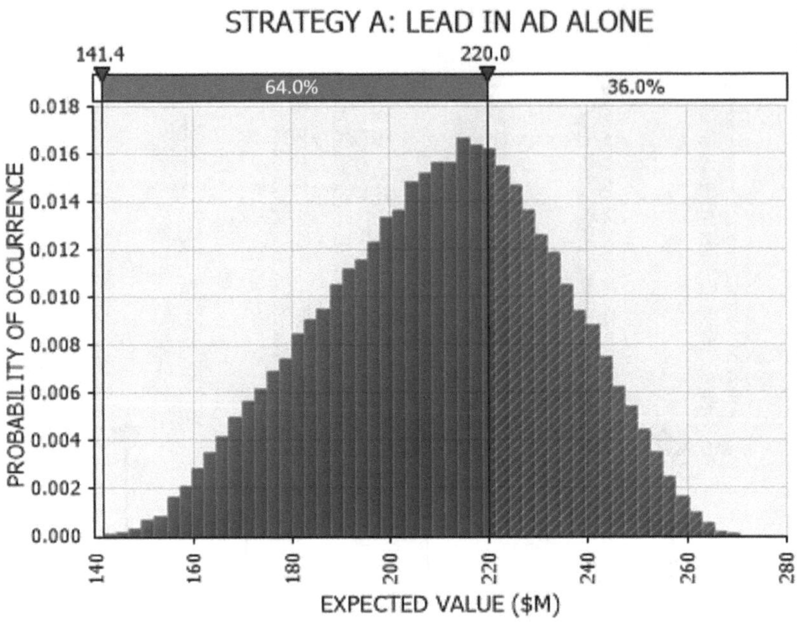

Statistics	LEAD AD ALONE
Cell	C8
Minimum	141.42
Maximum	270.65
Mean	210.00
90% CI	± 0.123
Mode	216.63
Median	211.43
Std Dev	23.57
Skewness	-0.1825
Kurtosis	2.4504
Values	100000
Errors	0
Filtered	0
Left X	141.4
Left P	0.0%
Right X	220.0
Right P	64.0%

Figure 3.8A Results for Strategy A of decision tree simulation.

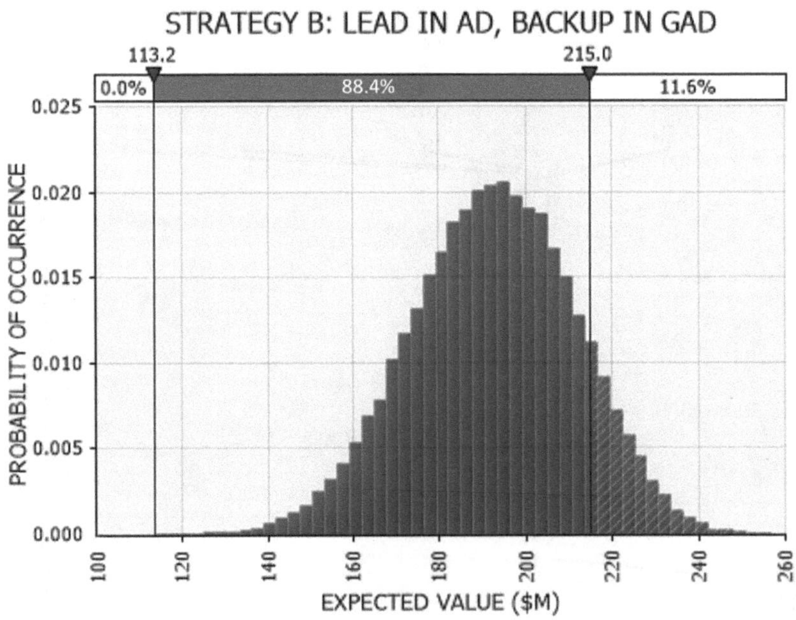

Statistics	LEAD AD, BACKUP ...
Cell	C20
Minimum	113.23
Maximum	256.61
Mean	192.33
90% CI	± 0.0986
Mode	195.93
Median	192.78
Std Dev	18.96
Skewness	-0.1297
Kurtosis	2.8350
Values	100000
Errors	0
Filtered	0
Left X	113.2
Left P	0.0%
Right X	215.0
Right P	88.4%

Figure 3.8B Results for Strategy B of decision tree simulation.

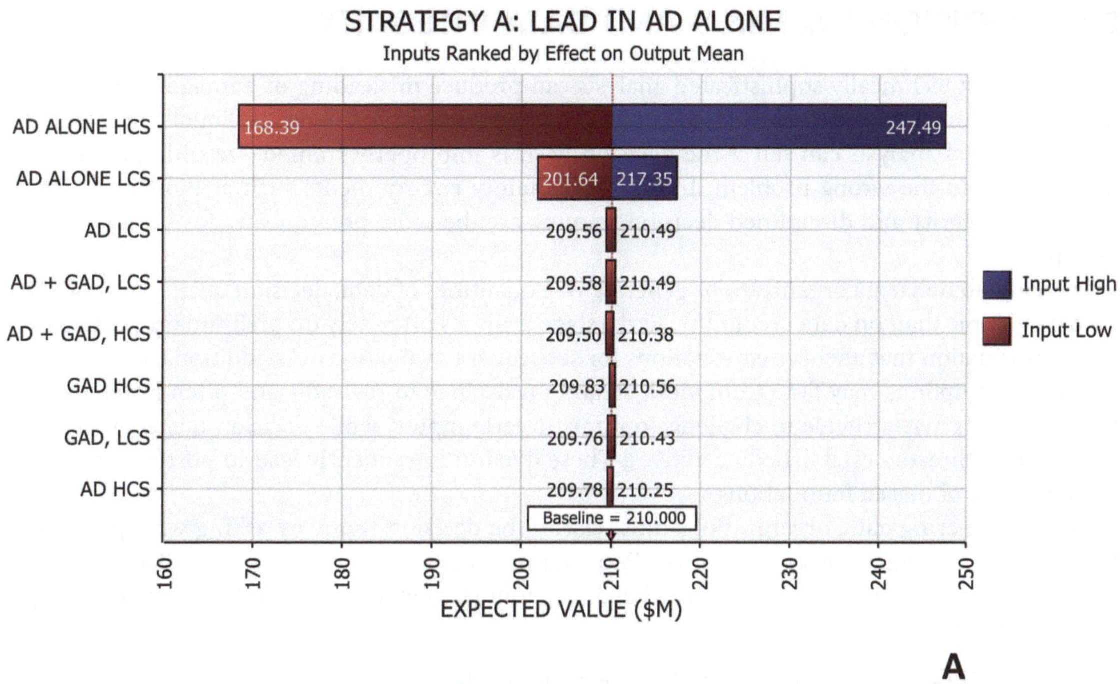

Figure 3.9A Tornado sensitivity analysis of commercial uncertainties to the EV of Strategy A.

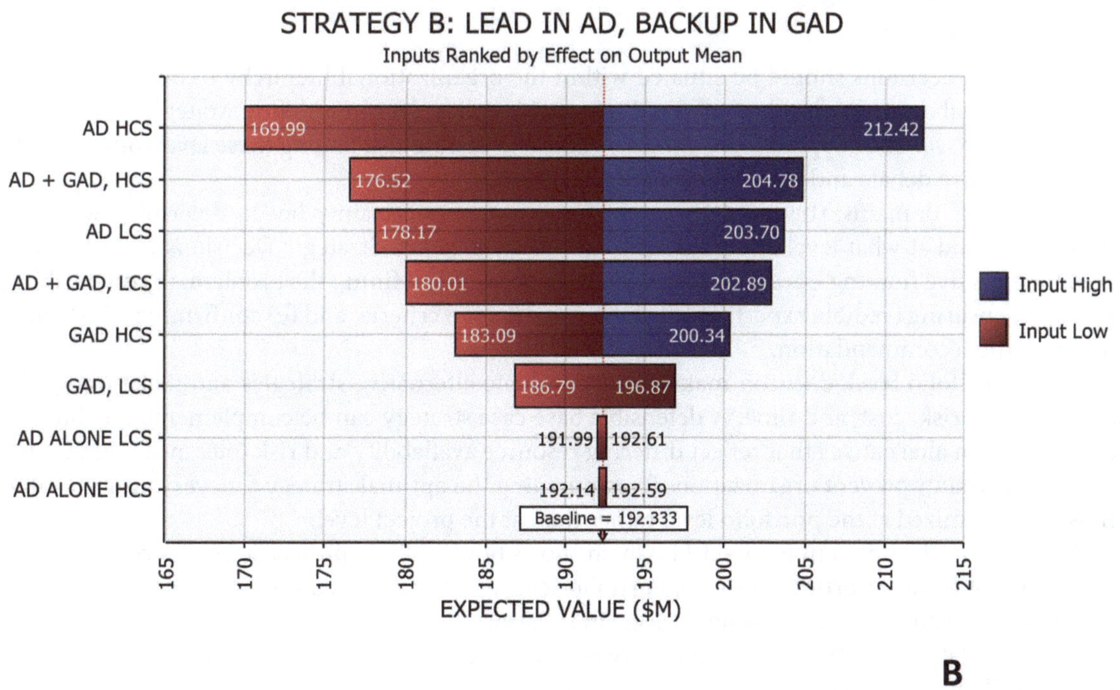

Figure 3.9B Tornado sensitivity analysis of commercial uncertainties to the EV of Strategy B.

3.5 DECISION FRAMING AND DATA INTEGRITY

Even the most technically sophisticated analysis can produce misleading or damaging recommendations if the underlying data are biased, logically inconsistent, or indefensible. Equally problematic, a well-executed analysis can fail if the decision itself is improperly framed—resulting in an optimal solution to the wrong problem. In high-uncertainty environments such as biopharma R&D, both data integrity and disciplined decision framing are therefore prerequisites for sound decision making.

Although biopharma organizations generate vast quantities of data, decision quality depends less on data volume than on data credibility. Early-stage analyses often rely on preliminary, partially filtered information that anchors expectations for subsequent evaluations. As additional data emerge, original assumptions may fade from view, creating resistance to revised—and often lower—valuations. Analysts may struggle to challenge optimistic early inputs, while decision makers may resist updates that threaten established narratives. These dynamics frequently lead to portfolio decisions built on weak or biased foundations.

Before gathering data, organizations must clarify the decision frame by asking what problem is being solved and what clarity is required. Poor framing leads to misdirected analysis regardless of technical rigor. Effective framing requires balancing four elements (Celona and McNamee, Strategic Decisions Group 2005):

- **Breadth**—Avoiding both excessive scope and overly narrow focus
- **Complexity**—Preventing analytical overload without oversimplifying material risks
- **Focus**—Ensuring attention is directed to unresolved, decision-relevant issues
- **Involvement**—Engaging appropriate subject matter experts (SMEs) without fostering bias or groupthink

Once framed, decisions should be situated within the organizational hierarchy of policy, strategic, and operational decisions, as illustrated in Figure 3.10. Policy decisions define strategic boundaries, which, in turn, constrain operational choices. Failure to distinguish among these levels often results in unproductive debate and misaligned analytical effort.

In high-risk domains, the objective is not to eliminate uncertainty but to determine what information—and at what level of confidence—is sufficient to act. Strategic Decisions Group (2005) recommends five framing questions to guide this process: (a) defining the problem, (b) establishing scope, (c) ensuring credible expertise, (d) clarifying decision criteria, and (e) confirming willingness to act on the recommendation.

At the portfolio level, decision makers must evaluate alternative strategies and make trade-offs among value, risk, cost, and time. A defensible base case strategy can be complemented by buy-up and buy-down alternatives that reflect differing resource availability and risk tolerance. Because resource constraints prevent organizations from pursuing the optimal strategy for every project, value must be maximized at the portfolio level rather than at the project level.

This logic is illustrated in Figure 3.11, which shows how multiple project strategies combine into a larger set of feasible portfolio options. Portfolio-level optimization requires selecting among these combinations rather than evaluating projects in isolation.

Ensuring data integrity under uncertainty requires prioritizing robustness over precision. *Good data* should satisfy three criteria:

- **Traceability**—Clear provenance, assumptions, and rationale
- **Logical consistency**—Internal coherence across estimates and models
- **Defensibility**—Justifiable reasoning based on evidence or analogs

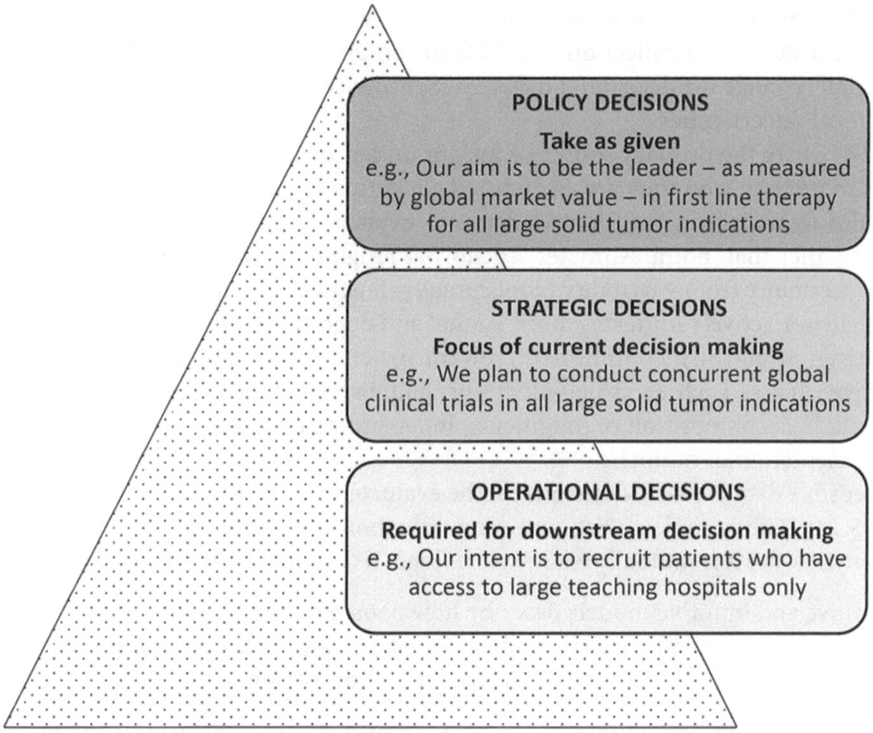

Figure 3.10 Hierarchy of policy, strategic, and operational decisions (Strategic Decisions Group 2005).

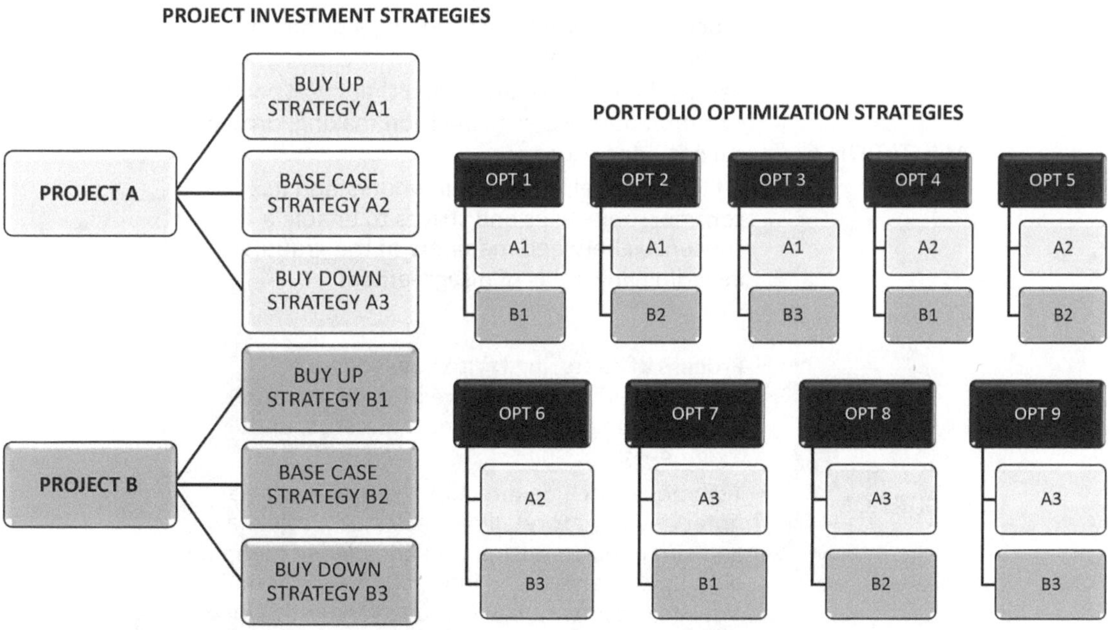

Figure 3.11 Portfolio decision making based on optimization from project investment strategies.

Robustness does not require precise estimates; excessive precision often leads to diminishing analytical returns. Instead, data collection should focus on decision-relevant information—inputs that would materially change the decision if different. Sensitivity tools such as tornado charts help identify these critical uncertainties.

Because SMEs are the primary source of insight under uncertainty, how their knowledge is elicited matters more than who they are. Structured elicitation techniques—distinct from informal facilitation—aim to transform tacit knowledge into explicit, testable inputs. Best practices include using ranges rather than point estimates, anchoring judgments in reference classes, and explicitly separating uncertainty from variability (conceptually illustrated in Figure 3.12).

A skilled elicitor actively mitigates motivational and cognitive biases (see Figure 3.13), including optimism driven by personal commitment, reward structures that support favorable estimates, and heuristic-driven biases such as availability, representativeness, and anchoring. Effective countermeasures include decomposition of judgments, independent estimates, iterative feedback, and systematic documentation of rationale.

With defensible data in place, decisions can be evaluated using different decision-making models (see Figure 3.14). These models—rational economic, bounded rational, cognitive, adaptive, intuitive, and analytic—are not mutually exclusive. In high-risk domains like drug development:

- Cognitive and intuitive models describe how people *naturally* think
- Rational economic and analytic models describe how people *ought* to decide
- Bounded rational and adaptive models describe how organizations *actually* decide over time

In practice, organizations blend them. Decision analysis does not replace intuition or experience; it externalizes and tests them, improving decision quality under uncertainty.

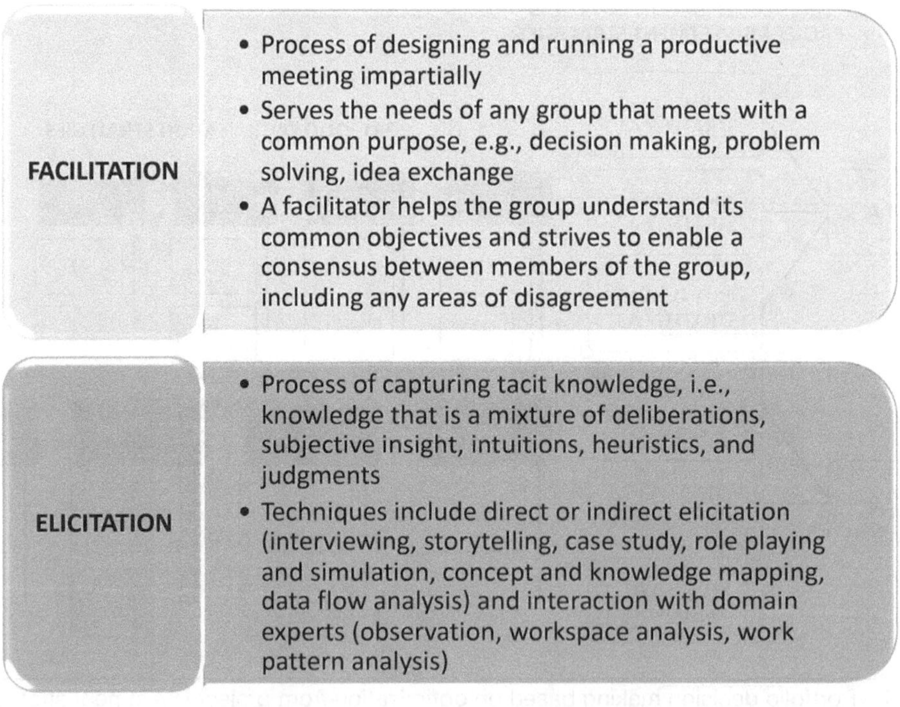

Figure 3.12 Facilitation and elicitation.

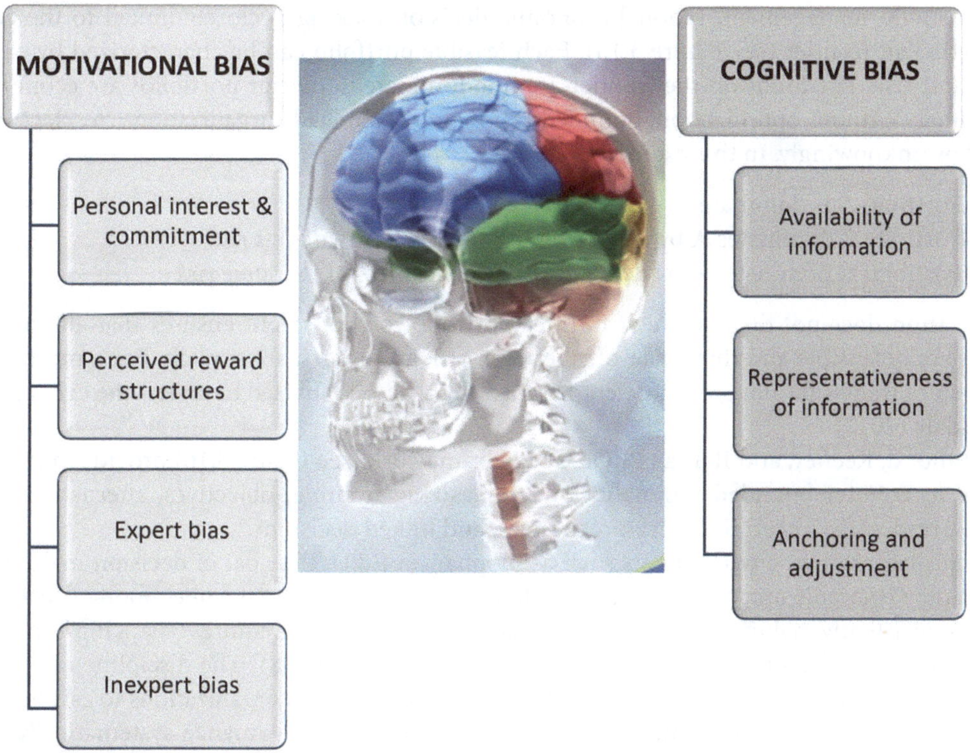

Figure 3.13 Motivational and cognitive biases.

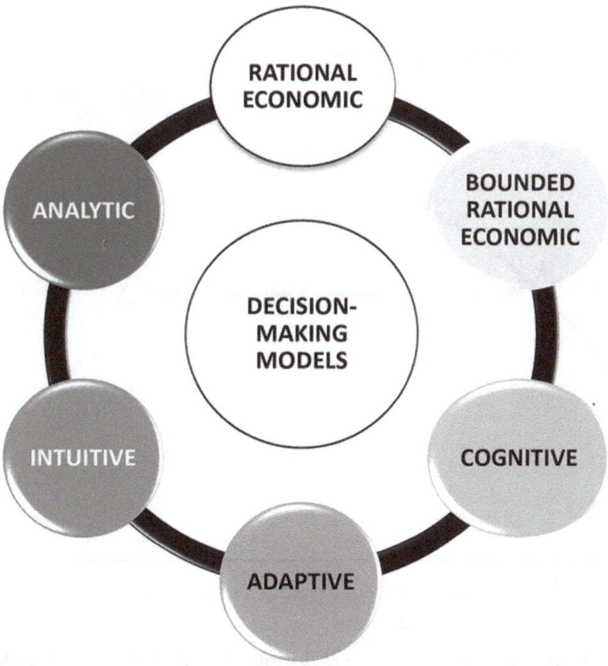

Figure 3.14 Decision-making models.

In portfolio management, rational economic decision making is closely linked to the concept of the efficient frontier (see Figure 3.15). Each feasible portfolio can be characterized by expected return and risk. Portfolios on the frontier are nondominated; interior portfolios are economically suboptimal. Without optimization, bounded rationality often leads organizations to select inferior portfolios unknowingly. In this example:

- Portfolio A is dominated
- Portfolio B dominates A by delivering the *same return with less risk*
- Portfolio C dominates A by delivering a *higher return for the same risk*

Optimization does not dictate which frontier portfolio to choose but ensures that all considered options are defensible and that trade-offs between risk and return are explicit. For a more detailed treatment of trade-offs along the efficient frontier, the reader is advised to review the excellent work of Ragsdale (2017).

Hammond, Keeney, and Raiffa's PROACT.url framework (see Figure 3.16) provides a structured method to counter bounded rationality by emphasizing framing, objectives, alternatives, consequences, trade-offs, uncertainty, risk preferences, and linked decisions.

In high-uncertainty environments such as biopharma R&D, the goal of decision making is not to guarantee favorable outcomes—an impossibility—but to maximize the likelihood of success over time by consistently making high-quality, defensible decisions. This requires balancing Type I errors (advancing failures) and Type II errors (terminating future winners). The discipline of structured framing, robust data, and explicit uncertainty management enables organizations to assess decision quality independently of outcomes and to improve portfolio performance systematically despite irreducible uncertainty.

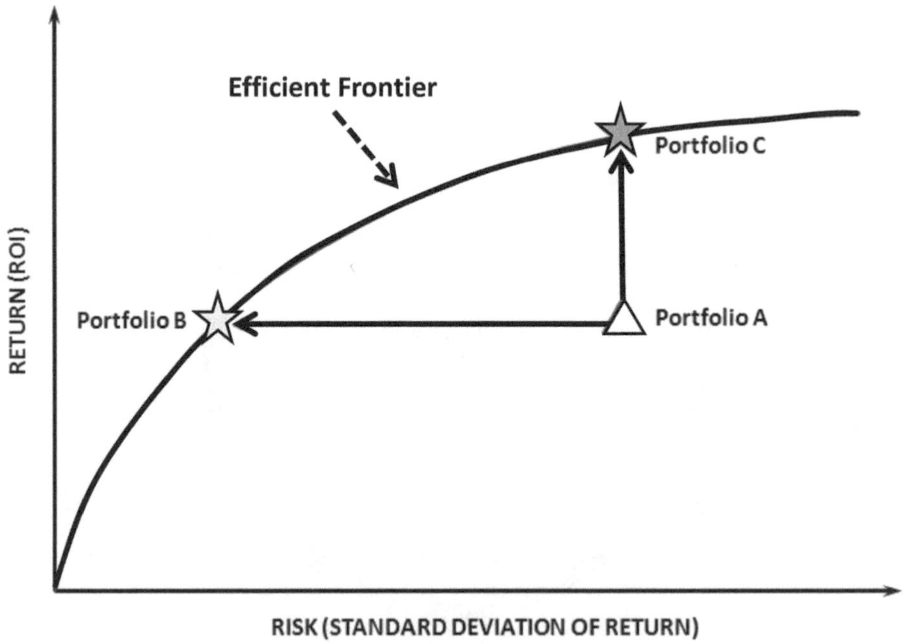

Figure 3.15 The efficient frontier and bounded rational economic decision making.

Pr	Work on the right decision **problem**, define its scope, and frame it well, i.e., determine what needs to be decided on. Avoid unwarranted assumptions and unnecessary complexity.
O	Determine what the **objectives** of the solution are, i.e., what needs to be accomplished and what issues (including conflicting goals) are most relevant to achieving the desired goals.
A	Create compelling, actionable, and defensible **alternative** solutions from which may emerge hybrid strategies; the best decision can be no better than the best alternative strategy.
C	Assess the **consequences** of each alternative solution (preferably in terms of value, risk, cost, and time) to help determine which alternative best meets the desired objectives.
T	Be prepared to make **trade-offs** between alternative solutions and conflicting objectives. Great hybrid solutions may exist but rarely does a perfect solution emerge.
u	Accept less-than-perfect knowledge and quantify the impacts of a range of possible **uncertainties**; like risk, uncertainty does not have to be an enemy of decision making.
r	Understand **risk attitudes/preferences/tolerances** to criteria that affect decision making. The greater the investment, longer the timeline, and higher the risk, there is less risk neutrality.
l	Consider how the current decision(s) could impact future, **linked decisions**. It is easier to uncouple the past from the future than it is the future from the present.

Figure 3.16 The PROACT.url framework for decision making (Hammond et al. 1999).

3.6 PORTFOLIO MANAGEMENT

In research-driven organizations—particularly in biopharma—portfolio management is typically framed as *project portfolio management* (PPM) to distinguish it from project management (PM). While PM focuses on execution efficiency at the individual project level, PPM addresses a fundamentally different question: how to allocate finite organizational resources across competing development opportunities to maximize enterprise value over time under uncertainty.

In biopharma, portfolio management can be rigorously defined as the continuous, data-driven governance of a collection of assets spanning discovery, clinical programs, life-cycle extensions, and enabling initiatives. This governance explicitly incorporates capital constraints, capacity limitations, and organizational tolerance for technical, regulatory, and commercial risk. In larger organizations, this function is often formalized as *enterprise project portfolio management* (EPPM), reflecting its role in linking R&D strategy, finance, and enterprise-level decision making.

Although portfolio management concepts originated in financial asset management, they have increasingly been adapted to nonfinancial portfolios characterized by long development timelines, high attrition rates, and asymmetric payoffs—hallmarks of biopharma R&D. In practice, discussing portfolios in terms of projects and programs provides a more operationally relevant framework, as projects serve as the primary vehicles for translating scientific hypotheses into future revenue.

Accordingly, the terms portfolio management, PPM, and EPPM are used interchangeably here, reflecting a pragmatic focus on R&D projects as the dominant value-creating assets.

3.6.1 The CREOPM™ Framework for R&D Portfolio Management

Across R&D-intensive organizations, portfolio decision making spans a continuum from highly centralized, personality-driven authority to formalized, committee-based governance supported by dedicated analytical teams. Regardless of structure, some form of portfolio management is almost always present. The critical question is not whether portfolio management exists, but how effectively it creates enterprise value—and how much value is lost through weak decision processes.

Even successful organizations often exhibit complacency, reinforced by beliefs such as reliance on seasoned judgment, skepticism toward formal analysis, or the perception that uncertainty renders structured portfolio management ineffective. A recurring observation is that the principal barrier to robust portfolio management is not technical complexity, but a failure to recognize its leverage as a value-creating capability. Portfolio management is frequently mischaracterized as administrative overhead rather than a strategic function. Dysfunction commonly manifests at two extremes: (1) portfolio groups that focus on data aggregation and reporting without analytical challenge and (2) those that produce extensive analyses disconnected from executive decision making. In both cases, the appearance of rigor masks a limited impact on decision quality.

To address these challenges, the CREOPM™ framework, developed by PPVC, is proposed as an end-to-end, decision-centric model for R&D portfolio management (see Figure 3.17). The framework emphasizes logical sequencing, interdependencies, and continuous feedback across the portfolio life cycle.

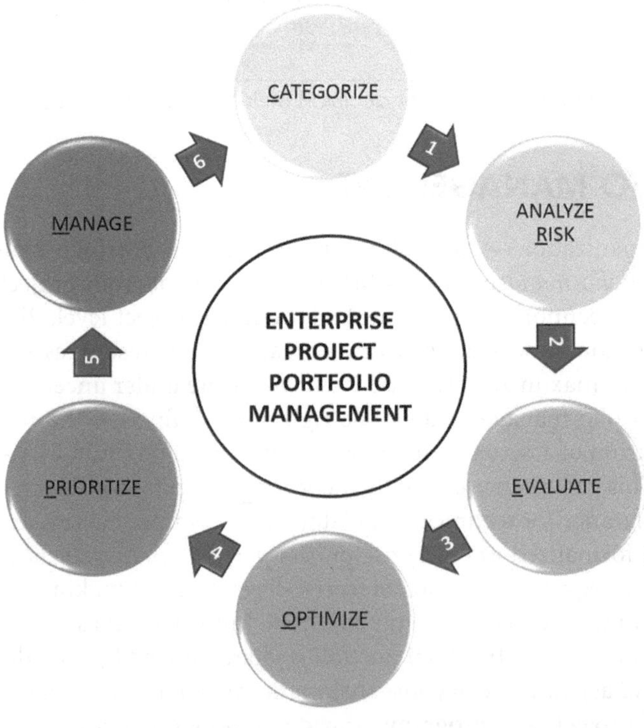

Figure 3.17 The CREOPM™ framework for portfolio management.

The CREOPM™ framework consists of six integrated components:

1. **Categorize**—Classify projects by strategic intent, development stage, risk profile, and value drivers
2. **Analyze risk**—Quantitatively and qualitatively assess technical, regulatory, commercial, and execution risks
3. **Evaluate**—Conduct integrated valuation using consistent economic, scientific, and strategic criteria
4. **Optimize**—Balance value, risk, and resource constraints at the portfolio level
5. **Prioritize**—Rank projects to guide funding, sequencing, and resource deployment
6. **Manage**—Continuously govern and adapt the portfolio as assumptions and evidence evolve

Together, these elements form a closed-loop framework that elevates portfolio management from reporting to a core strategic capability.

Categorize
Effective portfolio management begins with explicit classification of all projects into *must-do*, *may-do*, or *won't-do* categories. *Must do* initiatives are nondiscretionary obligations (e.g., regulatory commitments), *won't-do* initiatives no longer justify investment, and *may-do* projects represent discretionary choices competing for limited resources. This categorization defines the organization's true decision space and must be revisited dynamically as data and priorities evolve.

Analyze Risk
Rigorous portfolio decisions require quantitative risk assessment, particularly for technical uncertainty inherent in biopharma R&D. While operational risks may sometimes be mitigated through additional resources, risks that materially affect the probability of achieving value-defining milestones must be explicitly quantified. Qualitative labels (e.g., low/medium/high risk) offer limited decision support and undermine transparent trade-offs.

Evaluate
Portfolio value is built bottom-up from project-level valuations that integrate value, risk, cost, and time. The dominant metric in biopharma remains eNPV, which probability-weights future cash flows by technical and regulatory success. Robust evaluations treat uncertainty explicitly using ranges and probabilistic assumptions, distinguishing between risk (likelihood of success) and uncertainty (distribution of outcomes conditional on technical and regulatory success).

Optimize
Portfolio optimization identifies the combination of projects that maximizes value subject to constraints such as capital, capacity, risk exposure, and strategic balance. Optimization reveals trade-offs and opportunity costs and defines the efficient frontier of feasible portfolios. In practice, optimization is most effective when applied at the margin, guiding incremental investment decisions within existing commitments.

Prioritize
Prioritization ranks projects using explicit criteria and structured methods (e.g., MODA, AHP). While valuable for sequencing under operational constraints, prioritization alone rarely produces an optimal portfolio. Its greatest utility arises when applied within an optimized portfolio, where it informs execution order rather than strategic selection.

Manage

Portfolio management is sustained through integrated risk, resource, and stakeholder management. Risk is addressed through proactive planning and staged decision making; resources are deployed to minimize delays and bottlenecks; and stakeholder engagement ensures consistent governance and shared ownership of outcomes. Portfolio management is a collective, continuous process that synthesizes evolving information to support informed decisions.

Cost containment alone cannot compensate for persistent misallocation of resources. Frameworks such as CREOPM™ provide a disciplined mechanism for aligning investment decisions with strategy and risk tolerance. While portfolio management offers no immediate quick wins, its consistent application has demonstrated durable improvements in strategic alignment and long-term enterprise value across R&D-driven organizations.

3.6.2 Organizational Dynamics Underpinning EPPM

Achieving excellence in EPPM depends less on analytical sophistication than on an organization's willingness to invest deliberately in portfolio management as a scalable, enduring capability. This progression is best described through a capability maturity hierarchy, in which higher levels of maturity are associated with broader accountability, deeper functional integration, and greater influence on enterprise decision making (see Figures 3.18 and 3.19) (Letavec 2006).

Across all maturity levels, effective portfolio management rests on the coordinated execution of three foundational disciplines: project and program management, financial management, and resource management. These elements must be integrated through coherent process management

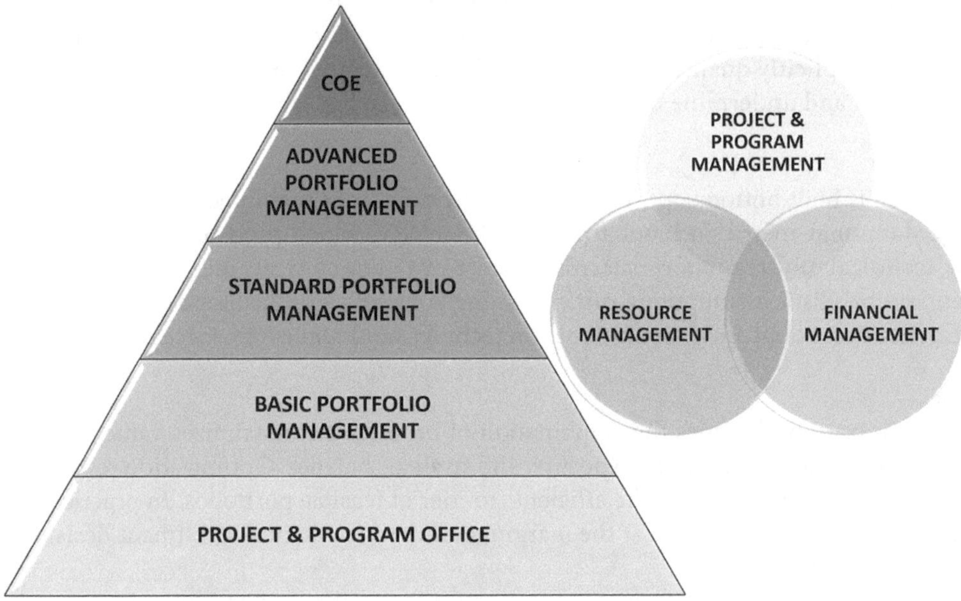

Figure 3.18 Project portfolio management capability maturity hierachy (Letavec 2006).

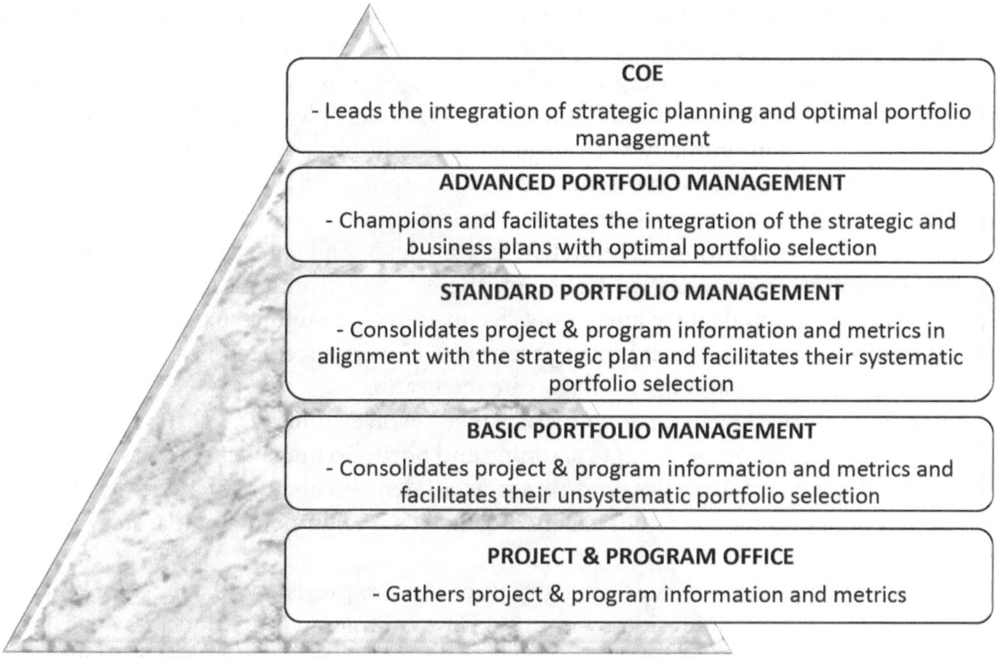

Figure 3.19 Overarching responsibilities across the PPM maturity hierarchy.

to ensure alignment, information flow, and logical interdependency. Without such integration, portfolio management remains fragmented and unable to support enterprise-level decisions reliably.

At the base of the maturity hierarchy, organizations typically establish a project and program office (PPO). At this stage, portfolio management is primarily informational, focused on collecting and reporting project metrics. Decision authority remains concentrated with senior leadership, while the PPO functions as a data provider rather than a governance or analytical partner. Data generation often outpaces validation and interpretation, increasing informational burden without materially improving decision quality. Portfolio decisions are episodic, judgment-driven, and weakly reproducible, particularly for marginal assets.

As maturity increases, organizations establish a basic *portfolio management office* (PMO) that consolidates timelines, costs, and high-level metrics. While recognized as the central source of portfolio information, its contribution remains largely descriptive. Inconsistent data ownership across portfolio management, finance, and resource functions frequently leads to multiple, conflicting views of the portfolio, undermining confidence and impairing governance. Portfolio reviews tend to align with annual planning cycles, and stage-gate decisions are rarely assessed in terms of cumulative portfolio impact.

A critical transition occurs when the enterprise strategic plan becomes the explicit foundation for portfolio analysis. At this stage, portfolio management assumes ownership of consolidated, validated enterprise data and begins translating strategy into actionable investment pathways. The function evolves from reporting historical data to framing forward-looking strategic options, supported by deliberate investment in decision analysis expertise and professional capability development.

In a standardized portfolio management configuration, portfolio management operates as a peer to finance and resource management, but portfolio selection remains driven primarily by prioritization, with finance typically imposing the effective budget constraint. While processes are more rigorous and transparent, true portfolio optimization is not yet fully realized. Portfolio reviews are conducted annually or semi-annually, with stage-gates providing project-level control.

Advanced portfolio management extends beyond consolidation and prioritization to explicitly maximize enterprise portfolio value. Using frameworks such as CREOPM™, portfolio construction begins with project categorization, quantitative risk analysis, and integrated evaluation, followed by formal optimization to generate portfolios on the efficient frontier. Alternative portfolios are developed under varying budgetary and resource constraints, enabling senior leadership to select among analytically sound, value-maximizing scenarios. Governance becomes more frequent, with quarterly portfolio updates and ongoing stage-gate integration.

At the highest maturity level, portfolio management evolves into a *Center of Excellence* (COE). The COE leads full integration of strategic planning and portfolio management, sets enterprise standards for methodology and analytics, and serves as a talent incubator for future leaders. Portfolio governance operates on a consistent cadence, reinforcing alignment from strategy formulation through execution.

Organizational placement and reporting structure remain persistent challenges. While portfolio management has historically resided within R&D or commercial functions, its enterprise-wide mandate often aligns best with strategic planning, provided long-term strategy is not routinely overridden by short-term considerations. Regardless of structure, centralized organizations tend to integrate portfolio management more effectively than decentralized enterprises, where success depends on sustained alignment among senior leaders.

In centralized organizations, where decision authority is clearly defined and standards, processes, and tools can be harmonized, portfolio management tends to integrate more smoothly into enterprise governance. In contrast, decentralized organizations—characterized by distributed authority, localized practices, and heterogeneous decision norms—present a more challenging environment. In such settings, the success of portfolio management depends less on any single function and more on sustained alignment among senior leaders across business units.

Standardizing portfolio management practices across decentralized enterprises can be extraordinarily difficult. Cultural norms, political dynamics, and legacy power structures can either enable or undermine the credibility and longevity of the function. As a result, organizational culture and internal politics play a decisive role in determining whether portfolio management becomes a durable strategic capability or remains a marginal, advisory activity with limited influence on enterprise outcomes (see Figure 3.20) (DuBrin 2007; Gido and Clements 2009).

Decentralization, geographic dispersion, and virtual operating models amplify challenges related to standardization, cultural resistance, and political dynamics. In such environments, portfolio management must establish credibility through analytical rigor while remaining sensitive to informal power structures, which often derive from control of resources, relationships, and tenure rather than formal authority (see Figure 3.21) (Lewis 2003).

Sustained success in portfolio management requires visible executive sponsorship. Without it, portfolio functions risk marginalization or repeated reinvention. The executive sponsor must

FUNCTIONAL	• Each unit or department has different activities and responsibilities • Orientation towards a specific set of activities creating skilled functional experts but de-emphasizes exchange of ideas and cooperation
PRODUCT	• Places responsibility for a product or product line with managers • Managers may not have sufficient authority to carry out responsibilities and spend more time coordinating activities and insufficient time planning
TERRITORIAL	• Units are divided on the basis of territory or geographical region • Important for adjustments to local conditions, markets, or resources and a need for training and development of managers with broad based abilities
MATRIX	• Functional structure combined with either a product or project structure where employees report to more than one supervisor • Can create confusion over who reports to whom
VIRTUAL	• Operates across space, time, and boundaries with people who communicate mainly through electronic means • With improved customer service comes loss of cost efficiencies
ADAPTIVE	• Flexible enough to adapt to environmental changes, trends, and shifts • Well suited to start-up ventures and organizations seeking to re-architect and reestablish themselves
MULTIPLE STRUCTURES	• Each division may have its own functional departments • Problems in coordinating policies, programs, and resources leading to increased management of overhead expenses and duplication of resources

Figure 3.20 Types of organizational structure (DuBrin 2007).

actively champion the function, navigate organizational politics, align stakeholders, clarify ownership boundaries, and ensure coherence across governance structures. Equally important is aligning portfolio recommendations with the organization's risk tolerance. Misalignment—most commonly, excessive risk aversion—can prevent analytically robust recommendations from being acted upon.

Common manifestations of organizational risk aversion include overemphasis on downside avoidance, inflation of adverse probabilities, suppression of uncertainty through single-point estimates, and decision avoidance. Addressing these behaviors through transparent communication, explicit articulation of risk appetite, and consistent executive support is essential for portfolio management to translate analytical insight into sustained enterprise value creation.

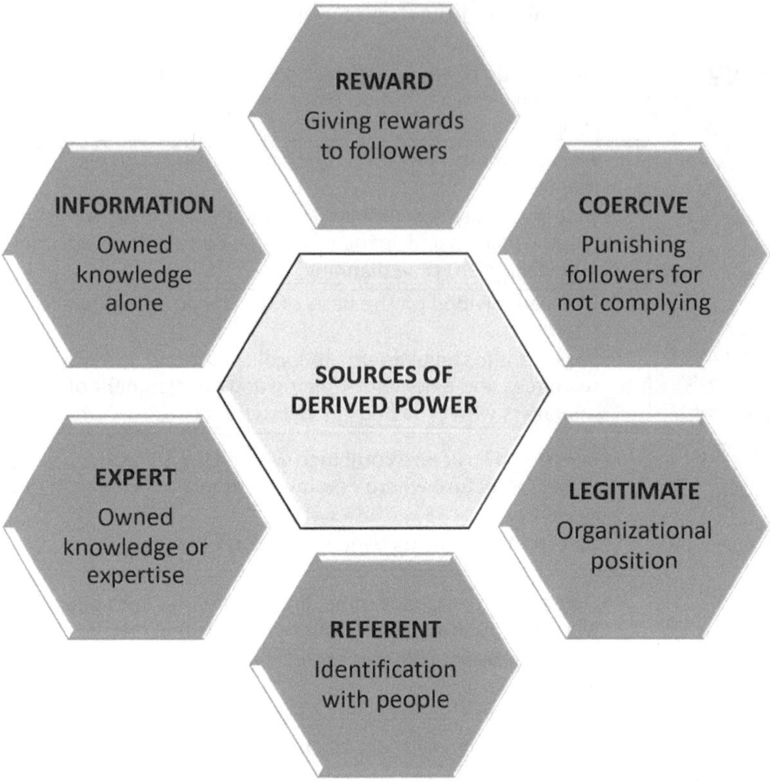

Figure 3.21 Sources of organizational power (Lewis 2003).

3.7 PORTFOLIO SELECTION

3.7.1 Prioritization

Project prioritization is among the most widely used—but frequently misunderstood—tools in portfolio management. Although often labeled *portfolio prioritization*, what is typically prioritized are *individual projects or programs*, not portfolios. Despite this conceptual limitation, prioritization is often treated as a surrogate for portfolio value maximization, a role it is not designed to fulfill.

Prioritization reflects a rational response to resource scarcity, seeking to rank projects based on their perceived contribution to organizational objectives such as financial performance, strategic alignment, market position, or customer impact. Decision attributes used for this purpose may be quantitative (e.g., NPV, IRR, and PV LRP sales), qualitative (e.g., strategic importance or competitive differentiation), or hybrid combinations.

When a single quantitative metric such as NPV (net present value) is used, prioritization is straightforward: projects are ranked and selected until a binding constraint is reached. However, rank consistency quickly breaks down when multiple quantitative criteria are applied, as projects rarely rank identically across metrics. Ambiguity increases further when qualitative attributes—ubiquitous in biopharma—are introduced, as strategic and patient value cannot be reduced to financial terms alone.

Utility theory provides a mechanism for integrating quantitative and qualitative attributes into a common evaluative scale. Multiple Objective Decision Analysis (MODA) (Kirkwood 1996) is a widely used approach that converts disparate attributes into utility scores and aggregates them using weighted sums. When applied within homogeneous R&D segments and with carefully defined attributes, MODA can provide a transparent and communicable ranking of projects. However, such rankings can misleadingly imply portfolio optimality, ignoring constraints, interdependencies, and uncertainty.

The Analytic Hierarchy Process (AHP) (Saaty 1980) offers an alternative by eliciting preferences through pairwise comparisons within a hierarchical structure. While AHP facilitates structured judgment and consistency checking, its scalability limitations and complexity restrict its practical use in large R&D portfolios.

In summary, prioritization techniques such as MODA and AHP are valuable for organizing information and supporting discussion, but they should not be conflated with portfolio optimization or value maximization.

3.7.2 Optimization

Optimization, rooted in mathematical programming, provides a fundamentally different paradigm for portfolio selection by identifying optimal portfolios under explicit objectives and constraints (Ragsdale 2017). Optimization problems consist of three elements: objectives, decisions, and constraints.

Objectives define what the organization seeks to maximize (MAX) or minimize (MIN) (e.g., expected portfolio value or development cost) and are expressed as mathematical functions. Decisions correspond to resource commitments, such as initiating or terminating projects, which are often costly or irreversible. Constraints represent practical limits, including budgets, full-time equivalent availability, manufacturing capacity, or regulatory throughput. Constraints may be hard (nonnegotiable) or soft (aspirational), and resources may be fungible or nonfungible.

Formally, optimization problems are expressed as:

$$\text{MAX (or MIN) } f_0(X_1, X_2, \dots, X_n)$$

subject to (less-than-or-equal-to; greater-than-or-equal-to; equal to):

$$f_i(X_1, X_2, \dots, X_n) \leq, \geq, = k_i$$

where decision variables are denoted as $X_1, X_2, \dots, X_n$ and a specified numerical limit is represented by the constant k.

The *objective function* of an optimization problem captures the criterion to be maximized or minimized and is expressed in the general form:

$$\text{MAX (or MIN)} = f(X_1, X_2, \dots, X_n)$$

More formally, an optimization problem can be written as:

$$\begin{array}{rl} \text{MAX (or MIN)} & f_0(X_1, X_2, \dots, X_n) \\ \text{subject to:} & f_1(X_1, X_2, \dots, X_n) \leq k_1 \\ & f_j(X_1, X_2, \dots, X_n) \geq k_j \\ & f_m(X_1, X_2, \dots, X_n) = k_m \end{array}$$

This formulation defines both the objective to be optimized and the set of constraints that must be satisfied. The purpose of optimization is to identify the values of the decision variables that yield the best possible outcome—maximum or minimum—while remaining entirely within the feasible region defined by the constraints (Ragsdale 2017).

Optimization evaluates projects collectively, rather than independently, enabling rigorous trade-off analysis under shared constraints. This distinction underpins its superiority over prioritization for advanced portfolio management.

3.7.2.1 Linear Programming

Linear programming (LP) applies when both the objective function and constraints are linear, implying proportional relationships between inputs and outputs. Despite this simplifying assumption, LP has proven highly effective for R&D investment planning and portfolio selection.

When decision variables must take integer values, the problem becomes an integer LP (ILP); when variables include both integer and continuous values, it is a mixed integer LP (MILP). Portfolio selection problems fundamentally rely on binary decision variables, where each project is either selected (1) or not (0), yielding a binary ILP (BILP):

$$\text{MAX (or MIN)} \sum_{i=1}^{n} c_i X_i$$

subject to:

$$\sum a_{ji} X_i \leq, \geq, = k_j$$

Here, $X_i \in \{0,1\}$, and coefficients c_i represent the contribution of each decision variable to the objective function and may correspond to marginal profit, EV, or cost associated with selecting each project.

As emphasized by Ragsdale (2017), optimization effectiveness depends more on problem formulation than on solution mechanics. A disciplined five-step process—defining scope, identifying decision variables, specifying objectives, defining constraints, and establishing bounds—reduces both analytical and organizational risk:

Step 1: Define the frame and scope of the problem: A clear understanding of what decision is being addressed—and what lies outside the boundaries of the analysis—is critical.

Step 2: Identify the decision variables: Decision variables represent genuinely discretionary choices. Projects that are nondiscretionary (e.g., regulatory commitments or mandated initiatives) should not be treated as variables.

Step 3: Specify the objective function: The objective function should be expressed as a linear combination of the decision variables and aligned with the organization's strategic and financial goals.

Step 4: Define the constraints: Constraints should be stated as linear combinations of the decision variables and categorized as either *hard* or *soft*. This distinction is crucial, as it allows decision makers to understand the consequences of violating soft constraints once an optimal solution is generated, in which case decision makers may knowingly choose a suboptimal solution that better aligns with operational or strategic preferences.

Step 5: Establish upper and lower bounds on decision variables: Investment boundaries, such as minimum or maximum funding levels by therapeutic area or business sector, should be explicitly defined. These bounds are particularly important when portfolio balance across risk, reward, and capability domains is a strategic objective. Without such constraints, optimization may yield solutions that are mathematically optimal but strategically destabilizing, such as starving core organizational capabilities in pursuit of short-term value maximization (Ragsdale 2017).

3.7.2.2 Goal Programming and Multiple Objective Linear Programming

Conventional linear programming approaches—including LP, ILP, MILP, and BILP—are based on the assumption that the constraints defining the decision space are *hard*, meaning they cannot be violated without material consequences. In many operational settings, such assumptions are appropriate; examples include fixed manufacturing capacity, regulatory limits, or legally binding budget ceilings. However, in portfolio management contexts—particularly in R&D-intensive industries such as biopharma—this rigidity can significantly limit the usefulness of optimization.

When numerous constraints are treated as inviolable, the resulting feasible region—the space of portfolio solutions that satisfy all constraints simultaneously—can become so restricted that few, if any, acceptable portfolio solutions exist. As described by Ragsdale (2017), this phenomenon is exacerbated when organizations attempt to address multiple, often competing, objectives simultaneously. Common portfolio aspirations—such as maximizing revenue, cash flow, and return on investment while simultaneously minimizing cost, risk, and time to market—can collectively shrink the feasible region to a narrow, fragile solution space.

Ironically, this tension is frequently ignored in practice when organizations rely on multi-attribute project prioritization techniques. Such approaches implicitly impose multiple objectives without explicitly assessing whether value is being created or destroyed by the resulting trade-offs. Optimization frameworks, by contrast, make these trade-offs explicit and therefore defensible.

Goal programming (GP) addresses this by minimizing weighted deviations from goal targets rather than optimizing a single objective. Deviational variables capture under- and overachievement relative to goals:

$$\Sigma a_{ji} X_i + d_j^- - d_j^+ = k_j$$

where d_j^- represents underachievement and d_j^+ represents overachievement of the j-th goal. These variables are constrained to be nonnegative and do not, by themselves, impose penalties.

The GP objective minimizes undesirable deviations:

$$\text{MIN}\Sigma(w_j^- d_j^- + w_j^+ d_j^+)$$

Instead of maximizing or minimizing a single measure, GP minimizes the deviations between achieved outcomes and predefined goal targets, weighted by their relative importance. The emphasis shifts from strict optimality to satisficing—identifying solutions that best align with organizational preferences across competing dimensions (Ragsdale 2017).

MINIMAX Objectives in GP

In situations where decision makers wish to minimize the *worst-case deviation* across all goals, a MINIMAX objective can be introduced. This involves defining a variable Q such that all deviational variables are constrained to be less than or equal to Q, and then minimizing Q. This approach ensures that no single goal is violated excessively and is particularly valuable when goals carry implicit lower or upper bounds (Ragsdale 2017).

Multiple Objective Linear Programming

A closely related approach is multiple objective linear programming (MOLP), which can be viewed as a specialized form of GP in which target values for objectives are not known a priori and must be derived during the analysis. MOLP explicitly addresses dominance relationships: a solution is dominated if another solution improves at least one objective without worsening any others.

As described by Ragdsale (2017), a typical MOLP process includes:

1. Defining decision variables and fixing (i.e., assigning an integer value of 1) nondiscretionary investments
2. Formulating each objective as a standalone LP problem
3. Identifying shared constraints
4. Solving the model separately for each objective (set at a weight of 100%) to determine its extreme (best) value
5. Treating these extreme values as goal targets
6. Introducing deviational variables and weights
7. Applying a MINIMAX objective to control deviations
8. Iterating until a satisfactory compromise solution is achieved

Chapter 7 provides applied case studies demonstrating how MOLP can be used to structure and select R&D portfolios under competing strategic and operational pressures.

3.7.2.3 Stochastic Optimization

Deterministic optimization assumes that project attributes are known with certainty—an assumption rarely valid in R&D. Most parameters are uncertain but can be characterized probabilistically. Despite this, organizations often rely on single-point *base-case* estimates that suppress uncertainty rather than model it.

Stochastic optimization incorporates uncertainty explicitly, typically through probability distributions (Shapiro et al. 2009). Unlike deterministic models, stochastic approaches evaluate decisions across a range of plausible futures. Stochastic programming is commonly applied to repetitive operational problems, such as logistics routing or inventory control, where decisions are made repeatedly and performance converges to long-run averages. In these settings, occasional suboptimal outcomes are acceptable because the law of large numbers governs aggregate performance.

Portfolio selection problems, however, often involve *one-off* or infrequently repeated decisions—such as committing to a specific configuration of R&D projects—where outcomes are path-dependent and irreversible. Each project carries unique technical, regulatory, and commercial uncertainties, frequently expressed as probabilities of technical and commercial success. In such contexts, reliance on expected value or expected utility alone can be problematic, as it implicitly assumes risk neutrality. Consequently, the subjective risk attitudes of decision makers must be explicitly acknowledged and incorporated into the modeling process.

Simulation-based approaches—particularly Monte Carlo simulation and Latin Hypercube Sampling (LHS)— are widely used in portfolio contexts. Simulation repeatedly evaluates a deterministic model:

$$Y = f(X_1, X_2, \dots, X_q)$$

using randomly sampled inputs to generate output distributions. LHS improves sampling efficiency by stratifying input distributions, often yielding more accurate estimates with fewer iterations.

Simulation outputs support decision-relevant metrics such as confidence intervals, downside risk, and probability-weighted performance. While simple simulations can be built in spreadsheets, enterprise portfolio problems typically require specialized tools such as @Risk Optimizer, as illustrated in Chapter 7.

3.8 THE FUTURE OF EPPM

Enterprise portfolio management delivers value gradually by improving decision quality, capital allocation, and strategic coherence, making sustained executive sponsorship essential. Its future depends on recognition as a distinct professional discipline rather than a derivative of financial asset management.

Academic institutions play a critical role in formalizing education in portfolio management, while organizations must integrate strategic thinking with decision analysis and operations research. Core requirements include mastery of enterprise decision competencies, holistic frameworks such as CREOPM™, and exposure to real-world portfolio case studies.

Ultimately, the future of portfolio management lies not in the mechanical application of tools, but in the disciplined integration of analytics, strategy, and organizational insight to support better decisions under uncertainty.

Section II

CREOPM™—A Holistic Portfolio Management Framework

4

PROJECT CATEGORIZATION

Categorization, in an organizational and portfolio management context, refers to the systematic classification and structuring of projects and programs into a mutually exclusive and collectively exhaustive set of categories. Each category is defined by a coherent set of attributes—such as regulatory obligation, strategic intent, risk profile, or resource criticality—that are consistently shared across all constituent initiatives. The purpose of such categorization is not merely descriptive; rather, it establishes a disciplined framework through which comparable assets can be evaluated, prioritized, governed, and resourced in a consistent and transparent manner.

Much like scenario analysis, strategic intent articulation, decision framing, and data integrity, project categorization represents a foundational enabler of effective portfolio decision making. While the aforementioned practices help leadership understand where the organization is headed and why, categorization provides the operational clarity required to determine how organizational objectives will be realized. In the absence of clear categories, decision makers are often forced to compare fundamentally dissimilar initiatives—such as regulatory commitments, platform investments, and exploratory research—using a single evaluative lens, thereby obscuring trade-offs and increasing the likelihood of suboptimal resource allocation.

An intuitive analogy can be drawn to the preparation for a journey. Prior to departure, individuals typically distinguish between items that are essential and nonnegotiable (e.g., legal identification, insurance documentation, payment instruments, and communication devices) and items that are optional or preferential (e.g., clothing selections, entertainment options, or discretionary accessories). Equally important is the deliberate exclusion of items that are unnecessary or inappropriate for the journey. This implicit triage—what must be taken, what may be taken, and what should not be taken—is instinctive at the individual level. Yet, paradoxically, organizations often struggle to impose the same discipline when determining which initiatives must be executed, which may be pursued opportunistically, and which should be explicitly excluded from consideration.

Within a robust portfolio governance framework, these distinctions are commonly formalized through three high-level categories: *must-do, may-do,* and *won't-do* initiatives. These categories are deliberately nonoverlapping. *Must-do* projects are largely nondiscretionary and mandatory; failure to execute them exposes the organization to unacceptable strategic, financial, regulatory, legal, operational, or reputational risk. In contrast, *may-do* and *won't-do* initiatives are largely discretionary. They compete for capital and organizational capacity based on strategic alignment, expected value, and risk-adjusted return, rather than obligation.

In the biopharma industry, the *must-do* category is dominated by externally imposed requirements and critical infrastructure commitments. Typical examples include regulator-mandated long-term carcinogenicity studies, post-marketing surveillance and pharmacovigilance trials required to maintain market authorization, and enterprise-wide information technology initiatives necessary to ensure compliance with HIPAA and related data privacy and reporting standards (see Figure 4.1).

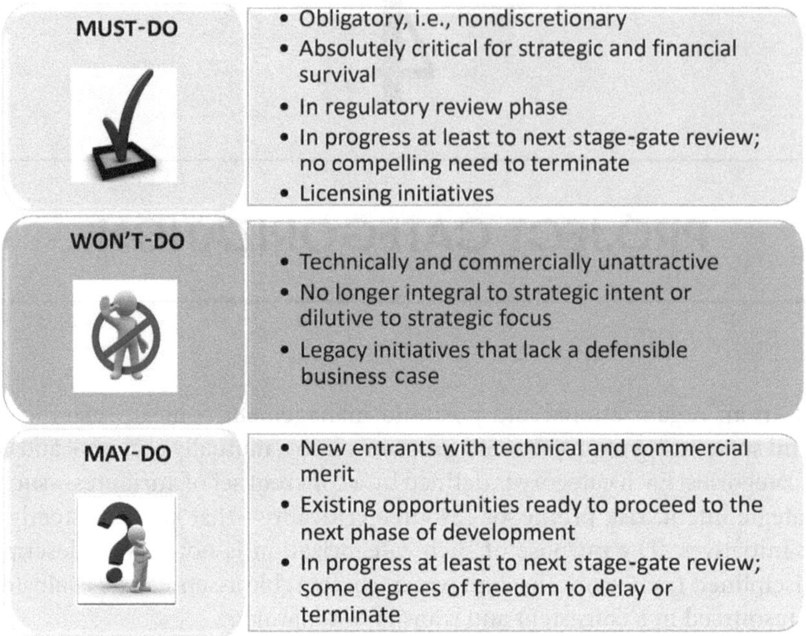

Figure 4.1 Project categorization.

These initiatives are not undertaken to create incremental shareholder value in the conventional sense; rather, they serve to preserve the organization's license to operate and protect existing value streams.

Importantly, classification as a *must-do* does not imply inflexible execution timing. In certain circumstances, organizations may elect to defer initiation or completion of mandated projects, particularly when capital constraints or competing priorities dominate near-term planning. Such deferrals, however, are explicit risk-management decisions and are often accompanied by predefined financial penalties, regulatory scrutiny, or negotiated remediation plans. As such, even when timing is discretionary, the obligation itself is not. Recognizing and explicitly categorizing these projects allows senior leadership to separate unavoidable commitments from strategic choices, thereby improving transparency, accountability, and the overall quality of portfolio-level decision making.

Certain classes of projects—particularly those that are central to the financial viability, strategic continuity, or long-term positioning of the enterprise—occupy a gray zone between strictly mandatory and discretionary investments. While these initiatives may not be legally or regulatorily obligatory in the narrow sense, they often constitute a de facto nondiscretionary subcategory because deferral, suspension, or termination would materially impair the organization's competitive position or threaten its future relevance. This is especially true for programs in advanced stages of development or those representing life-cycle extensions of existing products, where the cumulative investment, strategic signaling, and anticipated downstream value significantly outweigh the perceived benefits of delay or abandonment.

For projects already in execution, the presumption should generally favor continuation through the current development phase or until the next formal decision gate. Absent compelling evidence of value destruction—such as the emergence of unacceptable safety liabilities, insurmountable technical failures, or external developments that fundamentally undermine the commercial rationale—terminating an in-flight program prematurely often results in avoidable loss of learning, stranded investment, and erosion of organizational credibility. In some cases, a subset of initiatives is so tightly coupled to the organization's mission, values, or identity (e.g., a breakthrough diagnostic addressing a life-threatening condition) that they effectively become nondiscretionary, irrespective of short-term financial metrics.

Licensing agreements and broader business development transactions further reinforce the must-do category through contractual obligation. These arrangements typically impose explicit commitments to advance research and development activities in accordance with agreed-upon timelines and diligence standards. Failure to demonstrate good-faith performance—regardless of whether the organization is the sole sponsor or a development partner—can trigger financial penalties, contractual disputes, and reputational damage that may compromise future partnering opportunities. Consequently, such initiatives must be managed as mandatory commitments rather than optional investments.

A common and consequential failure mode in portfolio governance arises when organizations overinflate the must-do category by including projects whose strategic or financial value is merely asserted rather than demonstrably nondiscretionary. This behavior often reflects a desire to shield favored initiatives from rigorous comparison and trade-off analysis. An undisciplined portfolio is therefore characterized by an excessively large must-do category—one that overwhelms available financial and human resources and crowds out genuinely high-value discretionary investments that fall outside this protected class.

It is also critical to recognize that not all must-do projects possess intrinsic strategic or economic priority. Some initiatives are mandatory simply because they must be completed within a specified time frame, often for operational, regulatory, or compliance reasons. In these cases, the core decision question is not *whether* the project should proceed, but rather *how* it should be sequenced relative to other mandatory commitments. Unfortunately, many organizations conflate prioritization with funding eligibility by constructing a single rank-ordered project list and imposing a budgetary cutoff below which no further investments are sanctioned. Inevitably, this approach results in one or more mandated projects falling below the funding line, precipitating a range of reactive—and often problematic—responses.

These responses commonly include retroactively modifying the prioritization methodology to force mandated projects above the cutoff; applying arbitrary adjustment factors that artificially elevate their scores; overriding the formal ranking process by invoking undefined *other considerations*; or abandoning the prioritization framework altogether in favor of subjective managerial judgment. While these tactics may resolve immediate funding tensions, they undermine the credibility of the decision process and create a precedent that allows weak or poorly justified initiatives—often referred to as *pet projects*—to persist with minimal transparency or accountability.

Although most organizations aspire to make portfolio decisions grounded in data, structured analysis, and rational trade-offs, the absence of methodological discipline and neutral facilitation frequently allows decision making to devolve into a contest of influence. As discussed earlier, heuristics, advocacy, organizational politics, and personal incentives tend to dominate, with outcomes disproportionately shaped by those with the greatest authority or the strongest voice. The result is not merely suboptimal prioritization, but systemic erosion of portfolio value. Under resource constraints, it is neither feasible nor optimal to fund every technically sound or strategically attractive project; equally important, assembling a portfolio comprised entirely of *good* projects does not guarantee that the portfolio itself is optimal. In the end, project sponsors or detractors with the loudest voice or proverbial *biggest stick* generally get their way to the detriment of the value of the portfolio. As shown in Chapter 7, under constrained resource conditions, not every good project can be part of an optimal portfolio. However, as demonstrated in Chapter 8, the converse is also true: every good project can be part of a suboptimal portfolio.

Compounding these challenges is the persistence of projects that no longer merit continued investment. In virtually every organization, a fraction of the portfolio consists of initiatives for which the technical feasibility, commercial promise, or strategic relevance has materially deteriorated. Despite this, such projects continue to absorb scarce capital and talent—resources that could otherwise be redeployed to initiate new programs, accelerate higher-value assets, or strengthen core strategic initiatives. Particularly problematic are projects that were once compelling but failed to meet expectations despite extended development efforts. These initiatives often survive not because of objective merit, but because sponsors are reluctant to concede that the underlying rationale has eroded, or because of the psychologically powerful—but economically flawed—desire to justify past expenditures.

The difficulty organizations face in terminating projects in a timely and effective manner is multifactorial. Contributing factors include:

- The absence of clearly articulated decision criteria and success thresholds
- Weak or inconsistently applied stage-gate governance
- Disproportionate influence of politically powerful sponsors
- Failure to rigorously update and revalidate business cases
- Aversion to acknowledging failure
- Misinterpretation of sunk costs as recoverable investments
- Misaligned incentives and emotional bias
- Concerns about the potential demoralization of project teams

Collectively, these dynamics reinforce inertia and delay decisive action, ultimately diminishing portfolio performance and organizational agility.

Just as forward-looking organizations rely on scenario analysis and strategic planning to inform which initiatives should populate their portfolios, it is equally critical that they apply the same rigor and transparency to decisions regarding which projects should be explicitly excluded or removed. Portfolio optimization is as much an exercise in disciplined subtraction as it is in selective inclusion. Although achieving emotional and organizational detachment from prior investments can be challenging, the benefits of doing so are both tangible and immediate. Clear termination decisions sharpen strategic focus, reenergize project teams by eliminating ambiguity, and enable the redeployment of scarce capital and talent to initiatives with higher prospective value.

Following the full funding and resourcing of nondiscretionary must-do initiatives—and the deliberate reclamation of resources released from won't-do projects—the organization is positioned to concentrate on the discretionary may-do segment of the portfolio. This category represents the principal arena for value creation, where alternative combinations of projects can be systematically evaluated to determine which configurations most effectively advance enterprise objectives. May-do projects typically include early-stage internal programs, externally sourced opportunities, and in-flight initiatives approaching a formal end-of-phase or stage-gate review.

It is within the may-do category that senior leadership and portfolio governance bodies should devote the majority of their analytical and deliberative effort. Decisions in this space extend beyond binary go/no-go determinations and include questions of investment pacing, optionality, and staged commitment. In many cases, partial or time-limited funding may be strategically optimal, enabling the organization to acquire critical technical, regulatory, or commercial information before committing to full-scale development. As discussed in Chapter 7, the discretionary nature of may-do projects introduces degrees of freedom into the portfolio, allowing organizations to adapt dynamically to new information and evolving strategic priorities.

Once the boundaries of discretion are clearly defined, multiple portfolio strategies can be evaluated and compared, each reflecting different assumptions about risk tolerance, capital availability, and value realization horizons. These trade-offs are explored in depth in Chapter 7 and illustrated through applied case studies later in this text. However, this flexibility is frequently eroded when organizations misclassify discretionary initiatives as must-do projects. Such misclassification reduces the strategic maneuverability of the enterprise and often serves as a convenient shield for investments that lack robust justification but are supported by influential stakeholders.

Two questions commonly arise in discussions of project categorization:

1. How many projects should reasonably reside within the must-do versus the may-do categories?
2. How should available financial resources be apportioned between these two classes?

There is no universal prescription. The appropriate balance is contingent upon the organization's strategic objectives, financial posture, risk appetite, and time horizon. Enterprises that adopt a long-term perspective are generally better positioned to defend discretionary investments aimed at maximizing value over time. In contrast, organizations that operate under a persistent, and sometimes exaggerated, sense of urgency often prioritize immediate survival considerations. In such environments, an excessive number of projects are deemed indispensable, inflating the must-do category and paradoxically increasing the likelihood of portfolio value destruction.

At first glance, it may appear counterintuitive that reclassifying discretionary projects as nondiscretionary could reduce portfolio value. The explanation lies in a foundational principle of portfolio optimization: as the number of mandatory projects increases, the solution space of feasible project combinations contracts. This contraction reduces the likelihood of identifying portfolio configurations that lie on or near the efficient frontier for a given resource constraint. In other words, the more rigid the portfolio becomes, the fewer opportunities there are to build high-value investment combinations.

This principle can be illustrated with a simplified example. Consider must-do Project A requiring a $20M investment and generating a net present value (NPV) of $100M. Compare this with two discretionary projects—Projects B and C—requiring $9M and $10M, respectively, and generating NPVs of $50M and $60M. In aggregate, Projects B and C require less capital and generate greater total value than Project A. In isolation, the superiority of the B+C combination may be obvious. However, in a complex portfolio comprising dozens or hundreds of initiatives, such dominance relationships are rarely transparent without systematic analysis or computational support. Absent such tools, decision makers may inadvertently commit to suboptimal portfolios simply because the combinatorial alternatives are not readily apparent.

These dynamics underscore the importance of disciplined categorization and analytically supported decision making. The implications of constrained resources, combinatorial complexity, and efficient frontier analysis are examined further in Chapters 3 and 7, while the specific challenges of resource allocation and budgeting are addressed in Chapter 9.

5

RISK ANALYSIS

Some of the figures and tables in this chapter are too large or complex to properly format in printed form. Figures and tables marked with a "WAV" label can be downloaded from the Web Added Value Download Resource Center at www.jrosspub.com/wav for proper viewing.

5.1 CONCEPTUAL FOUNDATIONS OF RISK AND UNCERTAINTY IN BIOPHARMACEUTICAL R&D

Within biopharma research and development (R&D), risk is rigorously defined as the probability that an adverse event will occur that could compromise project viability, value, or timelines. Examples include unfavorable clinical outcomes, regulatory delays, manufacturing failures, or post-approval recalls. Quantitatively, risk is represented as a probability (p) bounded between 0 and 1, reflecting its likelihood of occurrence.

R&D risk is not monolithic and can be decomposed into two broad classes. The first includes risks that are meaningfully amenable to managerial intervention, where additional time, capital, or resources can reduce the probability of occurrence. These include operational execution risks such as delayed trial initiation, slow patient enrollment, or data quality deficiencies. The second class consists of risks that are largely resistant to mitigation, where incremental investment has little or no effect on outcomes. Fundamental biological uncertainty—particularly drug efficacy and unforeseen safety liabilities—dominates this category.

Closely related but conceptually distinct is uncertainty, which refers not to whether an event occurs, but to the range and distribution of outcomes conditional on its occurrence. Outcomes may be discrete (e.g., number of serious adverse events) or continuous (e.g., severity of toxicity). While both risk and uncertainty are probabilistic, risk emphasizes event frequency, whereas uncertainty emphasizes outcome dispersion.

In biopharma R&D, risks are best viewed as impediments to orderly progression toward regulatory approval and commercialization. These risks span the entire research and development continuum and collectively determine the probability that a project delivers its anticipated value.

Industry practice has produced relatively mature tools for identifying and managing operational risks, often characterized as temporal risks tied to schedule and budget adherence. Established project management techniques—risk registers, mitigation plans, and probability–impact matrices—are routinely applied to reduce such risks. By contrast, the systematic quantification and governance of intrinsic scientific risks, particularly efficacy and long-term safety, remain far less codified. These uncontrollable risks dominate value creation and destruction in biopharma yet resist straightforward mitigation, making portfolio-level frameworks for their aggregation and decision-weighting an active area of development.

Quantification and Scoring of Risk

Risk exposure may be assessed using quantitative or semi-quantitative methods, including weighted risk scoring frameworks adapted from project management. Individual risks are evaluated along two dimensions: probability of occurrence and magnitude of impact. A common implementation is the probability–impact matrix (PIM) shown in Table 5.1, which combines predefined likelihood and impact scales into a composite score.

While effective for operational risk prioritization, composite risk scores lack the mathematical rigor required for formal decision analytic tools such as decision trees, expected value calculations, or integrated portfolio comparisons. They highlight where risks reside but do not readily translate into coherent measures of aggregate portfolio risk.

From a decision-making standpoint, the primary value of risk analysis lies not in generating abstract scores but in elucidating how discrete risk elements contribute to the overall probability of project success. By decomposing risk in this manner, organizations can more clearly identify the dominant drivers of failure or delay and assess whether meaningful mitigation strategies exist. This perspective enables leadership to distinguish between risks that warrant active intervention and those that must be strategically accepted.

Despite significant advances in project-level risk management, many organizations across industries—and particularly within biopharma R&D—remain at an early stage of maturity with respect to portfolio-level risk analysis. While formal risk identification, scoring, and mitigation practices are often applied with reasonable effectiveness to individual programs, these approaches have proven far less successful when scaled to the portfolio as a whole.

A more analytically robust approach to risk analysis involves the explicit assignment of judgmental probabilities to represent the perceived likelihood of the occurrence of distinct risk elements, particularly those that are intrinsic to drug development and are largely uncontrollable (e.g.,

Table 5.1 PIM—a combination of probability of occurrence and impact.

	SEVERITY	VERY LOW	LOW	MEDIUM	HIGH	VERY HIGH
PROBABILITY		0.05	0.10	0.20	0.40	0.80
VERY HIGH 71-90%	0.90	0.05	0.09	0.18	0.36	0.72
HIGH 51-70%	0.70	0.04	0.07	0.14	0.28	0.56
MEDIUM 31-50%	0.50	0.03	0.05	0.10	0.20	0.40
LOW 11-30%	0.30	0.02	0.03	0.06	0.12	0.24
VERY LOW Up To 10%	0.10	0.01	0.01	0.02	0.04	0.08

demonstrating clinical efficacy). In practice, structured checklists of risk factors may be used to identify the specific scientific, regulatory, and operational hurdles that define success at each phase of development.

Once these phase-specific success criteria are articulated, the risks that could prevent attainment of each hurdle are probabilistically assessed. A joint probability of success is then calculated for each development phase, conditional on successful completion of prior phases. By linking these conditional probabilities across the full development pathway, an overall *probability of technical and regulatory success* (PTRS) can be derived for the project. This probabilistic framing provides a transparent and decision-relevant measure of project risk that is directly compatible with decision trees, valuation models, and portfolio optimization techniques.

It must be acknowledged, however, that both qualitative scoring systems and probabilistic approaches are susceptible to subjectivity, particularly when probabilities are elicited from subject matter experts (SMEs). Cognitive biases, overconfidence, anchoring, and framing effects can materially distort risk assessments if left unaddressed. While formal methods exist to reduce bias in probability elicitation—such as structured interviews, calibration training, and aggregation techniques—a detailed discussion of these methods lies beyond the scope of this text. Readers seeking a rigorous treatment of probability judgment are encouraged to consult Kirkwood (1996).

At the portfolio level, risk shifts from executional success to the probability of achieving strategic and financial objectives within resource constraints while managing outcome variability. Rigorous portfolio-level risk analysis provides an objective baseline for evaluating investments and assessing alignment with enterprise goals under uncertainty. Sensitivity and scenario analyses (see Chapters 3 and 6) further strengthen robustness, provided organizational risk tolerance is explicitly defined.

Clarifying the Language of Risk and Uncertainty

Statements such as "this project is a sure bet" conflate risk and uncertainty. No legitimate R&D investment is risk-free, probabilities cannot equal 1.0, and *success* must be explicitly defined. Maintaining a clear distinction—risk as event frequency, uncertainty as outcome range—is essential for rigorous decision making. Integrated risk and uncertainty assessment also enables organizations to capture upside and mitigate downside once events occur. Scenario planning supports the anticipation of developments such as competitive entry or regulatory change.

Just as financial portfolio risk depends on individual asset contributions, R&D portfolio risk depends on project-level risks shaped by the strategic pathway selected. These pathways are articulated through the target product profile (TPP), which is more accurately represented as a range rather than a single point target. This concept is illustrated in Table 5.2.

TPP ranges are justified because safety and efficacy manifest along continua and remain only partially controllable. Development choices influence outcomes, but biological variability imposes irreducible uncertainty. Defining minimum, base, and upside TPPs explicitly links the likelihood of attainment to the associated value. The minimum TPP reflects the lowest approvable and commercially viable profile, while the upside TPP captures outcomes exceeding the base-case and potentially surpassing the future standard of care.

The TPP is a living document guiding development and regulatory dialogue. In interactions with agencies such as the FDA, it aligns development intent with evidentiary requirements and helps reduce late-stage regulatory risk. Core TPP elements typically include dosage and administration, pharmacologic class, indications, safety information, and patient counseling considerations.

In early research, where data are insufficient to support detailed TPP ranges, a Disease Target Profile may be used (see Table 5.3) to frame disease biology, unmet need, and therapeutic hypotheses. This approach preserves strategic discipline while deferring detailed product profiling until

Table 5.2 Target Product Profile—minimum, base, and upside variations. SOC = standard of care.

ATTRIBUTES	MINIMUM	BASE	UPSIDE
SAFETY & TOLERABILITY	Upper GI disturbances; muscle pains; headaches; rash; sleep disturbances equal to leading competition	Upper GI disturbances; muscle pains; headaches; rash; sleep disturbances marginally better than leading competition	Upper GI disturbances; muscle pains; headaches; rash; sleep disturbances significantly better than leading competition
EFFICACY	Lower LDL by 20% - 40% Increase HDL by 5% - 9% Lower triglycerides by 10% - 20%	Lower LDL by 40% - 60% Increase HDL by 10% - 15% Lower triglycerides by 20% - 40%	Lower LDL by at least 60% Increase HDL by at least 15% Lower triglycerides by at least 40%
THERAPEUTIC BENEFIT	Reduction in coronary heart disease and stroke equal to current SOC	Reduction in coronary heart disease and stroke marginally better than current SOC	Reduction in coronary heart disease and stroke significantly better than current SOC
DOSAGE & ADMINISTRATION	20 – 40 mg. per os	10 – 20 mg. per os	5 – 10 mg. per os
TARGET POPULATIONS	Adults	Adults and late adolescents	Adults and adolescents

Table 5.3 Disease Target Profile.

ATTRIBUTES	BASE CASE
THERAPEUTIC AREA & DISEASE	Central Nervous System: Mild-to-Moderate Alzheimer's Disease
TARGET PATIENT POPULATION(S)	Aged ≥ 65 years; MMSE Score Between 12 - 22
UNMET MEDICAL NEEDS	Improving treatment options (including neuropsychiatric symptoms) beyond symptom management Enhancing early detection and diagnosis (including misdiagnosis) Reducing health disparities and providing greater support for patients and caregivers
UNIQUENESS OF THERAPY	Disease modifying mechanisms that intervene in pathological processes, such as amyloid plaque deposition and neurofibrillary tangle formation, that offer the potential to slow or halt disease progression Novel drug delivery systems (e.g., nanoliposomes and exosomes) that overcome the blood-brain barrier
DRIVERS OF USE	Reducing the rate of cognitive decline and improving quality of life of sufferers Interventional approaches to target the underlying pathology of the disease
COMPETITIVE LANDSCAPE	Companies targeting early intervention in the pathology of the disease beyond amyloid deposition and tau formation, including neuroinflammation, neuroprotection, gene therapy, and precision medicine
ECONOMIC COST OF DISEASE	Between $15,000 - $20,000 annually per patient (includes both direct and indirect costs)
KEY STRATEGIES FOR SUCCESS	Apart from interventional therapies, provide support for patients aimed at simplifying their living environment to reduce excessive stimulation, promoting a sense of safety, establishing predictable routines to reduce anxiety and help with orientation, and using visual cues to help with memory and orientation

uncertainty is reduced. Some organizations alternatively employ a Research Target Profile focused on preclinical performance criteria.

5.2 PROJECT RISK

A Structured, Six-Step Framework for Project Risk Analysis

The authors propose a disciplined six-step framework for project-level risk analysis, designed to support probabilistic decision making in capital-intensive, high-uncertainty environments such as biopharma R&D:

- **Step 1** establishes foundational alignment by ensuring that the business case, TPPs, and the clinical development plan (CDP) are current, internally consistent, and broadly shared. These documents define strategic intent and success criteria; misalignment materially undermines subsequent risk analysis.
- **Step 2** focuses on expert judgment elicitation. SMEs with scientific, clinical, regulatory, and operational expertise are engaged through structured individual and group interviews led by a trained facilitator. Advocates and skeptics are both deliberately included to surface divergent perspectives critical to balanced risk assessment.
- **Step 3** addresses systematic risk identification and classification. Risks are categorized into (a) largely uncontrollable intrinsic risks—primarily efficacy and safety—and (b) operational risks that may be mitigated but not eliminated within CDP constraints.
- **Step 4**, the analytical core, involves formal risk elicitation and quantification. Prepared SMEs assess, independently and collectively, the probability of overcoming each major intrinsic risk across R&D phases. Operational risks are excluded unless they cannot be neutralized through resources or time extensions.
- **Step 5** integrates elicited probabilities into a decision-analytic model. Decision trees capture joint, conditional probabilities of success by phase, supported by sensitivity analyses to identify dominant uncertainties.
- **Step 6** emphasizes iteration. As new data emerge, probabilities and decision trees are updated to maintain decision relevance throughout the project life cycle.

Step 4 is the most resource-intensive and is frequently abbreviated in practice, despite being critical to rigorous probability-based decision making. Elicitation externalizes SME tacit knowledge—intuition, heuristics, and experiential judgment—that is difficult to codify yet represents a vital organizational asset. To ensure decision relevance, elicited judgments must be validated through structured *playback* to SMEs, allowing clarification and refinement. Common elicitation techniques include structured interviews, facilitated storytelling, simulations, concept mapping, process analysis, and iterative questioning.

PPVC recommends a formal risk decomposition framework developed by Rang (2006) to support completeness and objectivity. This framework, shown in WAV Table 5.4, functions as a structured checklist of potential failure modes and uncertainties that may jeopardize a project's success. By systematically decomposing risk into technical, regulatory, and operational dimensions, the framework provides a common vocabulary and analytical structure for SME discussions. Each identified risk can then be assessed for its relevance to the asset under consideration and, where appropriate, quantified probabilistically. This structured approach reduces the likelihood of overlooked risks, mitigates cognitive bias, and enhances comparability across projects.

A frequently cited critique of formal risk decomposition and probabilistic risk analysis is their susceptibility to human bias. Skeptics argue that, rather than eliminating subjectivity, structured risk analysis may inadvertently institutionalize bias if the underlying judgments are systematically distorted. As introduced earlier (see Chapter 3), such biases influence not only how knowledge is expressed, but also how it is filtered, interpreted, and transformed prior to articulation. The roots of these distortions lie in both motivational and cognitive mechanisms which shape how experts access, process, and communicate their internal knowledge when participating in risk elicitation exercises.

Motivational bias arises when individuals, consciously or subconsciously, adjust their stated beliefs or probability assessments in response to incentive structures, organizational expectations, or perceived rewards and penalties. Under these conditions, experts may unintentionally recalibrate their assessments to support outcomes that are viewed as desirable or institutionally favored, rather than those that most accurately reflect their private state of knowledge.

By contrast, cognitive bias reflects systematic patterns in human information processing. These biases occur even in the absence of explicit incentives and arise from mental shortcuts, or heuristics, that individuals use to manage complexity and uncertainty. Such heuristics can lead to predictable distortions in judgment, including overconfidence, anchoring on early information, availability effects, and framing-dependent assessments.

Critically, both motivational and cognitive biases influence the transformation of underlying beliefs into stated judgments before those judgments are ever captured in a formal risk model. As a result, awareness of these biases and the deliberate design of elicitation processes to mitigate their impact is essential to preserving the integrity and credibility of risk decomposition and analysis.

Achieving proficiency in rigorous risk analysis and probabilistic quantification is neither trivial nor rapid. It demands sustained investment in analytical infrastructure, standardized processes, skilled facilitation, and organizational discipline. As a result, many biopharma organizations elect not to build fully bespoke risk-elicitation capabilities and instead rely on externally derived probabilities of success drawn from industry benchmarks, which are then selectively adjusted to reflect the perceived characteristics of individual projects. This pragmatic approach reflects both resource constraints and the recognition that large, cross-industry datasets can provide a credible empirical baseline against which internal judgments may be anchored. Benchmark-derived probabilities offer a degree of objectivity, comparability, and speed, particularly in early portfolio assessments or in organizations with limited internal historical data.

A widely cited and comprehensive source of such benchmarks has been developed through collaborative analyses by the Biotechnology Innovation Organization, Informa, and Quantitative Life Sciences. Their work aggregates historical development outcomes across a broad range of programs and systematically reports phase-specific attrition rates and probabilities of success. These data span 14 therapeutic areas and cover the full clinical development continuum, including Phase I, Phase II, Phase III, and regulatory approval, as summarized in Table 5.5.

While benchmark-based probabilities should not be viewed as substitutes for asset-specific risk analysis—particularly for differentiated or novel mechanisms—they nonetheless provide a valuable reference point. When applied thoughtfully and transparently, and when adjusted with explicit rationale, such benchmarks can serve as a defensible starting position for portfolio modeling, scenario analysis, and comparative investment decision making across heterogeneous R&D assets.

The use of industry-derived benchmarks for estimating the probability of success offers several practical advantages that explain their widespread adoption in biopharma portfolio analysis. Chief among these is their accessibility and ease of use. Benchmark datasets are typically organized along multiple dimensions—including therapeutic area, modality (e.g., small molecules, monoclonal antibodies), and broad mechanistic classes (e.g., DNA replication interference, receptor

Table 5.5 Industry benchmarks by therapeutic area and phase of development (Biotechnology Innovation Organization; Informa Pharma Intelligence; Quantitative Life Sciences 2021).

THERAPEUTIC AREA	PHASE I	PHASE II	PHASE III	REGULATORY
ALLERGY	56.4%	28.3%	64.7%	100.0%
AUTOIMMUNE	55.2%	31.4%	65.3%	94.1%
CARDIOVASCULAR	50.0%	21.0%	55.2%	82.5%
ENDOCRINE	43.3%	26.6%	66.2%	86.3%
GASTROENTEROLOGY	46.7%	34.2%	57.1%	90.9%
HEMATOLOGY	69.6%	48.1%	76.8%	93.1%
INFECTIOUS DISEASE	57.8%	38.4%	64.0%	92.9%
METABOLIC	61.8%	45.0%	63.6%	87.5%
NEUROLOGY	47.7%	26.8%	53.1%	86.7%
ONCOLOGY	48.8%	24.6%	47.7%	92.0%
OPHTHALMOLOGY	71.6%	35.5%	51.2%	91.1%
PSYCHIATRY	52.7%	26.8%	56.3%	91.2%
RESPIRATORY	55.9%	21.9%	64.5%	95.6%
UROLOGY	40.9%	15.0%	69.2%	84.6%

agonism)—allowing rapid alignment with a given asset. In addition, such benchmarks provide a common reference frame from which project-specific adjustments can be made, facilitating cross-project comparisons and enabling early-stage screening decisions.

Notwithstanding these benefits, reliance on industry benchmarks introduces a number of material limitations that constrain their utility for rigorous, decision-relevant risk analysis:

- The expanding volume of historical attrition data can foster a misleading perception of objectivity or *ground truth*, obscuring the heterogeneity and context dependence of underlying development outcomes.
- Benchmark data are rarely available at a sufficiently granular and defensibly statistical level to reflect specific disease subtypes or lines of therapy (e.g., refractory, second-line breast cancer), limiting their relevance for highly targeted development programs.
- Conventional benchmarks typically fail to distinguish between exploratory Phase IIa studies, which carry substantial scientific risk, and confirmatory Phase IIb studies, which are materially less risky but are often aggregated under a single Phase II success rate.
- Benchmark approaches do not permit explicit estimation of probabilities associated with individual risk drivers, such as the likelihood of achieving clinically meaningful efficacy.
- Many root causes of failure—such as suboptimal dosing strategies, inadequate statistical power, or flawed trial design—are either poorly documented or entirely absent from public datasets.
- The use of benchmarks tends to reinforce cognitive biases, particularly anchoring and insufficient adjustment, thereby discouraging departures from industry averages even when asset-specific information warrants deviation.
- Benchmark-based estimates are poorly suited to Bayesian updating, making it difficult to revise prior probabilities in a principled manner as new internal or external data emerge.

- Such benchmarks are not explicitly linked to the attributes of a defined TPP and therefore cannot be used to discriminate among different efficacy or safety outcomes relative to the current or future standard of care.
- They are ill-equipped to support program-level risk assessment, particularly in cases where individual project components exhibit positive correlation rather than statistical independence.
- Benchmarks do not readily enable evaluation of phase-specific risk implications of alternative development strategies, such as accelerated versus conservative clinical pathways.
- Finally, reported transitional success rates reflect the combined outcomes of both high-quality programs (representing sound strategic and operational decisions) and low-quality programs (reflecting poor decision making), thereby conflating decision quality with inherent scientific risk.

For project teams, the application of risk decomposition and structured elicitation to conduct phase-specific risk analysis across alternative development strategies is therefore of particular importance. Unlike benchmark-based approaches, this methodology enables teams to explicitly examine how different strategic choices alter the risk profile of an asset. For example, an organization may wish to understand the incremental risk it must accept in pursuit of an accelerated development and launch timeline, relative to a more conservative strategy that emphasizes risk reduction and evidence accumulation. By decomposing risk at the phase level and quantifying the probability of success associated with each major risk factor, the organization can make informed trade-offs between speed, uncertainty, and value.

Consider a scenario in which a project team has defined three strategic alternatives spanning early development, late-stage clinical execution, and regulatory submission for a given molecule and indication. While, in principle, a strategy matrix such as that illustrated in Table 5.6 could accommodate up to $3^3 = 27$ distinct strategic permutations, practical decision making typically focuses on a small subset of strategically coherent and actionable pathways. The three most compelling alternatives are therefore selected for detailed risk analysis, enabling leadership to compare their respective risk-reward profiles in a transparent and analytically defensible manner.

Table 5.6 Generation of strategic alternatives for clinical development and registration. SAD = Single Ascending Dose; MAD = Multiple Ascending Dose.

STRATEGIC ALTERNATIVE	STRATEGIC DECISION 1: EARLY DEVELOPMENT	STRATEGIC DECISION 2: LATE DEVELOPMENT	STRATEGIC DECISION 3: REGISTRATION
SLOW TO MARKET	*PHASE I*	*PHASE IIa THEN PHASE IIb THEN PHASE III*	*US ALONE*
<u>MOMENTUM TO MARKET</u>	**PHASE I WITH SAD**	**PHASE IIa & PHASE III CONCURRENTLY**	<u>US THEN EU SEQUENTIALLY</u>
FAST TO MARKET	<u>PHASE I WITH MAD</u>	<u>PHASE IIa THEN PHASE III</u>	**US & EU CONCURRENTLY**

Two contrasting examples illustrate the approach. A *Slow-to-Market* strategy emphasizes sequential phase progression and evidentiary robustness at the cost of longer timelines. A *Fast-to-Market* strategy compresses development by running phases in parallel and pursuing concurrent regulatory submissions, accepting higher early-stage risk. Decision-analytic modeling reveals how accelerated strategies shift unresolved risk forward in time (see Figure 5.1) while potentially accelerating value capture.

Evaluating the Evidentiary Basis for Risk Decomposition and Probabilistic Analysis

A critical and legitimate question arises when proposing risk decomposition and structured risk analysis as an alternative—or complement—to industry benchmark-based probabilities of success: What evidence exists that such an approach leads to better decisions? Before examining specific case studies, such as the Eli Lilly example, it is essential to establish the analytical criteria by which any risk analysis framework should be judged.

At a minimum, a process designed to support risk decomposition and probabilistic decision making must be grounded in empirical or statistical evidence demonstrating that it improves the quality, calibration, or reliability of decisions relative to simpler heuristics. Absent such evidence, methodological sophistication alone provides little justification for replacing established benchmarks. One useful way to assess the performance of a risk analysis methodology is through the concept of predictive calibration—that is, the degree to which predicted probabilities of success correspond to observed outcomes over a sufficiently large number of decisions. In the limiting case, a process that exhibits perfect calibration can be described as perfectly predictive. Graphically, such a process is represented by the 45-degree diagonal line shown in Figure 5.2. Along this line, average predicted probabilities are in full concordance with average realized outcomes. For example, if each set of projects within a given development phase is assigned a 60% probability of success, and across a large

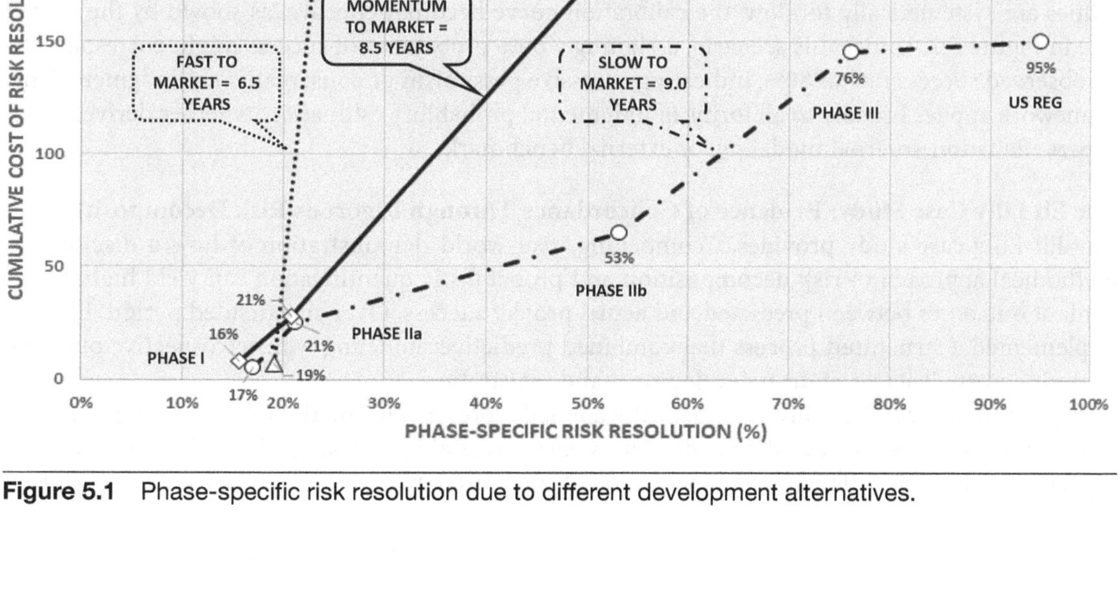

Figure 5.1 Phase-specific risk resolution due to different development alternatives.

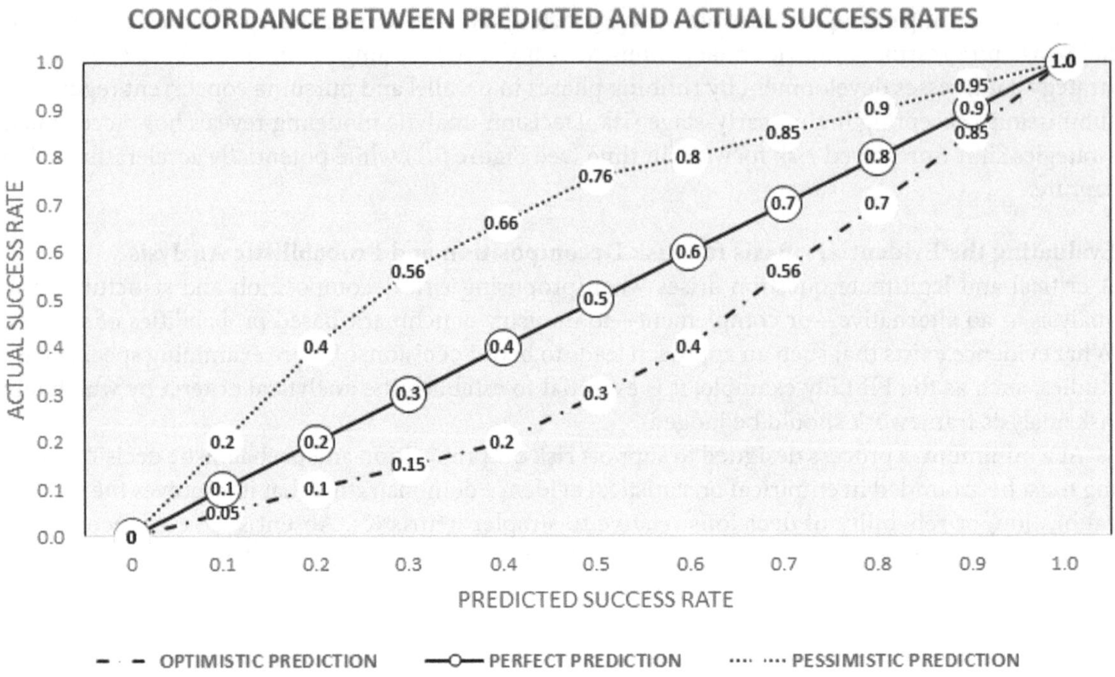

Figure 5.2 Comparing probability estimates with actual event occurrences.

sample, approximately 60% of those projects successfully progress, the prediction process displays perfect concordance between prediction and outcome. Importantly, this does not imply foreknowledge of which specific projects will succeed, but rather that the probability assignments are statistically consistent with observed frequencies.

Deviations from this diagonal line reflect systematic bias in probabilistic judgment. If predicted probabilities are consistently too high relative to observed outcomes, the resulting calibration curve becomes convex, as illustrated by the alternating dotted and dashed line in Figure 5.2. In this case, a stated 60% probability of success may correspond empirically to only 40% of projects achieving the desired outcome, reflecting overconfidence or optimism bias. Conversely, if predicted probabilities are systematically too low, the calibration curve becomes concave, as shown by the dotted line in Figure 5.2. Under this scenario, assigning a 60% probability of success might correspond to an observed success rate of 80%, indicating excessive pessimism or conservatism in judgment. This framework applies broadly to all forms of judgmental probability estimation, whether derived from expert elicitation, internal modeling, or external benchmarks.

The Eli Lilly Case Study: Evidence of Concordance Through Rigorous Risk Decomposition

The Eli Lilly case study provides a compelling, real-world demonstration of how a disciplined, methodical approach to risk decomposition and probabilistic quantification can yield highly concordant outcomes between predicted and actual project success. Over an extended period, Eli Lilly implemented a structured process that combined predictive modeling with retrospective outcome analysis across all stages of drug development and registration.

Beginning in 1997, Eli Lilly systematically applied a prescriptive methodology to assess project-level probabilities of success at four discrete stages: preclinical, Phase I, Phase II, and Phase III/regulatory submission. The process incorporated structured expert elicitation, risk decomposition aligned to TPPs, and decision-tree-based probabilistic modeling. The first public retrospective validation of this methodology was published in 2012, based on a database encompassing 730 observations across multiple therapeutic areas. The results demonstrated remarkable concordance between predicted probabilities of success and realized outcomes, particularly for probability estimates with a sufficiently large number of observations. For instance:

- Among 111 predictions of 60% probability, exactly 60% of projects ultimately succeeded, demonstrating perfect calibration.
- For 146 predictions of 80% probability, observed outcomes closely matched the predictions, reflecting near-perfect concordance.
- At the 50% probability level—the most challenging threshold to predict accurately—the observed success rate was slightly above 50% for 45 predictions, indicating robust predictive performance even at intermediate probabilities.

A notable deviation occurred at the 75% probability threshold, where 23 predictions yielded an actual success rate of approximately 96%, highlighting the inherent uncertainty in small sample sizes.

By 2025, the database had expanded to 1,394 probability judgments, enabling a more statistically robust evaluation of the methodology (Persinger and Andersen 2025). Even with the increased volume and diversity of projects, Eli Lilly's risk decomposition and elicitation framework continued to demonstrate exceptionally high concordance between predicted and actual success outcomes (see Figure 5.3).

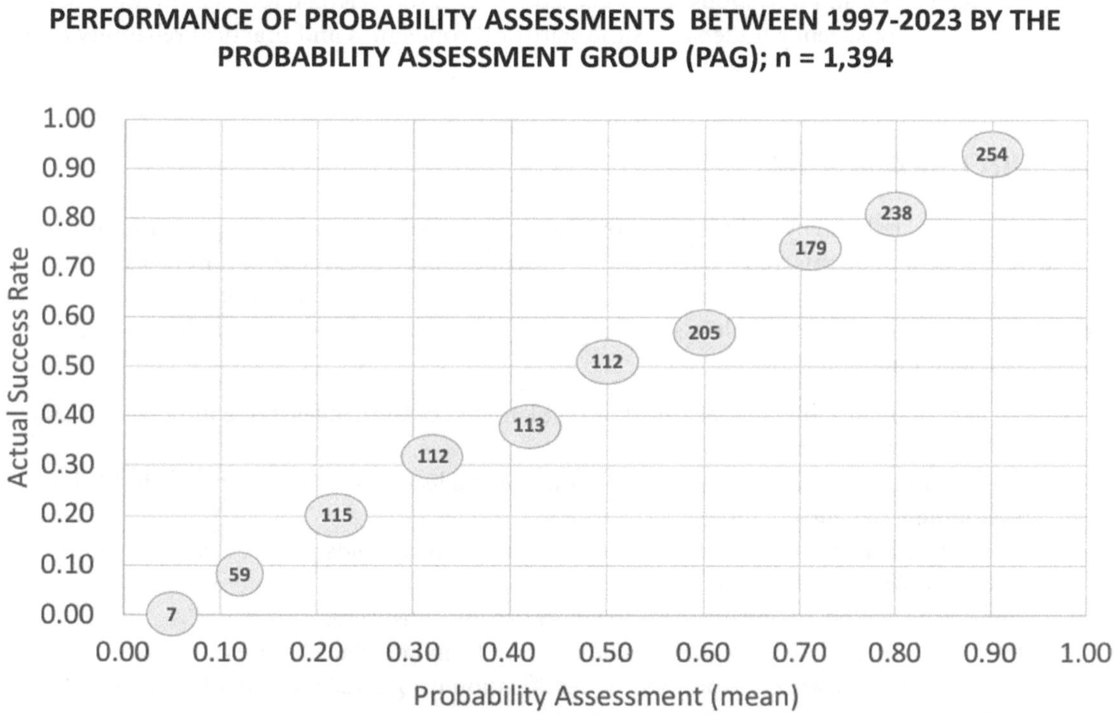

Figure 5.3 The Eli Lilly experience—concordance between probability estimates and actual event outcomes. Reproduced with permission (Persinger and Andersen 2025).

The Eli Lilly experience underscores several critical insights for biopharma R&D decision making:

1. **Protracted investment in methodology pays dividends:** Consistent application over decades allows the organization to refine elicitation protocols, calibrate expert judgments, and systematically incorporate historical outcomes into predictive models.
2. **Calibration improves with observation volume:** Concordance is strongest when probability estimates are supported by a sufficient number of observations, highlighting the value of large, longitudinal datasets.
3. **Structured risk decomposition enhances decision quality:** By decomposing projects into component risk factors and aligning probabilities to specific development stages and TPP attributes, organizations can achieve predictive accuracy far beyond what is typically attainable with generic industry benchmarks.
4. **Transparency and retrospective validation matter:** Publicly reporting retrospective concordance fosters accountability and provides an empirical foundation for confidence in decision-making processes.

Over more than two decades (1997–2025), Eli Lilly's structured approach to risk decomposition and probability assessment has demonstrated exceptionally high concordance between predicted and actual project outcomes. Across more than 1,394 probability-of-success judgments spanning preclinical, Phase I, Phase II, and Phase III/regulatory stages, the predicted probabilities consistently aligned with realized outcomes. Analysis of the data, irrespective of how it is stratified—by development phase or probability threshold—reveals that observed outcomes typically fall within ±5% of the predicted values, and deviations rarely exceed ±10% of the diagonal line representing perfect concordance. This consistency underscores the statistical robustness and practical reliability of the methodology.

5.3 PROGRAM RISK

While project-level risk can often be decomposed and quantified with precision, program- and portfolio-level risk cannot be inferred by aggregating individual project risks. Program risk is nonadditive and arises from structural dependencies, strategic objectives, shared resource constraints, and correlation effects that span multiple assets. Program risk may be defined across several overlapping dimensions, including:

- The likelihood of achieving technical or regulatory success in at least one indication pursued by a molecule or platform
- The probability of establishing proof-of-concept in a lead indication upon which subsequent development depends
- The probability of meeting economic and strategic thresholds, such as value creation and life-cycle objectives
- Exposure to capital, operational, and manufacturing constraints
- Downside risk, including negative value creation from late-stage failures
- Structural fragility arising from poor diversification or overreliance on a limited number of assets, disease areas, or biological hypotheses

Program Risk Under Technical Independence: Joint Outcome Analysis

Figure 5.4 illustrates a development program comprising two technically independent projects—distinct molecules targeting mild cognitive impairment (MCI) and Alzheimer's disease (AD). The

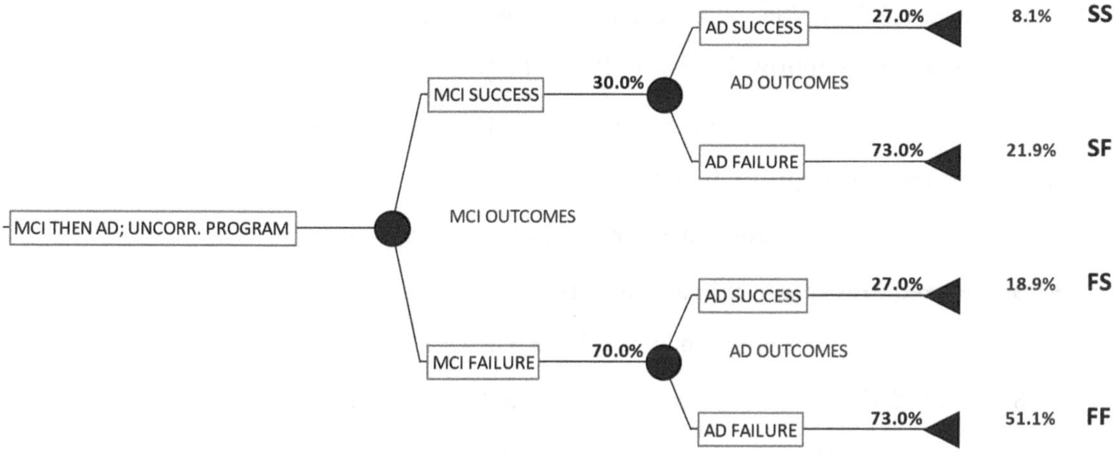

Figure 5.4 A program comprised of two independent, technically uncorrelated projects, each with binary outcomes.

projects are uncorrelated: the probability of AD success (27%) is unchanged regardless of the MCI outcome.

The MCI-AD program, consisting of two technically independent projects, gives rise to four mutually exclusive and collectively exhaustive joint outcomes. Under technical independence, joint probabilities are obtained as the simple product of the marginal probabilities of success and failure for each indication.

Let:

- p(MCI success) = 0.30 and p(AD success) = 0.27

The resulting joint outcome space is as follows:

1. **Success in MCI and success in AD (SS)**

$$p(SS) = 0.30 \times 0.27 = 0.081$$

2. **Success in MCI and failure in AD (SF)**

$$p(SF) = 0.30 \times 0.73 = 0.219$$

3. **Failure in MCI and success in AD (FS)**

$$p(FS) = 0.70 \times 0.27 = 0.189$$

4. **Failure in MCI and failure in AD (FF)**

$$p(FF) = 0.70 \times 0.73 = 0.511$$

As expected, the four outcomes sum to unity, confirming that the joint probability space is complete.

Crucially, program-level risk cannot be interpreted without an explicit and operational definition of *success*. Depending on how success is defined—scientifically, strategically, or economically—the same underlying probability structure yields materially different conclusions regarding program attractiveness and risk exposure.

Several reasonable definitions of success may be considered:

1. **Success requires approval in both MCI and AD**

$$p(\text{Program Success}) = p(SS) = 0.081$$

2. **Success requires approval in at least MCI**

$$p(\text{Program Success}) = p(SS) + p(SF) = 0.30$$

3. **Success requires approval in at least AD**

$$p(\text{Program Success}) = p(SS) + p(FS) = 0.27$$

4. **Success is defined as any technical success in the program**

$$p(\text{Program Success}) = p(SS) + p(SF) + p(FS) = 0.489$$

Under the most permissive definition—success of any type—the probability of total program failure is therefore:

$$p(FF) = 0.511 = 1 - p(\text{Success of any type})$$

This simple example illustrates a fundamental principle of program risk analysis: identical project-level probabilities can produce markedly different program-level risk profiles depending solely on how success is defined. As such, ambiguity in success criteria can lead to inconsistent or misleading interpretations of risk at the governance and portfolio levels.

Introducing Technical Dependence: Weak Positive Correlation

Now consider a structurally similar program involving the same two indications—MCI and AD—but with technical dependence between the projects. Specifically, assume a weak positive correlation, such that success in MCI modestly increases the likelihood of success in AD, while failure in MCI modestly decreases it. As shown in Figure 5.5:

- $p(\text{AD success} \mid \text{MCI success}) = 0.40$ and $p(\text{AD success} \mid \text{MCI failure}) = 0.21$

Because the conditional probability of AD success varies depending on the outcome in MCI, the projects are no longer independent. This dependency structure fundamentally alters the joint probability distribution and, consequently, the interpretation of program risk under each definition of success.

The implications of even weak correlation are profound: program risk can increase or decrease materially without any change in marginal project probabilities, underscoring why portfolio and program risk cannot be inferred from project-level metrics alone.

As shown in Figure 5.5, the same two-project program—targeting MCI and AD—now exhibits weak positive technical correlation. In this case, success in MCI modestly increases the probability of success in AD, while failure in MCI modestly decreases it. This dependence alters the joint probability structure relative to the independent case.

Given:

- $p(\text{AD success} \mid \text{MCI success}) = 0.40$ and $p(\text{AD success} \mid \text{MCI failure}) = 0.21$

The four mutually exclusive and collectively exhaustive joint outcomes are:

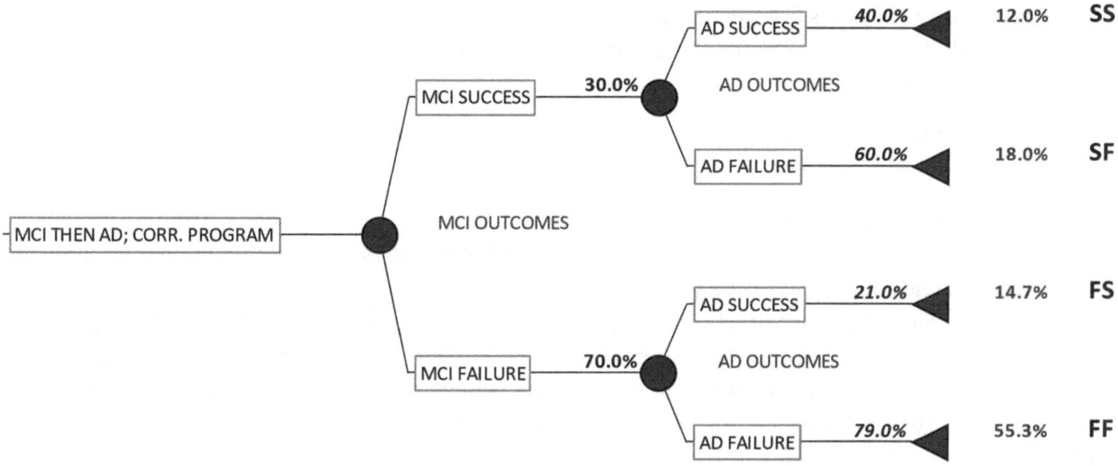

Figure 5.5 A program comprised of two dependent, positively correlated projects, each with binary outcomes.

1. **Success in MCI and success in AD (SS)**

$$p(SS) = 0.30 \times 0.40 = 0.120$$

2. **Success in MCI and failure in AD (SF)**

$$p(SF) = 0.30 \times 0.60 = 0.180$$

3. **Failure in MCI and success in AD (FS)**

$$p(FS) = 0.70 \times 0.21 = 0.147$$

4. **Failure in MCI and failure in AD (FF)**

$$p(FF) = 0.70 \times 0.79 = 0.553$$

As in the independent case, program risk remains undefined without an explicit definition of success. Using the previous joint probability space, several reasonable interpretations emerge:

1. **Success requires approval in both diseases**

$$p(\text{Program Success}) = p(SS) = 0.120$$

2. **Success requires approval in at least MCI**

$$p(\text{Program Success}) = p(SS) + p(SF) = 0.300$$

3. **Success requires approval in at least AD**

$$p(\text{Program Success}) = p(SS) + p(FS) = 0.267$$

4. **Success is defined as any technical success within the program**

$$p(\text{Program Success}) = p(SS) + p(SF) + p(FS) = 0.447$$

Accordingly, the probability of total program failure is 0.553.

A comparison of Figures 5.4 (technical independence) and 5.5 (weak positive correlation) highlights a critical insight for program-level decision making. Positive correlation simultaneously increases tail outcomes:

- The probability that both projects succeed increases from 8.1% to 12.0%
- The probability that both projects fail increases from 51.1% to 55.3%

Thus, correlation does not simply raise or lower overall success—it amplifies extremes, increasing both upside concentration and downside exposure. This effect is invisible when risk is assessed solely at the individual project level.

As will be explored quantitatively in Chapter 6, positive correlation among linked projects has a first-order impact on program valuation, risk concentration, and capital efficiency. Such linked decision making is common in practice, particularly when (a) pursuing multiple diseases with a single molecule, (b) developing multiple lines of therapy within the same disease, and (c) sequencing indications contingent on prior proof of concept.

5.4 PORTFOLIO RISK

As with the program level, portfolio risk cannot be evaluated without an explicit definition of risk. At the portfolio level, risk is inherently multidimensional and extends beyond technical success probabilities. Common portfolio-level risk dimensions include:

- Probability of achieving economic and strategic objectives (e.g., revenue, value creation)
- Probability of exceeding critical constraints (e.g., R&D budget, manufacturing capacity)
- Likelihood of launching a sufficient number of new molecular entities (NMEs) within a defined horizon
- Degree of balance and diversification across diseases, modalities, mechanisms, and platforms

Competing Definitions of Portfolio Risk

Consider two alternative portfolios—a base-case and a high-risk portfolio—represented by overlapping probability density functions in WAV Figure 5.6. Because the distributions overlap, there is neither deterministic dominance nor first-order stochastic dominance. Absent such dominance, neither portfolio can be labeled *riskier* without reference to organizational objectives and risk tolerance.

Consequently, the question "Which portfolio is riskier?" depends on how risk is defined:

- **Variance-based risk:** If risk is defined as dispersion around the mean, the high-risk portfolio is riskier by construction due to its higher variance
- **Downside-focused risk:** If risk is defined by downside exposure—for example, probability of negative net present value (NPV) or value at risk—either portfolio may appear riskier depending on the threshold and tail behavior
- **Goal-attainment risk:** If risk is defined as failing to meet strategic goals (e.g., ≥X NMEs or revenue targets), the portfolio with less probability mass above the goal threshold is riskier
- **Resource-constraint risk:** A portfolio may be riskier if it has a higher probability of exceeding fixed constraints (e.g., peak R&D spend), regardless of expected value

These perspectives underscore a central principle: portfolio risk is not an intrinsic property of a distribution but a function of preferences given constraints and objectives. Consequently, portfolio selection requires explicit success criteria, quantified downside exposure, and alignment with stated risk tolerance.

A close examination of the distribution of value outcomes in Figure 5.7 highlights the trade-offs. The base-case portfolio (broken curve) spans approximately \$0M–\$10,125M with a standard deviation (SD) of \$1,510M. The high-risk portfolio (solid curve) spans \$0M–\$13,524M with a higher SD of \$2,039M, indicating greater volatility. If risk is defined narrowly as the probability of absolute loss, both portfolios are equivalent, as each has a lower bound of \$0M. However, when success is defined relative to performance thresholds, conclusions diverge. For example, when success is defined as exceeding a portfolio value of \$800M, the high-risk portfolio offers a >13% probability of success, versus ~7% for the base case. Conversely, if success is defined as avoiding outcomes below \$4,000M, the base-case portfolio is preferred, with <13% probability below the threshold compared with >21% for the high-risk portfolio. These results demonstrate that portfolio preferences reverse depending on whether upside optionality or downside protection is more important.

Because annual R&D budgets are among the most binding constraints, portfolio risk is often evaluated in terms of spending volatility. As shown in Figure 5.8, the high-risk portfolio exhibits a broader annual budget distribution (\$0M–\$2,650M; SD \$399M) than the base-case portfolio (\$0M–\$2,038M; SD \$296M).

If the annual budget ceiling is \$1,500M, the base-case portfolio has a <13% probability of exceeding it, while the high-risk portfolio exceeds it by 17%. However, the high-risk portfolio also has greater downside flexibility, with a 39% probability of spending less than \$1,000M, versus 29% for the base case. Thus, portfolios that are riskier in terms of overspend may simultaneously offer greater capital optionality under failure scenarios.

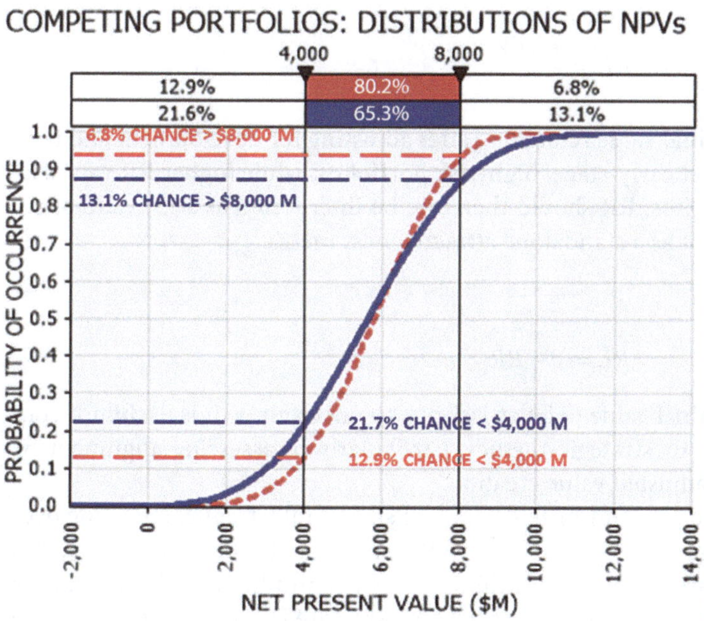

Statistics

	BASE CASE/...	HIGH RISK/...
Cell	'IRS Port Ri...	'IRS Port Ris...
Minimum	0	0
Maximum	10,124.94	13,524.11
Mean	5,752.92	5,661.93
90% CI	± 7.85	± 10.61
Mode	5,925.48	5,112.46
Median	5,796.33	5,622.77
Std Dev	1,509.50	2,038.84
Skewness	-0.1273	0.1110
Kurtosis	2.7612	2.7711
Values	100000	100000
Errors	0	0
Filtered	0	0
Left X	4,000	4,000
Left P	12.9%	21.6%
Right X	8,000	8,000
Right P	93.2%	86.9%

Figure 5.7 Portfolio risk defined in terms of financial targets.

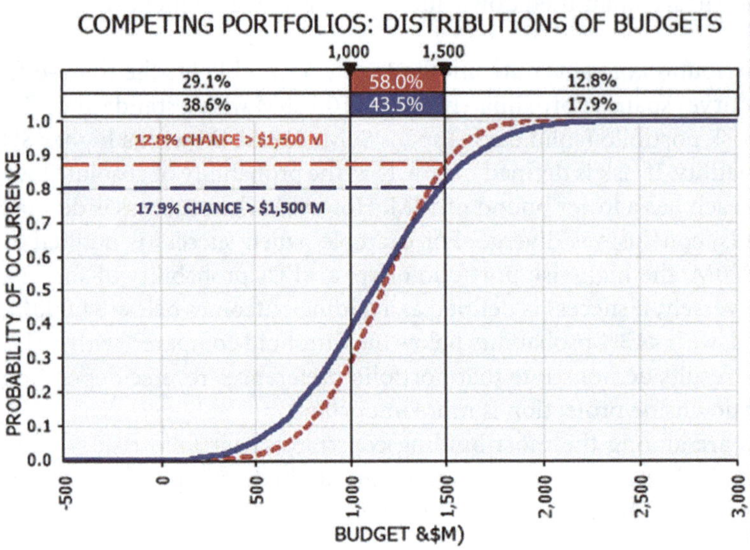

Statistics	BASE CASE/...	HIGH RISK/...
Cell	'IRS Port Ri...	'IRS Port Ris...
Minimum	0	0
Maximum	2,037.79	2,649.73
Mean	1,160.02	1,124.73
90% CI	± 1.54	± 2.08
Mode	1,171.42	1,219.09
Median	1,167.83	1,118.53
Std Dev	295.83	399.44
Skewness	-0.1548	0.0903
Kurtosis	2.7878	2.7692
Values	100000	100000
Errors	0	0
Filtered	0	0
Left X	1,000	1,000
Left P	29.1%	38.6%
Right X	1,500	1,500
Right P	87.2%	82.1%

Figure 5.8 Portfolio risk defined in terms of budget targets.

Portfolio Risk Metrics: Risk Ratio and Risk Load

To support consistent portfolio comparisons, two complementary metrics are proposed:

Risk Ratio (RR)

$$RR = \frac{eNPV}{NPV}$$

RR represents the proportion of nominal value remaining after adjusting for development and regulatory risk. Low RR values often reflect early-stage, high-risk portfolios, while higher RR values are typical of later-stage, de-risked portfolios. RR should therefore be interpreted as a measure of risk exposure relative to nominal value, not as a standalone attractiveness metric.

Risk Load (RL)

$$RL = 1 - RR$$

RL quantifies the fraction of the nominal value to be *at-risk* due to uncertainty. It is useful for comparing therapeutic areas, modalities, or strategic themes, particularly in assessing alignment between investment intensity and risk-adjusted value creation.

However, portfolios optimized solely to minimize RL may be structurally fragile, reflecting overconcentration in mature or late-stage assets. Both RR and RL are strongly influenced by portfolio maturity; early-stage portfolios are expected to exhibit lower RR and higher RL than mature portfolios with similar NPV. Accordingly, RR and RL should be compared across portfolios or subportfolios of comparable maturity. When combined with absolute value, budget exposure, and strategic objectives, these metrics provide a disciplined framework for balancing risk-adjusted value creation, capital efficiency, and long-term portfolio sustainability.

5.4.1 Portfolio Balance as a Function of Organizational Risk Tolerance

If organizational risk tolerance is the primary driver of portfolio risk management, a temporally balanced portfolio represents the aspirational end state of effective governance. In practice, however, portfolio balance is neither absolute nor universal. Like value, it is contextual and preference-dependent: what constitutes balance for one organization may be misaligned with another's strategic intent.

Accordingly, any meaningful discussion of portfolio balance must begin with an explicit organizational definition of *balance* and identification of the dimensions that matter most. Without this clarity, balance remains a qualitative aspiration rather than an operational construct.

A balanced portfolio reflects deliberate alignment across several interdependent dimensions:

- **Strategic intent:** Balance must be evaluated relative to a clearly articulated future state, including therapeutic focus, modality mix, geographic footprint, and competitive positioning
- **Financial orientation:** Multi-year financial objectives—revenue growth, margins, and capital efficiency—define the economic boundaries for portfolio choices
- **Diversification:** Balance requires intentional decisions about where to compete and where not to compete, spanning diseases, mechanisms, platforms, and competitive intensity, while avoiding excessive concentration risk
- **Risk tolerance:** The acceptable mix of early-, mid-, and late-stage assets must be consistent with leadership's tolerance for uncertainty and outcome variability
- **Risk–reward trade-offs:** Higher expected rewards typically require acceptance of higher risk; governance must therefore define acceptable risk limits relative to potential reward
- **Resource allocation:** Balance is constrained by the organization's ability to commit financial capital, scientific talent, clinical capacity, manufacturing readiness, and managerial attention

Together, these elements underscore that portfolio balance is dynamic rather than static, continuously recalibrated as information and external conditions evolve.

Organizations pursue balance as a pragmatic response to the tension between near-term performance and long-term value creation. Sustained innovation depends on near-term cash flow, yet excessive emphasis on short-term objectives—often reinforced by incentive systems—can bias portfolios toward lower-risk, later-stage assets, eroding long-term pipeline strength. Achieving balance, therefore, requires explicit alignment among strategic intent, multi-horizon financial objectives, and executive incentives that reward sustainable, risk-adjusted value creation rather than short-term results alone.

In practice, portfolio balance is often conceptualized as a deliberate mix of higher-risk, higher-reward projects—typically earlier-stage or mechanistically differentiated—and lower-risk, lower-variance projects closer to commercialization. These categories contribute differently to near-term performance and future optionality. Visual tools such as three-dimensional bubble charts are commonly used to map projects across value, risk, cost, and time dimensions (see Figure 5.9).

While useful for communication, such visualizations do not, in themselves, confer balance. True balance emerges from scenario-informed decision making that explicitly recognizes interdependencies among risk, time, value, and resources.

Importantly, balance is not an intrinsic property of project volume but a function of alignment with strategic and financial goals. For example, an organization focused on maximizing near-term growth (e.g., one- to three-year revenues) will rationally prioritize late-stage or near-commercial assets with high probability of near-term returns. Conversely, an organization emphasizing long-term

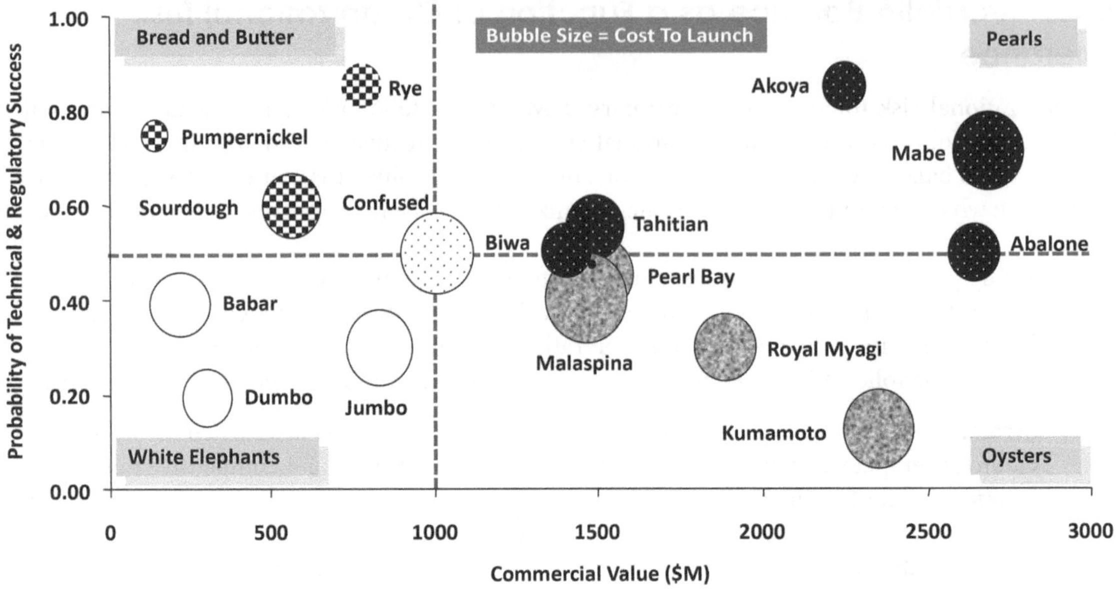

Figure 5.9 Portfolio diversification—representation of assets across value, risk, and cost dimensions.

growth (e.g., seven- to ten-year horizons) will place greater value on discovery and early clinical assets that provide differentiation and pipeline renewal. Thus, the same portfolio may be judged balanced or unbalanced depending entirely on the temporal orientation of strategic objectives.

5.4.2 Quantifying Pipeline Risk in Achieving Annual Product Launch Targets

The *launch profile* is a commonly used visualization in biopharma to depict the expected cadence of annual product launches across the R&D pipeline. When properly constructed, it provides a useful communication tool for conveying pipeline momentum and perceived growth potential to internal and external stakeholders. However, naïve interpretation of launch profiles can create a misleading sense of productivity by implicitly assuming equal probabilities of success across projects and development stages—an assumption inconsistent with historical R&D outcomes.

A more analytically robust approach applies the phase-specific PTRS, derived from internal benchmarks or external industry data. Risk-adjusting the launch profile transforms it from a descriptive artifact into a quantitative decision-support tool, enabling estimation of expected launches on a probability-weighted basis. This allows leadership to evaluate whether credible launch output aligns with long-range plan (LRP) revenue objectives, shifting the focus from nominal milestone counts to expected commercial contribution.

Consider the hypothetical company *Innovations 'R' Us*, with 24 active projects spanning preclinical through regulatory stages, including 10 NMEs. The unadjusted launch profile suggests clustering of launches in year five and apparent gaps elsewhere. Without risk adjustment, such patterns can be overinterpreted. Applying PTRS reveals that apparent peaks and troughs often reflect optimistic scheduling rather than credible revenue drivers.

For example, although five nominal launch candidates appear in year five, applying PTRS (assuming independence) yields an expected 1.2 launches in that year (see Figure 5.10). Across the full 10-year horizon, the portfolio of 24 projects delivers an expected 7.05 launches. This risk-adjusted

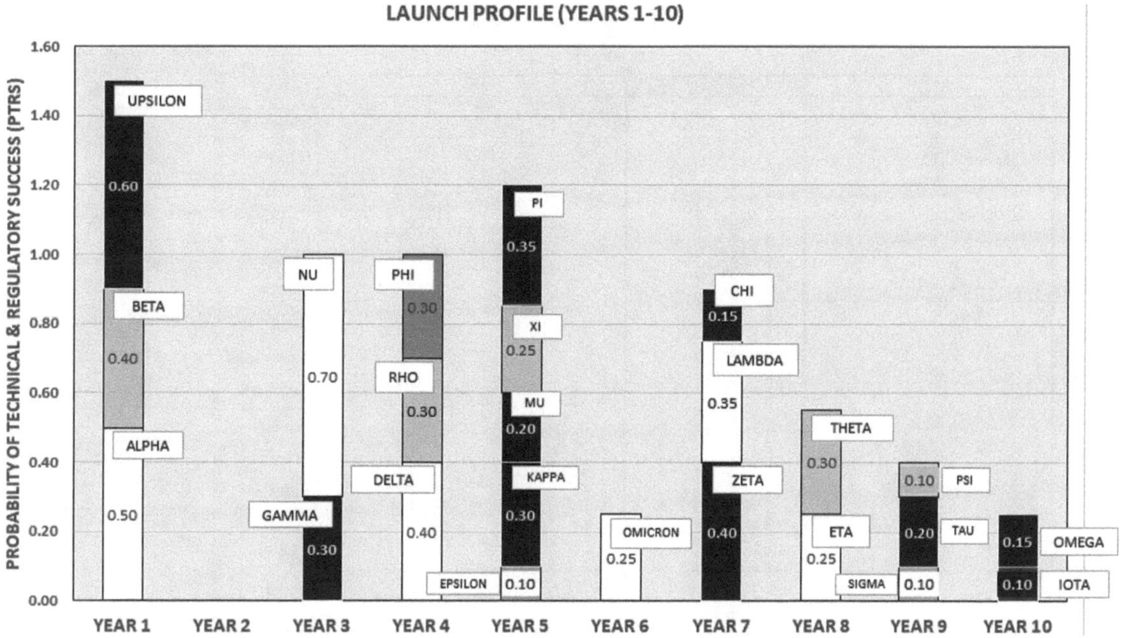

Figure 5.10 Launch profile of 24 pipeline projects (preclinical to regulatory).

launch profile shows expected NME launches in six of nine eligible years, with approximately one launch expected in years three and four. Even projects with high PTRS (e.g., 0.70) retain meaningful failure risk; treating them as deterministic *sure bets* systematically understates portfolio risk.

To capture this launch uncertainty explicitly, the analysis is extended to a stochastic framework using simulation. Each project's PTRS is modeled as a Bernoulli random variable, generating simulated portfolio realizations across thousands of iterations. This produces empirical distributions for total launches, annual launches, zero-launch years, and tail outcomes. The resulting distribution (see Figure 5.11A) centers near the deterministic mean, with a mode of seven launches (18.8%). There is a 71.3% probability of achieving between six and ten launches, but a meaningful 23.3% probability of five or fewer launches, highlighting nontrivial downside risk. Upside optionality exists, with a 5.4% probability of more than ten launches (maximum observed: 17), which may strain downstream capabilities.

The stochastic distribution therefore enables leadership to move beyond heuristic statements such as, "We expect around seven launches," toward a more disciplined discussion of:

- How much downside launch risk the organization is willing (and able) to absorb
- Whether upside outcomes can be operationally supported
- How sensitive financial and resource plans are to deviations from the modal outcome

Isolating the NME subset (see Figure 5.11B) reveals greater fragility. The mode is three launches (28.1%), with a 53.0% probability of achieving three to five NMEs and a 44.5% probability of two or fewer. This asymmetry underscores the vulnerability of innovation-driven growth and the need for explicit NME-focused risk management.

Year-specific simulations further expose short-term volatility (see WAV Figures 5.12A and B). In year 5, there is a 75.5% probability of one to five launches, but also a 24.5% probability of zero launches. The most likely outcome is a single launch (41.4%), despite multiple nominal candidates.

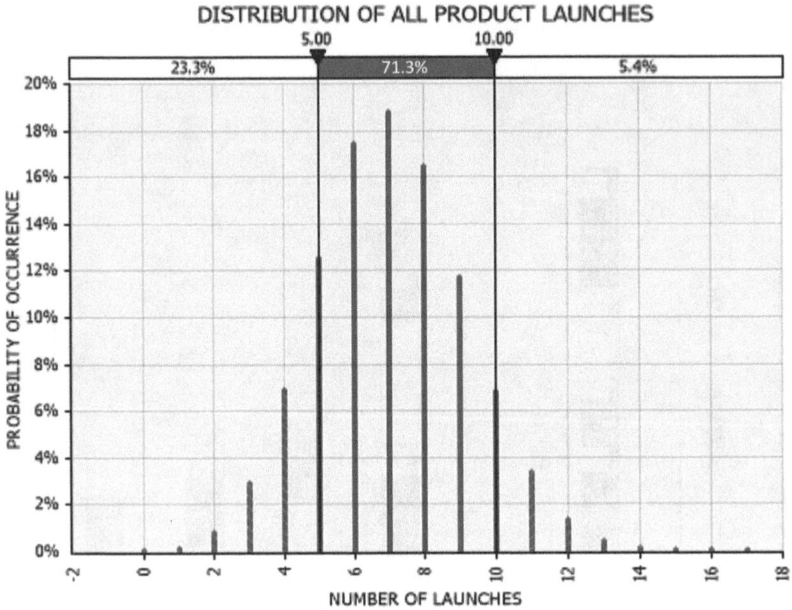

Statistics	
	TOTAL/Bernoulli
Cell	'Launch Profile 2'!D27
Minimum	0
Maximum	17
Mean	7.050
90% CI	± 0.0109
Mode	7
Median	7
Std Dev	2.099
Skewness	0.1678
Kurtosis	2.9565
Values	100000
Errors	0
Filtered	0
Left X	5.00
Left P	23.3%
Right X	10.00
Right P	94.6%

A

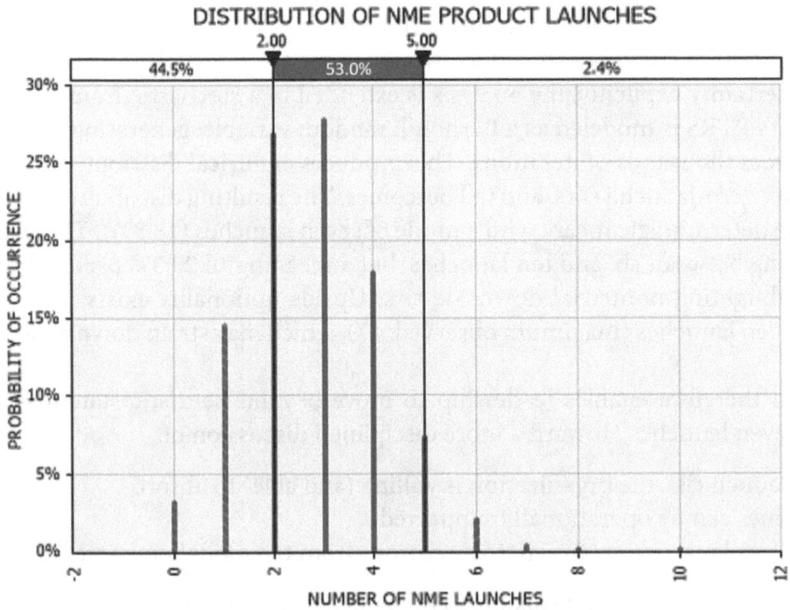

Statistics	
	NME TOTAL/Berno ...
Cell	'Launch Profile 2'!D41
Minimum	0
Maximum	10
Mean	2.750
90% CI	± 0.00697
Mode	3
Median	3
Std Dev	1.340
Skewness	0.2808
Kurtosis	2.8951
Values	100000
Errors	0
Filtered	0
Left X	2.00
Left P	44.5%
Right X	5.00
Right P	97.6%

B

Figure 5.11A and B Distribution of all (A) and NME (B) product launches over 10 years.

Similar dynamics apply to NMEs, where there is a 36.4% probability of zero NME launches and a mode of one (44.4%), reinforcing the risk of innovation gaps.

Simulation can also be applied to pipeline supplementation decisions. For example, adding nine preclinical projects, each with a PTRS of 11%, yields a 65.1% probability of one to six launches but a 34.9% probability of zero launches, despite a naïve expectation of ~1 launch (see WAV Figure 5.13A). Increasing to 14 projects raises the probability of at least one launch to 80.5%, reducing the zero-launch probability to 19.5% (see WAV Figure 5.13B). Alternatively, increasing PTRS to 16% for nine projects yields results comparable to those shown in Table 5.7.

Parenthetically, without employing simulation analysis, these probabilities of product launches can be calculated using binomial formulae in Excel. For example, the probability of exactly one launch from a portfolio of nine projects each with a PTRS of 11.0% can be calculated as:

- =(BINOM.DIST(1,9,0.11,FALSE)

while the probability of securing at least 1 product launch from the same portfolio can be calculated as:

- =1-BINOM.DIST(0,9,0.11,TRUE)

Accounting for technical dependencies between projects

Thus far, all R&D projects in the pipeline have been considered to be probabilistically independent of each other. Stated otherwise, an overarching assumption has been made that the eventual success or failure of some products through the different phases of R&D to launch does not probabilistically impact the eventual success of any other projects in the R&D pipeline. To be sure, this assumption is, more often than not, fraught with problems as some projects—through their underlying mechanisms of action (MoAs)—bear technical similarities to others. Because of this, it is reasonable to infer that success or failure in such projects can have a judgmental bearing on the success or failure of others.

Table 5.7 Probability of achieving one or more product launches from a pipeline of independent projects.

Average of Active Substances Across All TAs	Phase I	Phase II	Phase III	Reg.
Probability of Success to Next Phase	50%	33%	73%	92%
Probability of Success to Launch	11%	22%	67%	92%
Average # of Projects to Achieve 1 Launch	9	5	2	2
Probability of Achieving 1 Launch From 9 Phase I Projects each with a PTRS of 11%	39%			
Probability of Achieving >=1 Launch From 9 Phase I Projects each with a PTRS of 11%	65%			
Average # of Phase I Projects each with a PTRS of 11% to Attain 80% Probability of >= 1 Launch	14			
Average PTRS of 9 Phase I Projects to Attain 80% Probability of >= 1 Launch	16%			

To illustrate this point, let's examine an R&D pipeline that has its research projects in oncology bucketed into three distinct MoAs (see WAV Table 5.8):

- Chimeric antigen receptor-T Cell (CAR-T) therapies such as Kymriah (Cetuximab) for the treatment of relapsed or refractory B-cell acute lymphoblastic leukemia
- Tyrosine Kinase (TK) inhibitors such as Gleevec (Imatinib) for the treatment of chronic myeloid leukemia
- Monoclonal antibody (MAb) therapies such as Herceptin (Trastuzumab) for the treatment of early-stage and metastatic breast cancer.

Potential and risk-adjusted sales for each molecule over years 1–10 are shown along with annual sales targets. For example, in year 10, aggregated potential and risk-adjusted sales are $15,715M and $4,716M, respectively, while the target is $7,000M.

To account for intra-bucket correlations, a correlation matrix is set up for the CAR-T bucket of projects, as shown in Table 5.9, reflecting the strength of positive relatedness among these projects.

It is assumed that CAR-T molecules are highly correlated (Spearman coefficients between 0.60 and 0.80), TK Inhibitor projects are moderately correlated (Spearman coefficients between 0.40 and 0.60; table not shown), and MAb entities are weakly correlated (Spearman coefficients between 0.20 and 0.50; table not shown). We are now positioned to analyze the differences in (a) product output and (b) sales attainment between these three buckets of projects under two scenarios: (a) probabilistic independence (no correlations) and (b) probabilistic dependence (positive correlations ranging from weak to strong).

Figure 5.14A and B show the distributions of output for CAR-T molecules when (a) no correlations and (b) high positive correlations between molecules are assumed. We can deduce the following:

- Without correlation, there is a 2.4% likelihood that 0 product launches will occur, and with positive correlation, there is a 20.0% likelihood that 0 launches will occur
- Without correlation, there is a 97.6% probability that at least 1 launch will occur; the corresponding estimate with positive correlation is 80.0%
- The modes are 3 product launches without correlation and 0 product launches with positive correlation

Table 5.9 Correlation matrix for CAR-T projects—highly positively correlated with Spearman coefficients between 0.60 and 0.80.

CORR	CAR-T-1	CAR-T-2	CAR-T-3	CAR-T-4	CAR-T-5	CAR-T-6
CAR-T-1	1	0.80	0.75	0.70	0.65	0.60
CAR-T-2		1	0.80	0.75	0.70	0.65
CAR-T-3			1	0.80	0.75	0.70
CAR-T-4				1	0.80	0.75
CAR-T-5					1	0.80
CAR-T-6						1

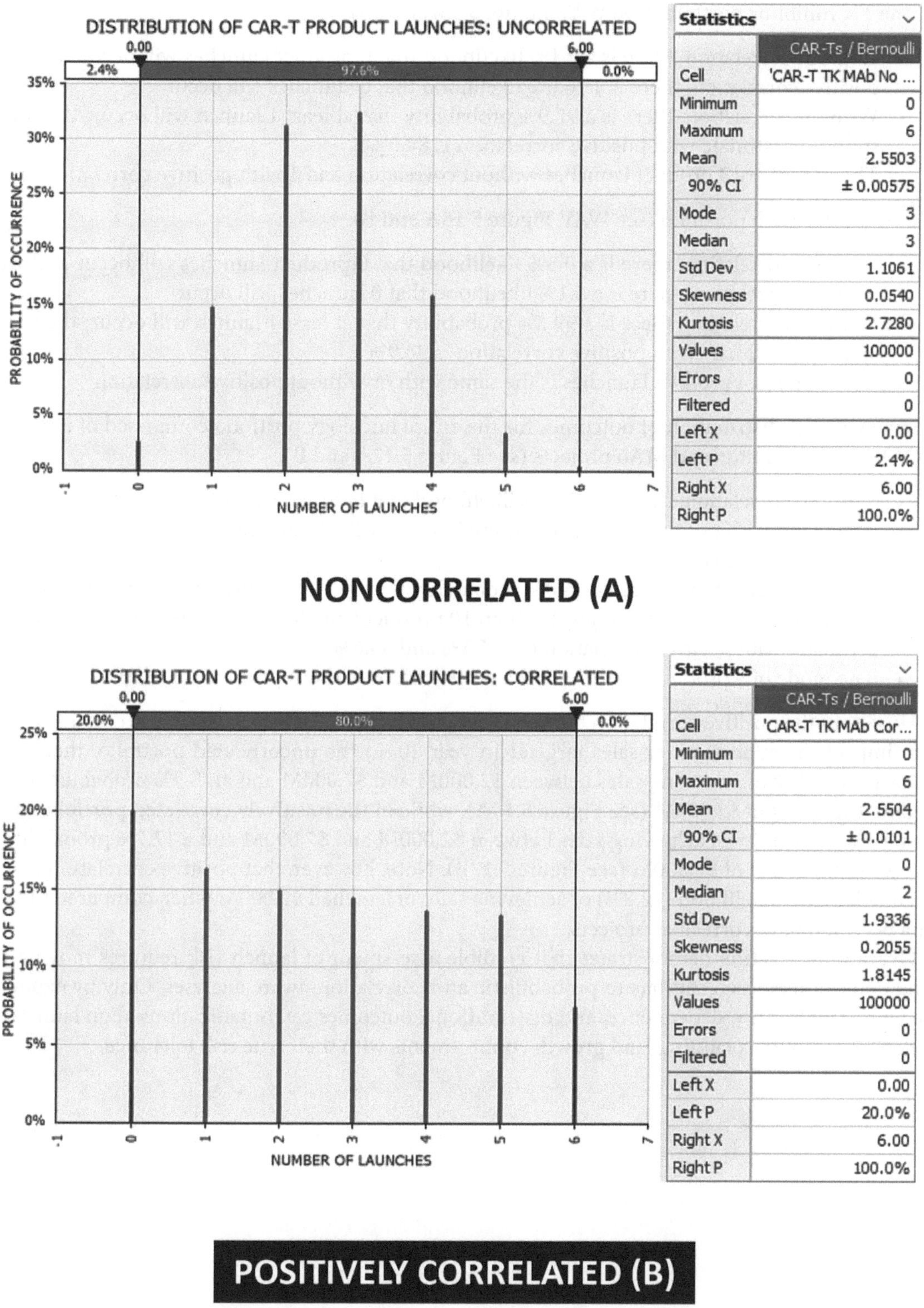

Figure 5.14A and B CAR-T portfolio—likelihood of securing at least 1 launch from 6 noncorrelated (A) and positively correlated (B) projects.

For the TK Inhibitor portfolio (see WAV Figure 5.15A and B):

- Without correlation, there is a 2.1% likelihood that 0 product launches will occur, and with positive correlation, there is a 16.0% likelihood that 0 launches will occur
- Without correlation, there is a 97.9% probability that at least 1 launch will occur; the corresponding estimate with positive correlation is 84.0%
- The modes are 3 product launches without correlation and 1 with positive correlation

Finally, for the MAb portfolio (see WAV Figure 5.16A and B):

- Without correlation, there is a 0.3% likelihood that 0 product launches will occur and with positive correlation, there is a 5.1% likelihood that 0 launches will occur
- Without correlation, there is a 99.7% probability that at least 1 launch will occur; the corresponding estimate with positive correlation is 94.9%
- The mode of 4 product launches is the same with or without positive correlation

In reviewing the distribution of outcomes for the entire oncology portfolio comprised of 3 buckets of CAR-T, TK Inhibitor, and MAb projects (see Figure 5.17A and B):

- Without correlation, there is a 72.1% likelihood that between 6 and 10 product launches will occur, and with positive correlation, there is a 50.3% likelihood that between 6 and 10 product launches will occur
- Without correlation, there is a 4.0% probability that 5 or fewer product launches will occur and a 23.9% probability that greater than 10 product launches will occur; the corresponding estimates with positive correlation are 16.1% and 33.6%
- The mode of 9 product launches is the same with or without positive correlation

How does the positive correlation between these three classes of molecules in the oncology portfolio impact the attainment of sales targets? In year 10, in the uncorrelated portfolio, there is an 88.7% probability of achieving sales between $2,000M and $7,000M and an 8.9% probability of exceeding the target of $7,000M (see Figure 5.18A), while in the positively correlated portfolio, there is a 69.4% probability of achieving sales between $2,000M and $7,000M and a 17.7% probability of exceeding the target of $7,000M (see Figure 5.18B). Note, however, that positive correlation results in a much higher likelihood (12.8%) of achieving sales of less than $2,000M when compared to 2.5% for a portfolio of uncorrelated projects.

Overall, this section demonstrates that credible assessment of launch risk requires moving beyond deterministic expectations to probabilistic and correlation-aware analyses. Only by explicitly modeling uncertainty, dependence, and distributional outcomes can organizations align launch expectations, resource planning, and growth commitments with their true risk tolerance.

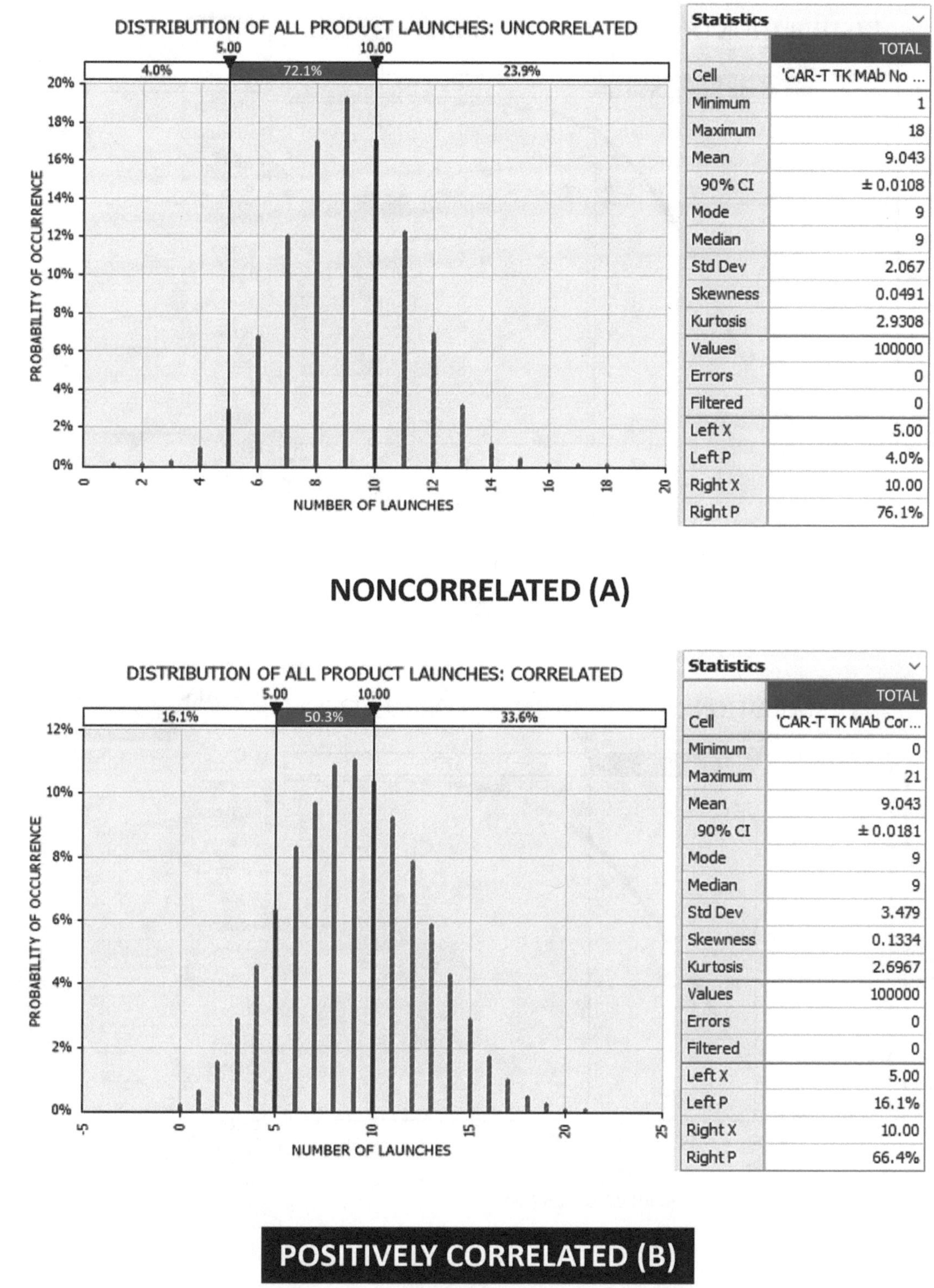

Figure 5.17A and B Oncology portfolio—likelihood of securing multiple launches from 3 buckets of noncorrelated (A) and positively correlated (B) projects.

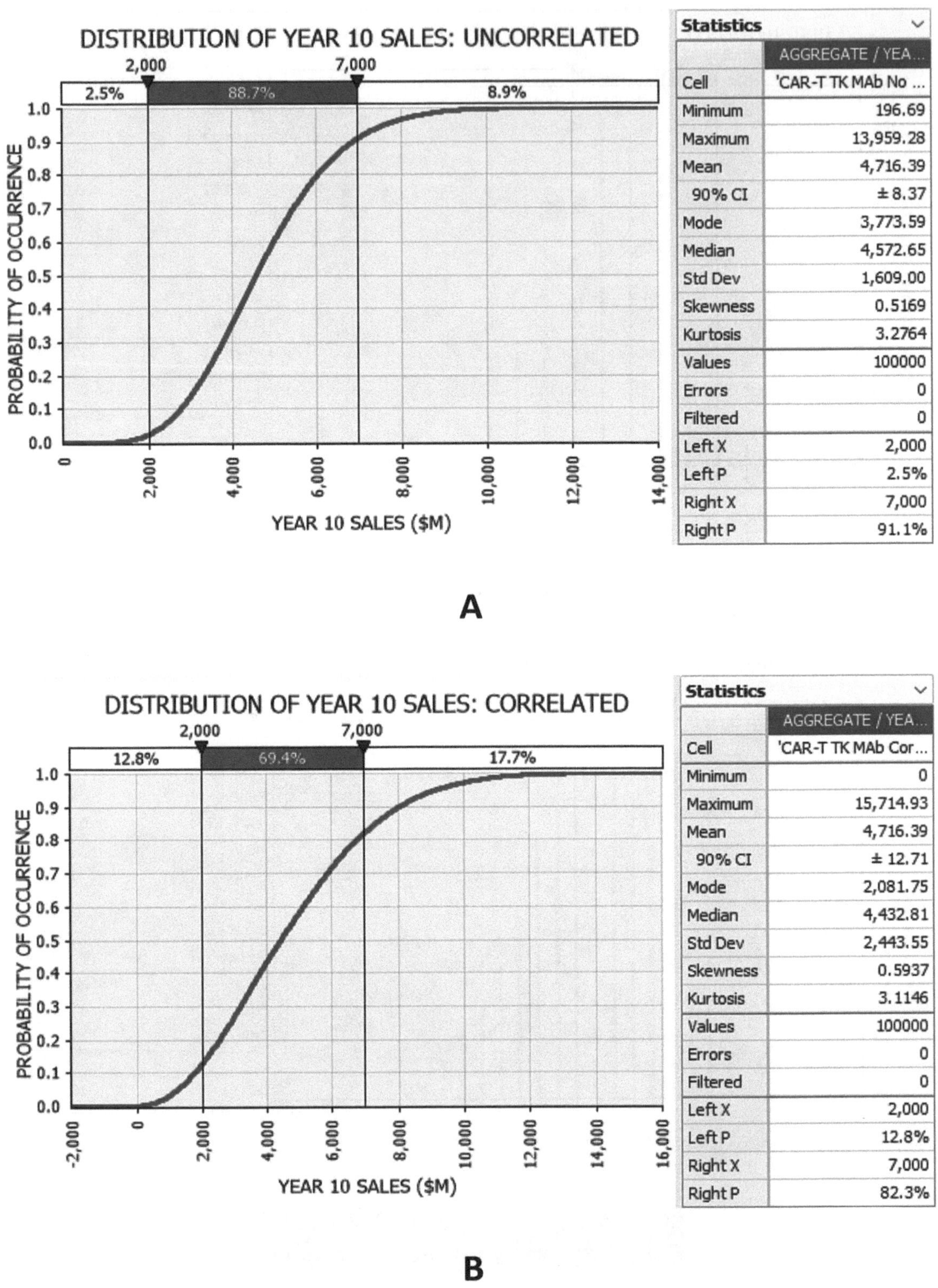

Statistics	AGGREGATE / YEA...
Cell	'CAR-T TK MAb No ...
Minimum	196.69
Maximum	13,959.28
Mean	4,716.39
90% CI	±8.37
Mode	3,773.59
Median	4,572.65
Std Dev	1,609.00
Skewness	0.5169
Kurtosis	3.2764
Values	100000
Errors	0
Filtered	0
Left X	2,000
Left P	2.5%
Right X	7,000
Right P	91.1%

Statistics	AGGREGATE / YEA...
Cell	'CAR-T TK MAb Cor...
Minimum	0
Maximum	15,714.93
Mean	4,716.39
90% CI	± 12.71
Mode	2,081.75
Median	4,432.81
Std Dev	2,443.55
Skewness	0.5937
Kurtosis	3.1146
Values	100000
Errors	0
Filtered	0
Left X	2,000
Left P	12.8%
Right X	7,000
Right P	82.3%

Figure 5.18A and B Likelihood of attaining the year-10 sales target assuming no intra-bucket project correlations (A) and positive intra-bucket project correlations (B).

5.4.3 Quantifying Portfolio Risk in Attaining Annual Financial Growth Targets

While launch probability provides an important measure of pipeline robustness, the more consequential question is whether the portfolio can deliver against annual sales growth commitments. This assessment is anchored in the LRP, typically spanning 10 years, in which year 1 reflects realized revenues escalated forward using an assumed compound annual growth rate (CAGR) that embodies management's growth ambitions and external guidance.

LRP revenues derive from two structurally distinct sources: the *base business* (BB) and the *R&D pipeline*. The BB consists of marketed products free of technical and regulatory risk but exposed to substantial commercial uncertainty, including competition, pricing pressure, and loss of exclusivity. In contrast, pipeline assets carry both commercial uncertainty and significant technical and regulatory risk until they transition into the BB. This asymmetry is central to portfolio risk assessment. A defensible LRP must reconcile two competing realities:

1. **Short-term growth continuity**—which is largely sustained by the BB and near-launch assets
2. **Long-term growth sustainability**—which depends on a diversified, risk-adjusted pipeline capable of replenishing revenues as BB products mature

Failure to explicitly incorporate PTRS, program correlation, and commercial uncertainty systematically overstates future revenues. Consequently, leading organizations increasingly rely on stochastic simulation to translate pipeline and commercial risk into probabilistic revenue outcomes. With this structure sufficiently positioned, leadership can assess the magnitude of growth gaps, the likelihood that the portfolio will meet annual targets, and the extent of pipeline or BB supplementation required.

With this LRP in place, an organization is positioned to address the following:

- What is the magnitude of annual gaps between growth aspirations and expectations?
- How likely is it that the current portfolio, in aggregate, will meet annual growth expectations?
- What is required to supplement the pipeline and BB in order to increase the likelihood of meeting annual growth expectations?

An example of the components of an LRP is shown in WAV Table 5.10, in which we can see the following:

- The BB target is set at \$5.00B in year 1 with a CAGR of 5.0% until year 10 when it reaches \$7.76B
- The target for the late clinical development (LCD; Phase IIb and Phase III/Regulatory) component of the pipeline is set at \$5.00B in year 2 (no sales from the LCD are possible until year 2) with a CAGR of 20.0% until year 10 when it reaches \$21.50B
- The target for the early clinical development (ECD; Phase I and Phase IIa) component of the pipeline is set at \$5.00B in year 6 (no sales from ECD are possible until year 6) with a CAGR of 30.0% until year 10 when it reaches \$14.28B
- The target for the preclinical development (PCD) component of the pipeline is set at \$3.00B in year 8 (no sales from PCD are possible until year 8) with a CAGR of 40.0% until year 10 when it reaches \$5.88B
- The combined target for the BB and pipeline is set at \$5.00B in year 1 with a CAGR of 21.6% until year 10 when it reaches \$49.42B

A consolidation of these annual growth targets is shown in Figure 5.19.

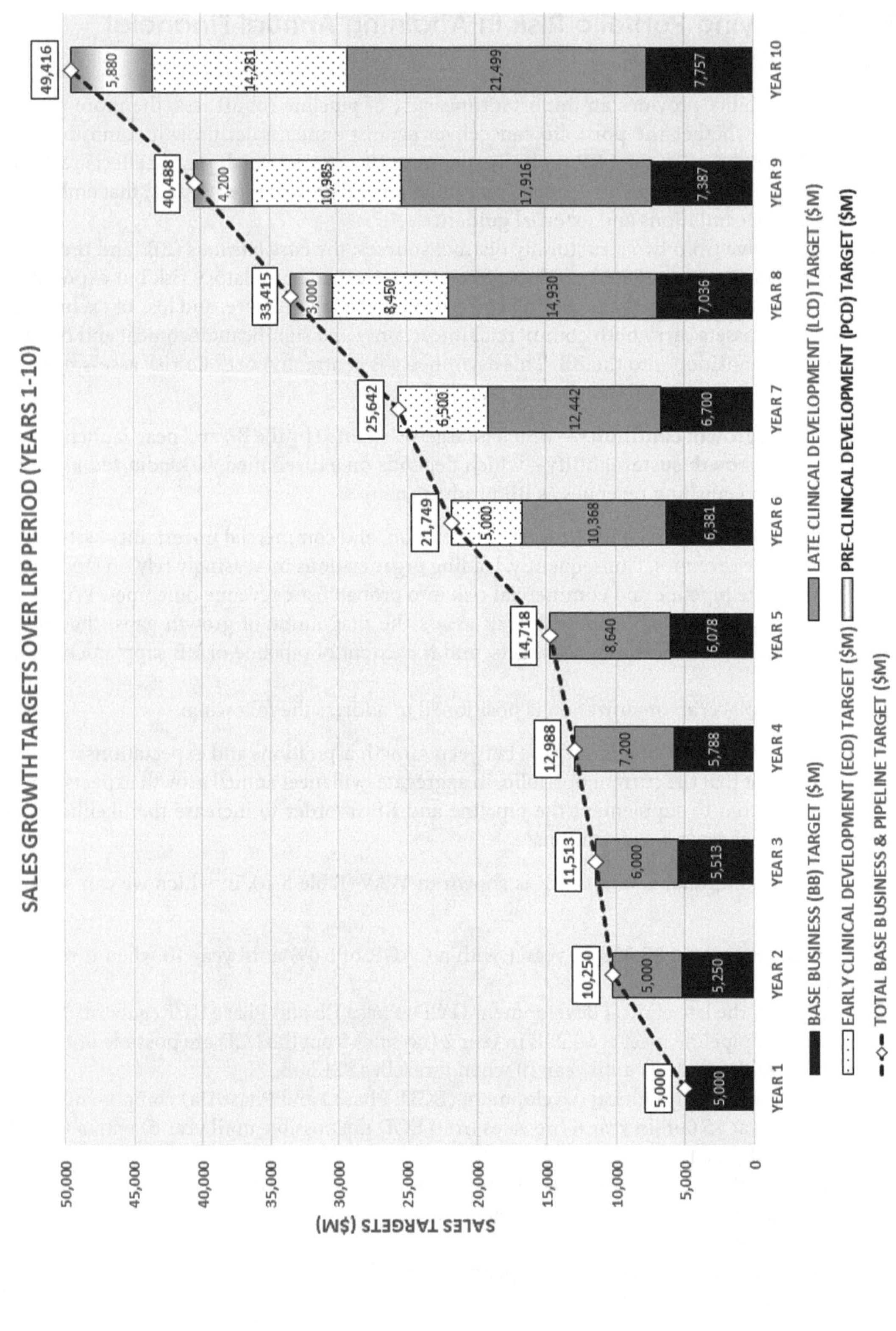

Figure 5.19 Consolidation of annual sales targets of the long-range plan.

Forecasted potential and risk-adjusted sales from each of these five components of the LRP reveal the following (see WAV Table 5.11):

- The BB is forecasted to grow by a CAGR of −9.7% from $5.00B in year 1 to $2.00B in year 10. This is due to the patent expirations of several *mature products*. Because there is no technical or regulatory risk in the BB, annual risk-adjusted sales remain the same as potential sales.
- The LCD pipeline is forecasted to grow by a CAGR of 14.7% from $9.71B in year 2 to $29.01B in year 10. On a risk-adjusted basis, the LCD pipeline is calculated to grow by a CAGR of 15.0% from $4.56B in year 2 to $13.95B in year 10.
- The ECD pipeline is forecasted to grow by a CAGR of 29.0% from $21.36B in year 6 to $59.16B in year 10. On a risk-adjusted basis, the ECD pipeline is calculated to grow by a CAGR of 26.8% from $3.77B in year 6 to $9.73B in year 10.
- The PCD pipeline is forecasted to grow by a CAGR of 23.4% from $37.27B in year 8 to $56.74B in year 10. On a risk-adjusted basis, the PCD pipeline is calculated to grow by a CAGR of 23.4% from $4.48B in year 8 to $6.82B in year 10.

On a non-risk-adjusted basis, aggregate revenues appear highly robust, growing from $5.00B to $146.92B by year 10 and exceeding all LRP targets (see Figure 5.20). However, once technical and regulatory risks are incorporated, shortfalls emerge as early as year 2 and widen over time, reaching nearly $17B by year 10 (see Figure 5.21). This divergence underscores a critical governance insight: potential does not equate to deliverability. The widening delta between potential and risk-adjusted sales serves as an early warning signal of structural growth risk, reinforcing the necessity of risk-adjusted portfolio analytics as a foundational element of LRP construction and strategic decision making.

Given these gaps, analysis begins with the BB, which establishes the revenue floor. Although free of development risk, BB revenues are subject to commercial uncertainty and are best represented as probabilistic distributions rather than point estimates. It is important to note that, absent clear conventions, commonly used descriptors such as *low, base*, and *high* forecasts are inherently ambiguous, inviting misinterpretation and encouraging inconsistent decision making.

Using the P50 forecast, BB shortfalls appear beginning in year 7 and become pronounced in years 8–10 (see Figure 5.22).

BB uncertainty is modeled via stochastic simulation using the Program Evaluation and Review Technique distributions parameterized by minimum, most likely, and maximum estimates (see WAV Table 5.12). In year 1, for example, the BB distribution yields an expected value of ~$4.9B, lower than the deterministic forecast due to asymmetric downside risk. By constructing annual BB revenues in this manner, the simulation model produces not only an expected value but also a full distribution of possible outcomes for each year. This probabilistic foundation is essential for evaluating the likelihood of meeting annual sales targets, assessing downside exposure, and determining the extent to which pipeline-derived revenues must compensate for BB uncertainty in order to satisfy the LRP.

A tornado analysis indicates that uncertainty in products Gamma, Epsilon, and Alpha dominates year-1 BB variability (see Figure 5.23).

Simulation results show that even when forecasts equal targets, the probability of exceeding them may be well below 50% due to asymmetric downside exposure. For example, the probability of exceeding the year-1 BB forecast of $5.0B is only 30.4%. By year 7, the probability of exceeding the BB target falls to 0.2%, and by year 10 it is 0% (see Figure 5.24).

If sales growth targets are not intended to function merely as unattainable stretch objectives—or as clarion calls to action—then a fundamental question arises: Why should an organization continue to embed such targets within its LRP? The authors argue that while growth targets should clearly

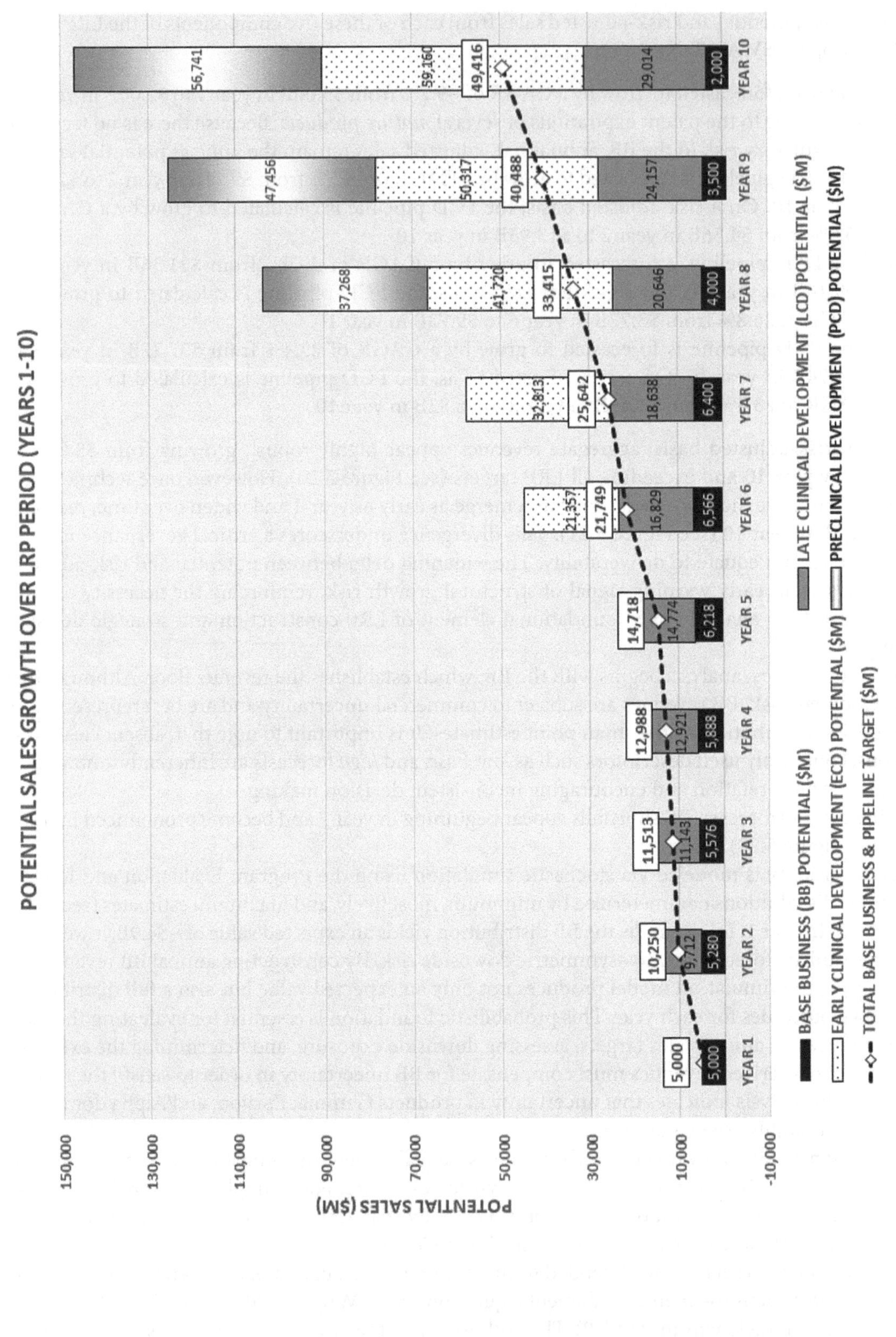

Figure 5.20 Potential annual sales of components of the LRP.

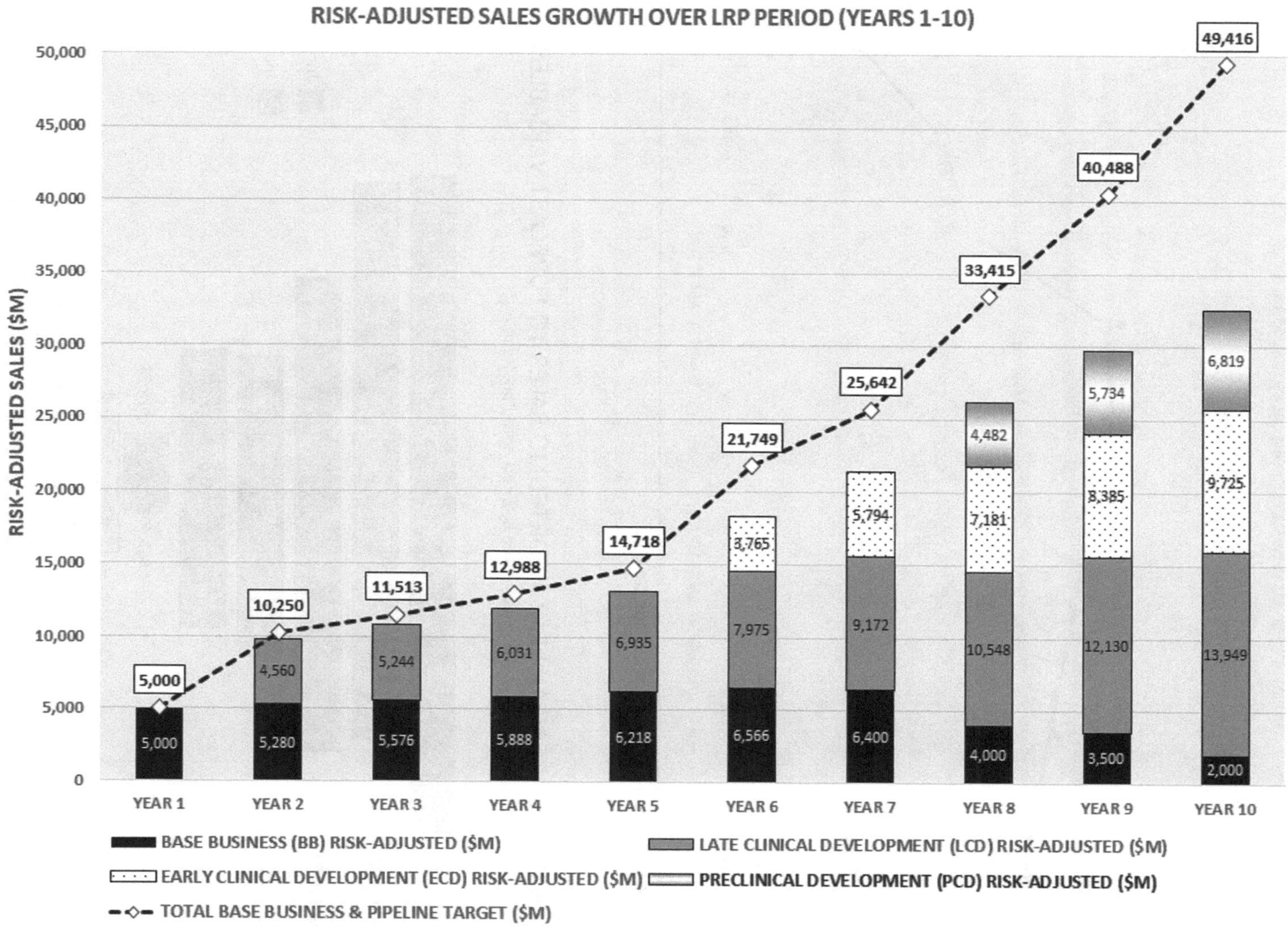

Figure 5.21 Risk-adjusted annual sales of components of the LRP.

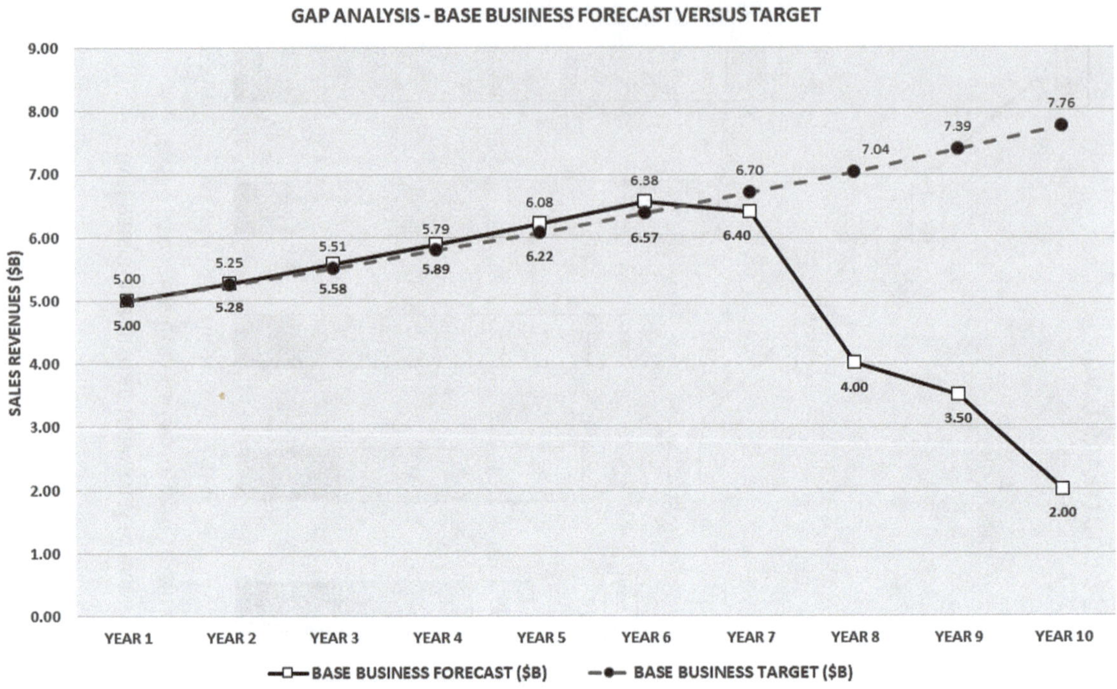

Figure 5.22 Annual BB forecast versus target.

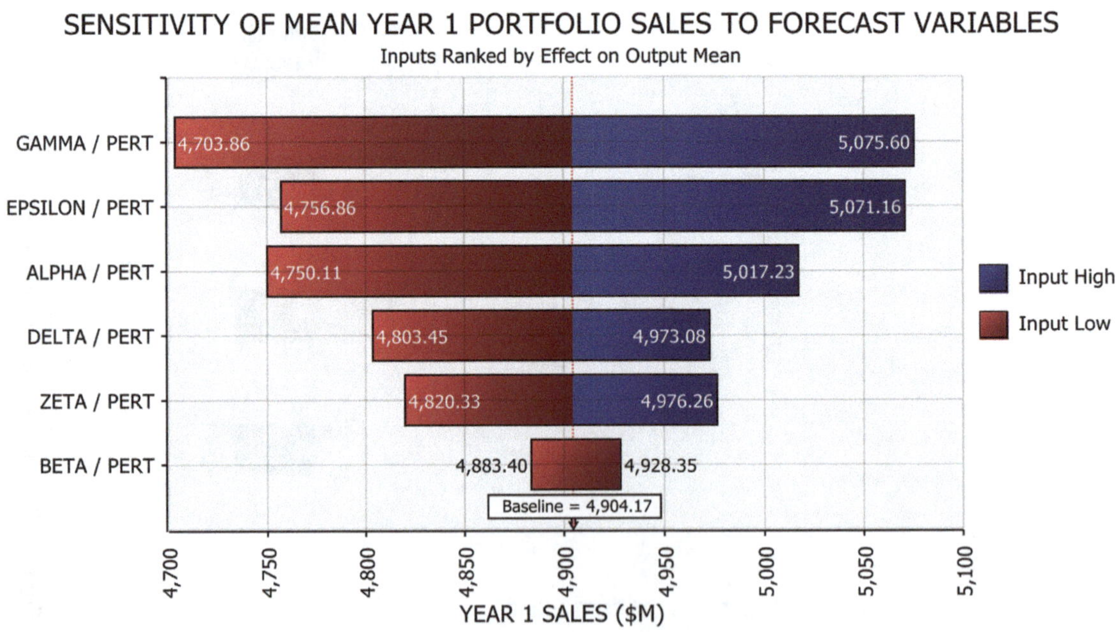

Figure 5.23 Tornado sensitivity analysis of year-1 aggregate mean sales of the BB product portfolio.

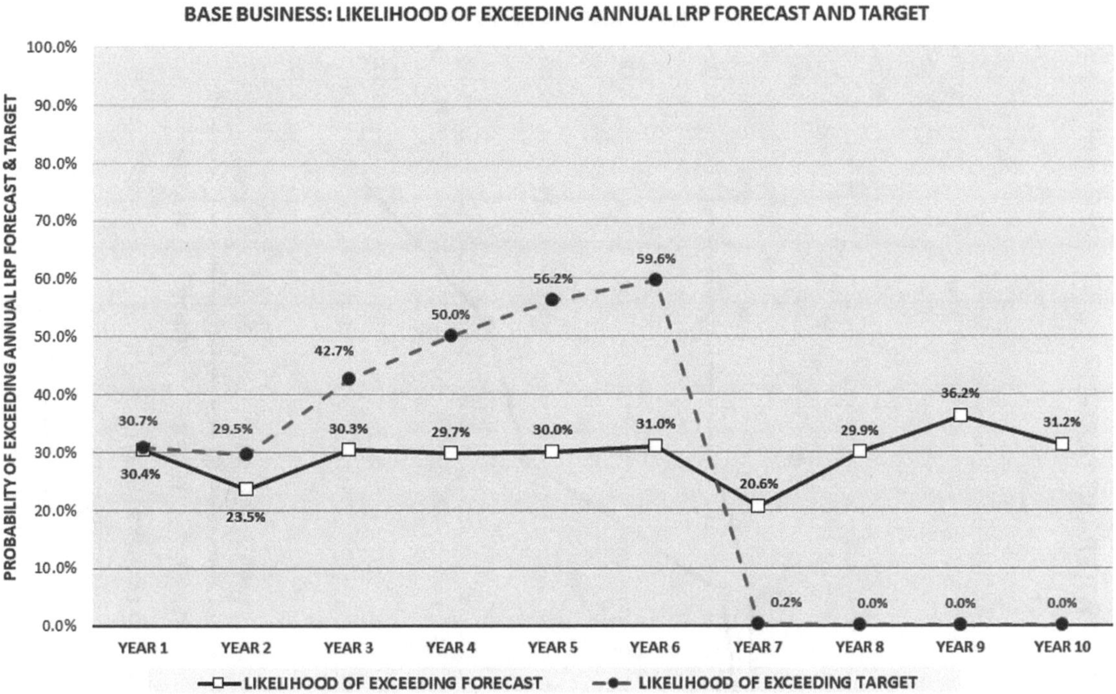

Figure 5.24 Probability of exceeding annual BB sales forecasts and targets over the LRP period.

embody aspirational ambition, they must also retain a credible connection to achievable outcomes if they are to serve as effective decision-making anchors. On this basis, we advance the concept of a *risk-tolerant* LRP, defined as one in which the probability of exceeding annual growth targets reflects a meaningful likelihood of success, rather than a near-zero chance of attainment. In probabilistic terms, targets that are exceeded only under extreme upside scenarios offer limited strategic utility; they neither inform capital allocation nor provide actionable signals for portfolio management.

To operationalize this concept, the authors propose a probability-based heuristic whereby annual growth targets are set such that the likelihood of exceeding them falls within a 60%–70% range. This interval represents a pragmatic balance between ambition and realism: sufficiently demanding to drive performance, yet achievable with reasonable confidence given the inherent technical, regulatory, and commercial uncertainties of the portfolio.

In this case, growth targets in years 7 through 10 must be recalibrated downward in order to restore a risk-tolerant probability of attainment (see WAV Table 5.13). When adjusted accordingly, the probability of exceeding annual targets converges toward the proposed 60%–70% range, thereby re-establishing alignment between strategic intent and portfolio-derived reality (see Figure 5.25).

The same probabilistic logic is applied to pipeline LCD, ECD, and PCD components. Simulation models for each component (see WAV Tables 5.14, 5.15, and 5.16) incorporate PTRS and commercial uncertainty.

The impact of both risk and commercial uncertainty from each of the eight LCD projects on the simulated mean ($4,636M) of the LCD pipeline is shown in Figure 5.26. From the tornado sensitivity analysis, the risks of seven of the eight projects account for the vast majority of the sensitivity of the mean pipeline value in year two.

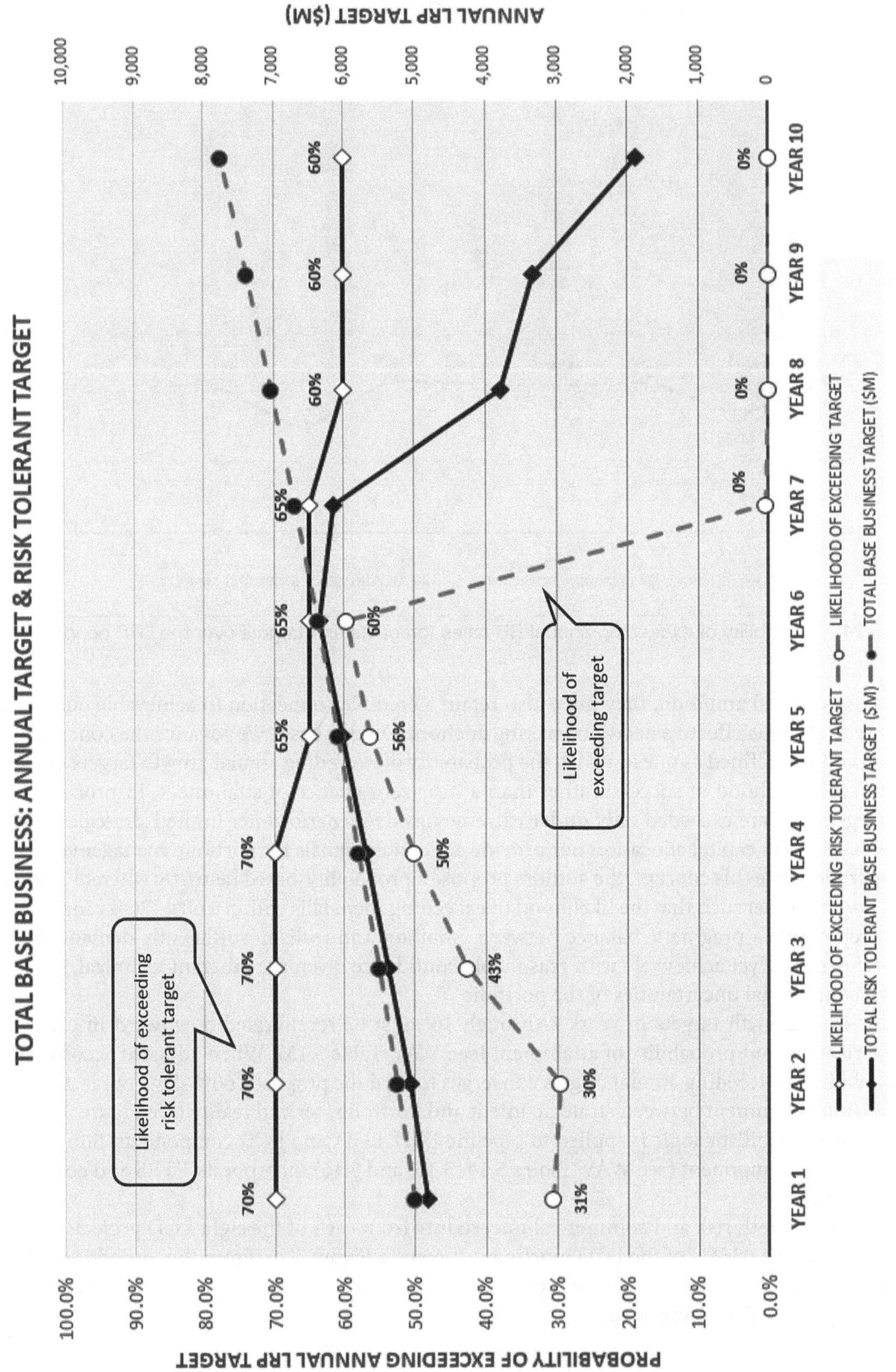

Figure 5.25 Impact of risk-tolerant target sales growth expectations to attainment of the LRP.

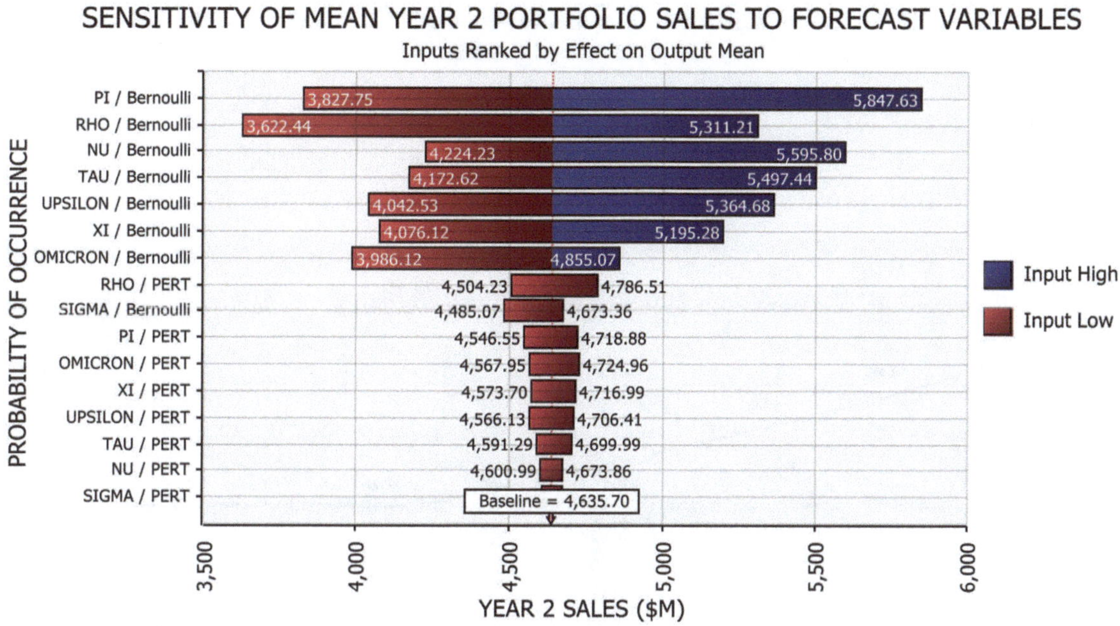

Figure 5.26 Tornado sensitivity analysis of the impact of projects in the late clinical development pipeline to simulated mean sales in year 2.

For LCD, the probability of exceeding targets declines from 42.8% to ~20% over the LRP (see Figure 5.27). ECD probabilities peak at ~38% before declining (see Figure 5.28), while PCD probabilities range from ~57% to ~49% (not shown).

Having quantified the probability of meeting annual sales growth targets for (a) BB and (b) the individual LCD, ECD, and PCD components of the R&D pipeline, it is both logical and necessary to extend this analysis to the total enterprise portfolio, comprising the combined contributions of marketed products and pipeline assets. Such an integrated view provides the most decision-relevant assessment of whether the organization's LRP is achievable in aggregate.

An integrated portfolio simulation combining BB and all pipeline assets is constructed in WAV Table 5.17. The probability of exceeding total portfolio targets peaks at 40.7% in year 2 and declines steadily to 8.5% by year 10 (see Figure 5.29).

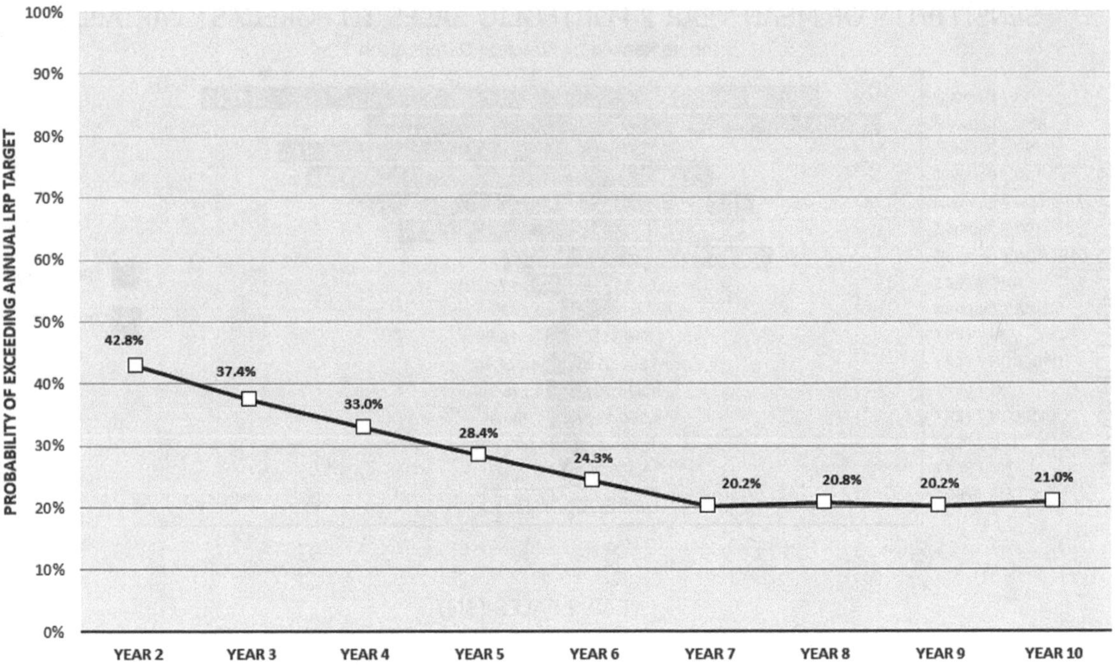

Figure 5.27 Probability of exceeding annual *late* clinical development pipeline targets over the LRP period.

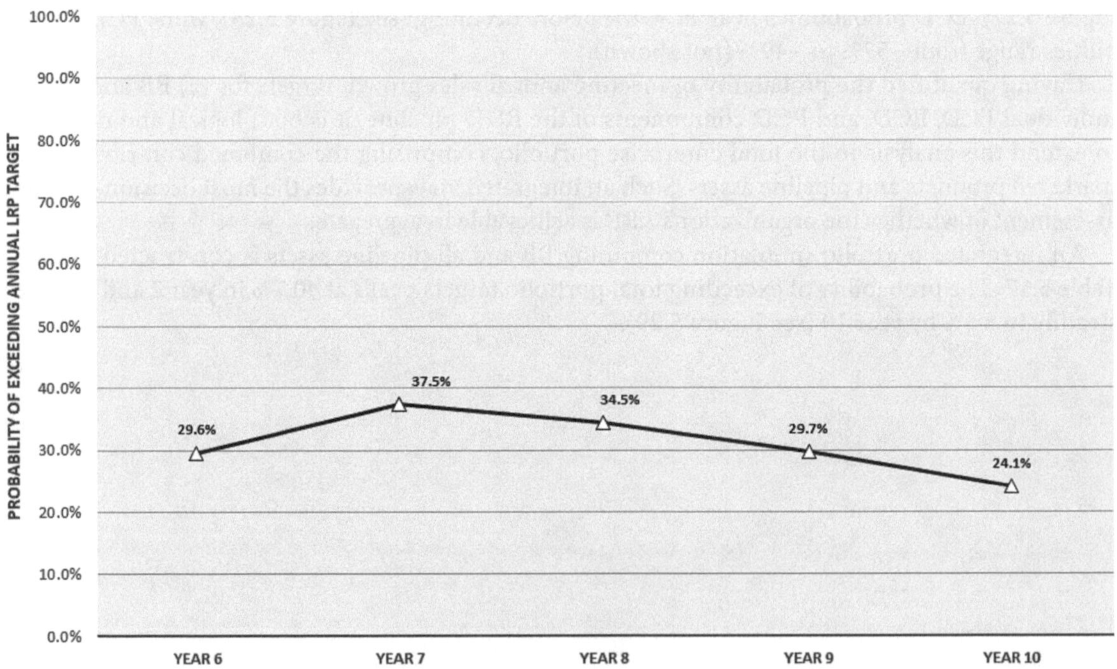

Figure 5.28 Probability of exceeding annual *early* clinical development pipeline targets over the LRP period.

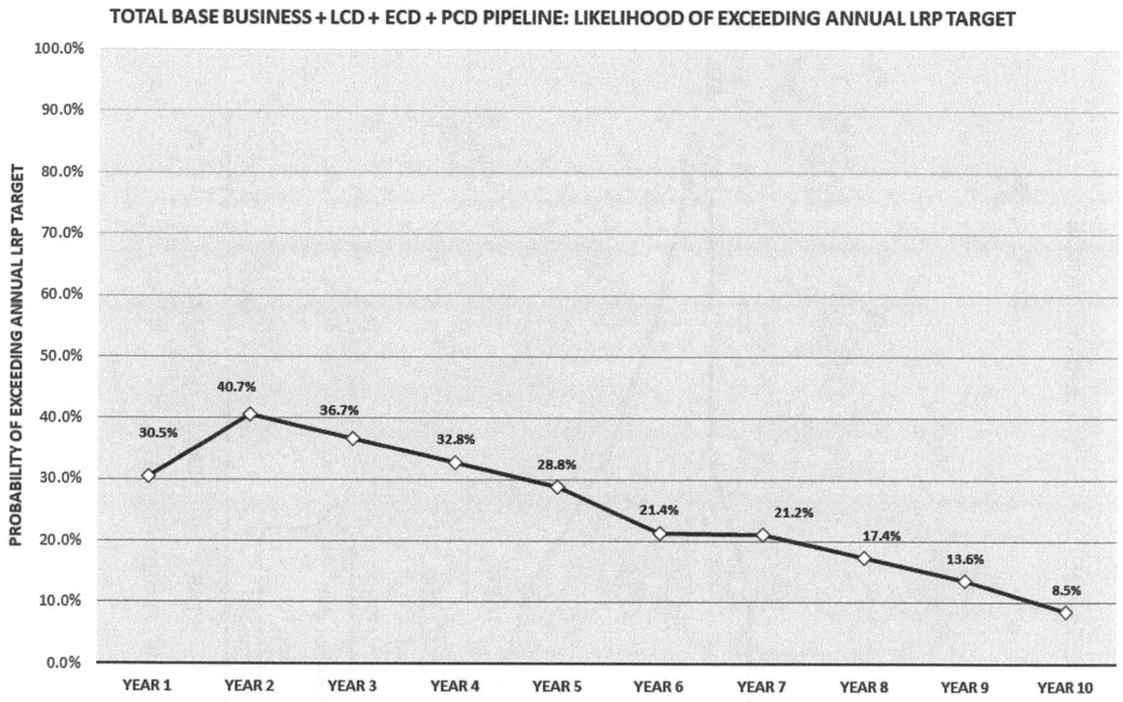

Figure 5.29 Probability of exceeding annual total portfolio targets over the LRP period.

The extent to which executive decision makers should be comfortable with an LRP that exhibits less than a 50% probability of achieving annual sales targets in the first five years—and less than a 25% probability in the latter five years—must be questioned directly. Again, applying the reasonable assumption that executive leadership would view a 60%–70% probability of target attainment as consistent with enterprise risk tolerance, annual sales targets should be systematically recalibrated downward (see WAV Table 5.18). Doing so raises the likelihood of success to a level that exceeds that of a random binary outcome (i.e., better than a coin toss), as illustrated by the upper solid line in Figure 5.30.

Importantly, experience suggests that realistic, probabilistically defensible targets act as a catalyst for concrete management action—including portfolio supplementation, life-cycle management, business development, and commercial investment prioritization—whereas chronically unattainable targets tend to obscure true performance gaps and undermine organizational credibility. In this sense, a risk-tolerant LRP does not lower ambition; rather, it translates ambition into decision-relevant accountability, enabling meaningful gap identification and executable mitigation strategies.

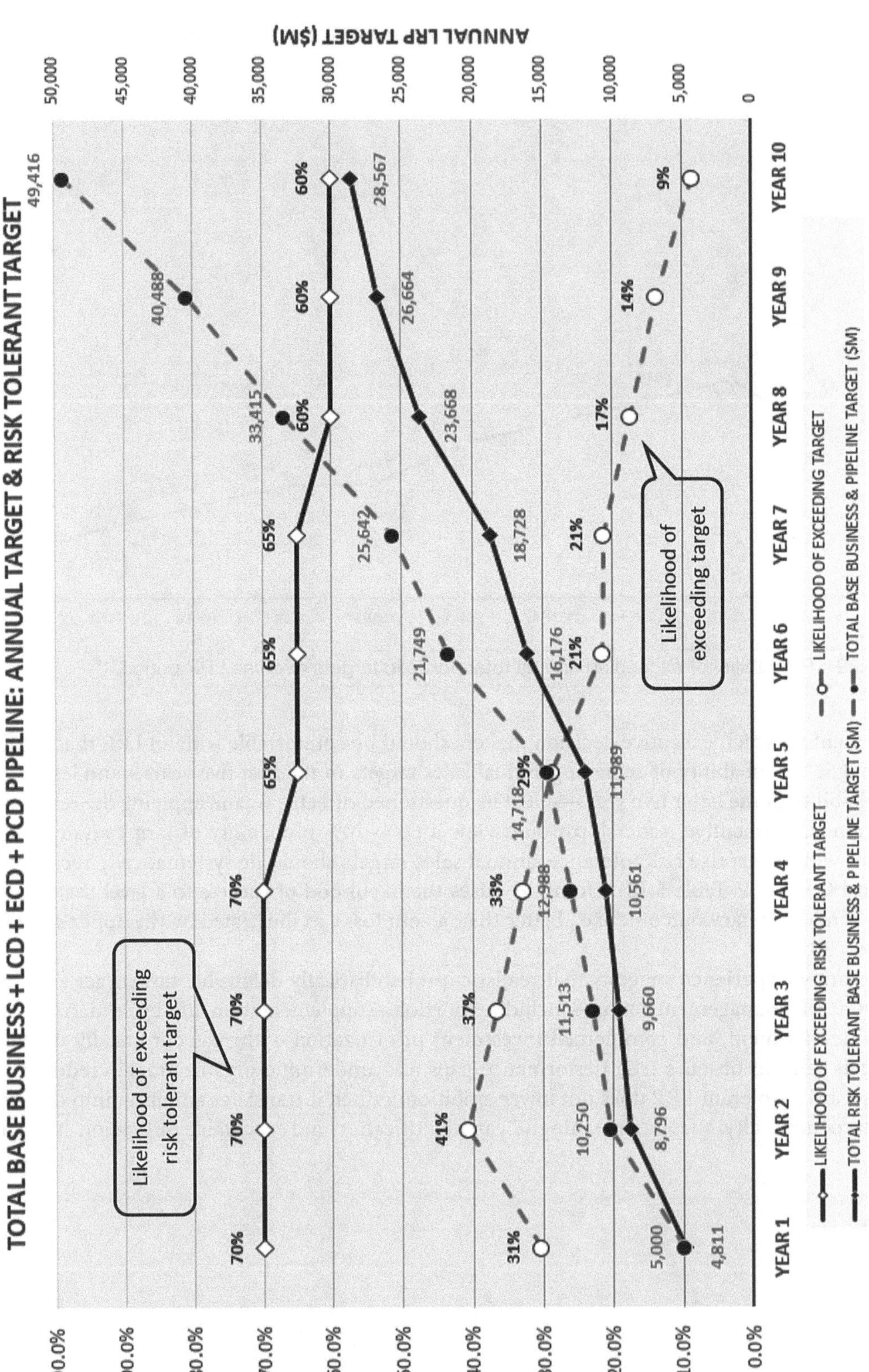

Figure 5.30 Annual portfolio sales targets and the probability of their attainment over the LRP period.

6

INTEGRATED EVALUATION

Some of the figures and tables in this chapter are too large or complex to properly format in printed form. Figures and tables marked with a "WAV" label can be downloaded from the Web Added Value Download Resource Center at www.jrosspub.com/wav for proper viewing.

6.1 KEY FINANCIAL VALUATION CONCEPTS

Expected Monetary Value

Expected monetary value (EMV), also referred to as expected value (EV) or integrated value, is a probabilistically adjusted measure that combines technical and regulatory risk along with commercial uncertainty. In decision analysis, EMV represents the risk-adjusted, rolled-back value of a decision tree.

For multi-stage drug development programs, EMV should not be estimated by applying a single aggregate probability of success to a commercial forecast. Instead, it is calculated by recursively rolling back terminal node values in a decision tree, incorporating stage-specific probabilities of success, expected value inflows (e.g., future cash flows), and value outflows (e.g., development and commercialization costs). This approach captures the sequential resolution of risk and preserves managerial flexibility inherent in biopharma research and development (R&D).

Across the biopharma industry, net present value (NPV) remains the primary metric for evaluating economic attractiveness, with internal rate of return (IRR) and return on investment (ROI) used as supplementary diagnostics. When development risk and commercial uncertainty are explicitly incorporated, expected net present value (eNPV) becomes the preferred measure. eNPV integrates four dimensions of value creation: (1) probabilistic commercial benefit, (2) phase-specific technical and regulatory risk, (3) capital requirements across development and commercialization, and (4) time, reflected through discounting and development duration. Probabilistic cash-flow modeling also enables calculation of risk-adjusted performance metrics such as expected IRR (eIRR) and expected ROI, facilitating more consistent comparisons under uncertainty.

From a capital allocation perspective, productivity metrics provide additional insight. The productivity index (PI = NPV / cost to launch or CTL) and its risk-adjusted analogue (ePI = eNPV / eCTL) are particularly useful for marginal investment decisions in constrained R&D budgets. However, these metrics are phase-dependent: early-stage assets may exhibit an attractive expected productivity index (ePI) due to limited capital exposure, while later-stage assets may appear less productive as cumulative investment increases and residual risk declines. PI and ePI must therefore be interpreted in the context of development maturity.

A common question is what constitutes a *good* NPV, eNPV, PI, or ePI. While financial theory suggests that projects with positive NPV or eNPV warrant consideration, this criterion is insufficient. Positive eNPV does not ensure capital efficiency relative to the weighted average cost of capital (WACC), nor does standard NPV reflect staged decision making or flexibility—namely, limitations that motivate real options analysis. Meaningful comparisons, however, should be made between real options valuations and eNPV, not deterministic NPV.

Rate-based metrics such as IRR and eIRR, despite known limitations, often provide clearer signals of capital productivity than PI or ePI. In practice, an eIRR exceeding WACC is a stronger indicator of value creation than a merely positive eNPV. When probabilistic valuation is applied, ePI is preferred over deterministic PI.

Ultimately, the quality of an ePI or eNPV is best assessed relative to other projects at the same stage of development. Cross-phase comparisons are misleading, as early-stage assets are penalized for unreduced uncertainty while late-stage assets benefit from risk already retired. Effective portfolio governance therefore evaluates value and productivity on a like-for-like, phase-specific basis.

6.2 ANATOMY OF A DECISION TREE

Before valuing projects with decision tree analysis and quantifying EV, it is essential to establish the conceptual basis of decision trees. They are not merely computational tools, but formal representations of the sequential logic through which value is realized in drug development.

Decision trees are structured chronologically from left to right, reflecting the temporal sequence of decisions, uncertainties, and outcomes that must be resolved before a drug's economic potential is realized. Nodes represent either managerial decisions or uncertain events, while branches denote mutually exclusive paths contingent on prior outcomes. This structure aligns naturally with biopharma R&D, where investment is staged, uncertainty is progressively resolved, and value depends on passing successive technical and regulatory milestones.

Consider two competing investments, Project A and Project B, which differ in phase-specific risk, cost structure, and commercial potential. Development risk is typically divided into controllable and uncontrollable components. Operational risks (e.g., enrollment delays or supply chain issues) are generally excluded from formal probabilities (p) of success because they can often be mitigated or nullified through additional investment. Instead, phase-specific risk is defined in terms of scientific and regulatory uncertainty: the probability of technical success (PTS) and the probability of regulatory success (PRS). When combined across phases, these yield the cumulative probability of technical and regulatory success (PTRS) to launch. While different conventions exist for expressing probabilities of success—POS, PTS, p(Success)—these terms are used interchangeably for phase-specific probabilities.

Phase-specific development costs are entered into the decision tree as discounted after-tax cash outflows for research, development, and registration, using the organization's WACC. Commercial value is treated separately as the NPV of post-launch cash flows only, referred to as the commercial value given success (CVGS). CVGS excludes preapproval development costs but includes all post-launch expenditures. In cases where post-marketing studies are required, these costs may be treated either as R&D extensions or as life-cycle commercial costs; when approval is contingent on such studies, they are typically included in CVGS.

The core inputs to the decision tree—phase-specific probabilities, discounted development costs, and CVGS for each alternative—are summarized in Table 6.1 and depicted in Figure 6.1. Together, they enable transparent comparison of Project A and Project B by explicitly linking risk resolution, capital deployment, and downstream value creation.

All after-tax cash flow inputs in Table 6.1 are derived from a structured income statement spanning the period from the present through loss of exclusivity (LOE), with additional years included to capture post-LOE erosion.

From Figure 6.1, the rolled-back eNPV for *Invest In A* is $259.5M, exceeding *Invest In B* at $222.0M. Under risk-neutral assumptions, *Invest In A* is therefore the economically dominant choice. The calculation of these values is illustrated through the phase-specific probabilities, path NPVs, and exit costs summarized in the *Invest In A* example, culminating in a rolled-back eNPV of $259.5M.

Although decision trees are constructed left-to-right, eNPVs are calculated right-to-left via rollback (see Figure 6.2). But what do the various numbers mean, and how are they (including the rolled back eNPVs) calculated by the decision tree? Let us therefore examine some of the various calculated data points for *Invest In A*:

- PTRS resulting in high commercial success = 17.6% = (0.65 * 0.90 * 0.30)
- PTRS resulting in base commercial success = 23.4% = (0.65 * 0.90 * 0.40)
- NPV of path resulting in high commercial success = $695M = $(800 – 5 – 100)M
- NPV of path resulting in base commercial success = $545M = $(650 – 5 – 100)M
- *p*(clinical success and regulatory failure) = 6.5% = (0.65 * 0.10)
- *p*(clinical failure) = 0.35
- Exit cost at clinical failure = –$100M
- eNPV at commercial node, assuming clinical and regulatory success = $515.0M
- eNPV at clinical node = $259.5M

The rollback calculations are shown in Figure 6.2. For example:

- First rollback: (0.30 * 695) + (0.40 * 545) + (0.30 * 295) = 515.00

For portfolio budgeting, aggregate risk-adjusted cost should be used rather than assuming 100% project success. Risk-adjusted cost can be derived from decision trees by assigning zero value to successful commercial outcomes (see Figure 6.3A and B), as illustrated for *Invest In A* and *Invest In B*.

The authors propose a six-step process for building robust decision trees (see Figure 6.4), emphasizing internal consistency, transparency, and traceability. Importantly, the decision tree output should be viewed as a prescriptive recommendation—not a mandate. Decision makers may

Table 6.1 Data requirements for evaluating strategic alternatives A and B.

	PTS	PRS	PV CLINICAL COST ($M)	PV REGISTRATION COST ($M)	CVGS HIGH ($M)	CVGS BASE ($M)	CVGS LOW ($M)
ALTERNATIVE A	0.65	0.90	-100.0	-5.0	800	650	400
ALTERNATIVE B	0.55	0.95	-60.0	-5.0	750	500	400

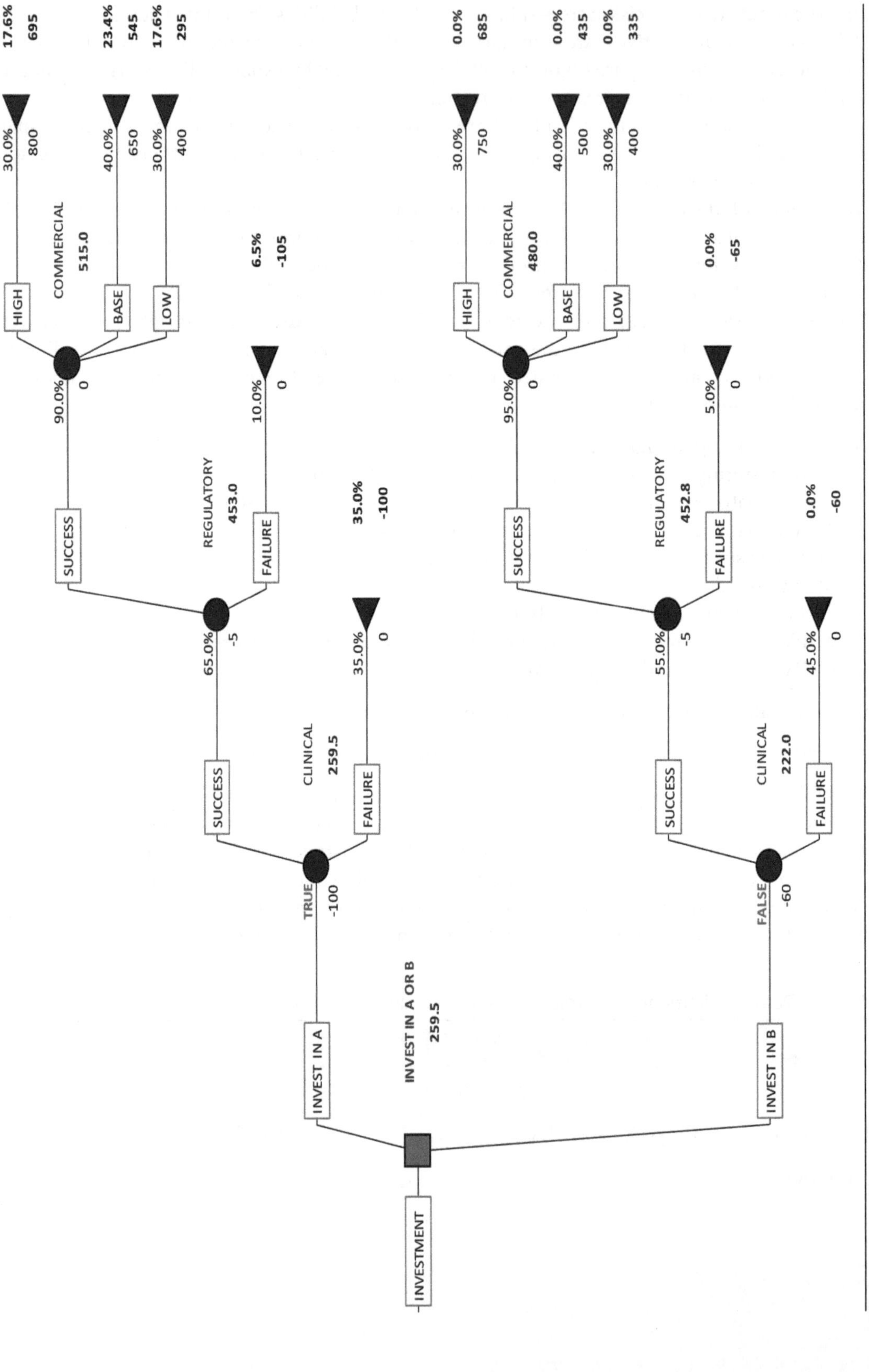

Figure 6.1 Decision tree—evaluation of strategic alternatives on the basis of eNPV.

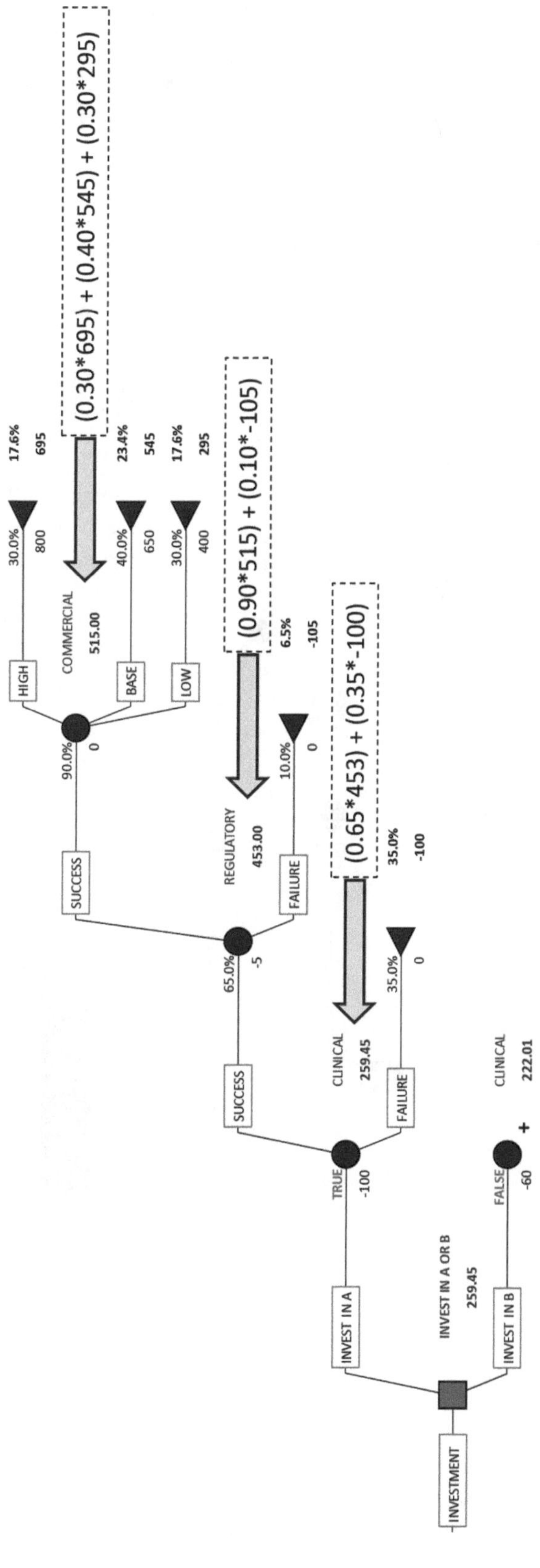

Figure 6.2 *Invest In A* decision tree—evaluation of rolled back eNPVs.

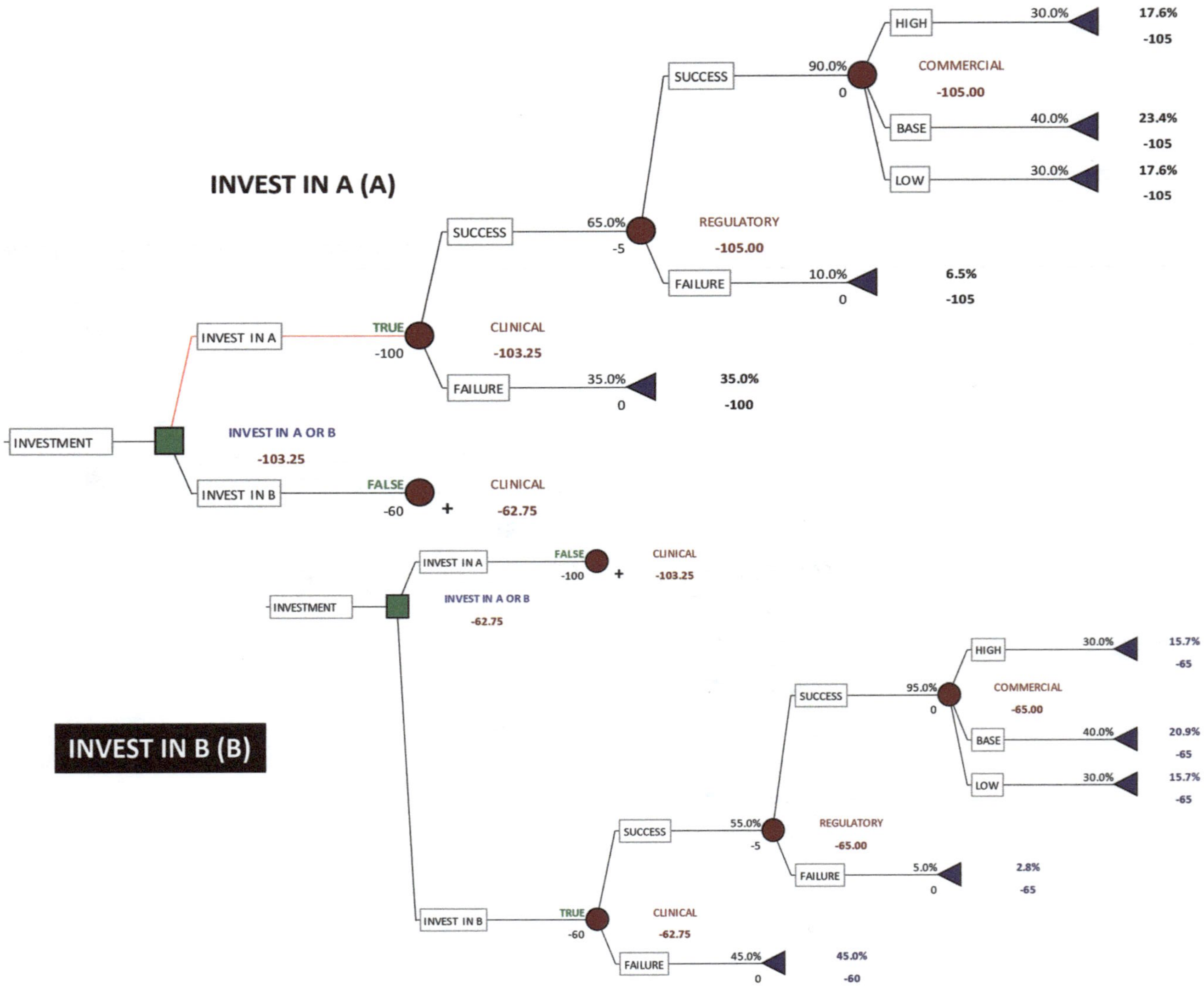

Figure 6.3A and B Decision tree—calculation of risk-adjusted cost.

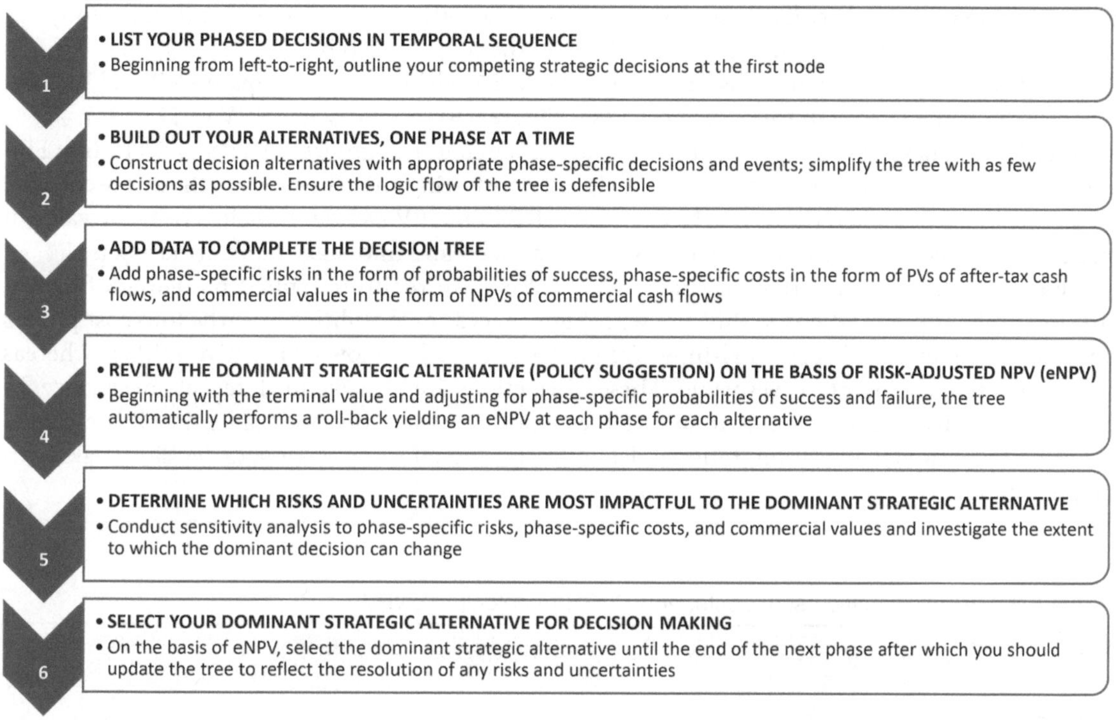

Figure 6.4 Building decision trees—a recommended process.

rationally choose a dominated alternative due to capital constraints, risk aversion, or strategic considerations. In such cases, the decision tree's value lies in explicitly quantifying the opportunity cost (foregone eNPV) of that choice. When qualitative factors materially influence decisions, Multiple Objective Decision Analysis (MODA; see Chapter 8) is recommended.

Unfortunately, shortcut approaches to generate an *approximate eNPV* persist. In most cases, such approximations rely on one of two methodological shortcuts:

- Applying a cumulative PTRS as a scalar multiplier to a deterministic NPV
- Adjusting for development risk and uncertainty by applying progressively higher WACCs to discount future cash flows

While both approaches are conceptually flawed, they differ in the extent to which they can be addressed constructively. The former practice—multiplying PTRS by NPV—can be shown analytically to misrepresent both risk resolution and value timing and is therefore addressed directly in this chapter. The latter approach—embedding risk through differential discount rates—remains highly contentious. Despite its persistence in some finance circles, the use of multiple WACCs to represent project- or phase-specific risk is inconsistent with decision-theoretic principles and sufficiently debated in the literature that a full treatment is beyond the scope of this book.

The PTRS × NPV shortcut, illustrated using Table 6.2, yields an *approximate eNPV* of $69.0M, whereas the correct decision tree calculation (see Figure 6.5) produces an eNPV of $21.39M.

Reordering phase-specific probabilities without changing the aggregate PTRS of 0.115 (see Figure 6.6) drives the eNPV negative (−$10.11M), demonstrating that shortcut methods obscure the economic impact of when risk is resolved. Because PTRS × NPV can never yield a negative value, it structurally precludes outcomes that are common in early- and mid-stage biopharma assets.

This example reinforces that accurate eNPV estimation requires explicit modeling of phase-specific probabilities and costs. Projects with high success probabilities in early, low-cost phases but low probabilities in late, capital-intensive phases tend to yield low or negative eNPVs, whereas the reverse sequencing preserves value. These distinctions are decisive for internal governance, licensing, and M&A decisions. Accurate eNPV estimation is therefore not academic—it is essential for disciplined capital allocation, rational deal making, and effective stage-gate decision making in biopharma R&D.

Table 6.2 Phase-specific risks, costs, and CVGS to calculate eNPV.

	PHASE I	PHASE II	PHASE III	REGISTRATION	AGGREGATE
p(SUCCESS)	0.90	0.25	0.60	0.85	0.115
COST ($M)	-10	-35	-100	-5	-150
CVGS ($M)					750
'APPROXIMATE eNPV' ($M)	= 0.115*((750)+(-10-35-100-5)) = 0.115*600				**69.00**

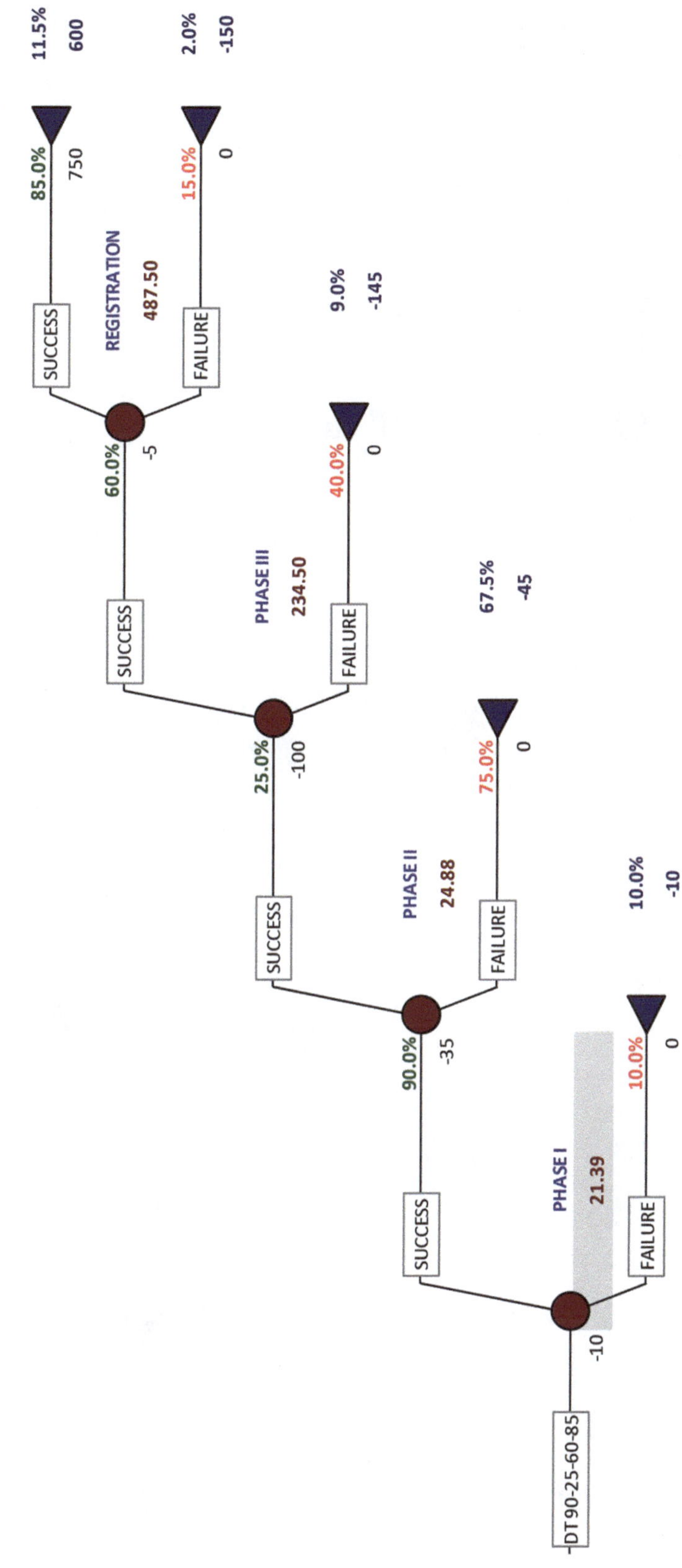

Figure 6.5 Decision tree calculation of eNPV (I).

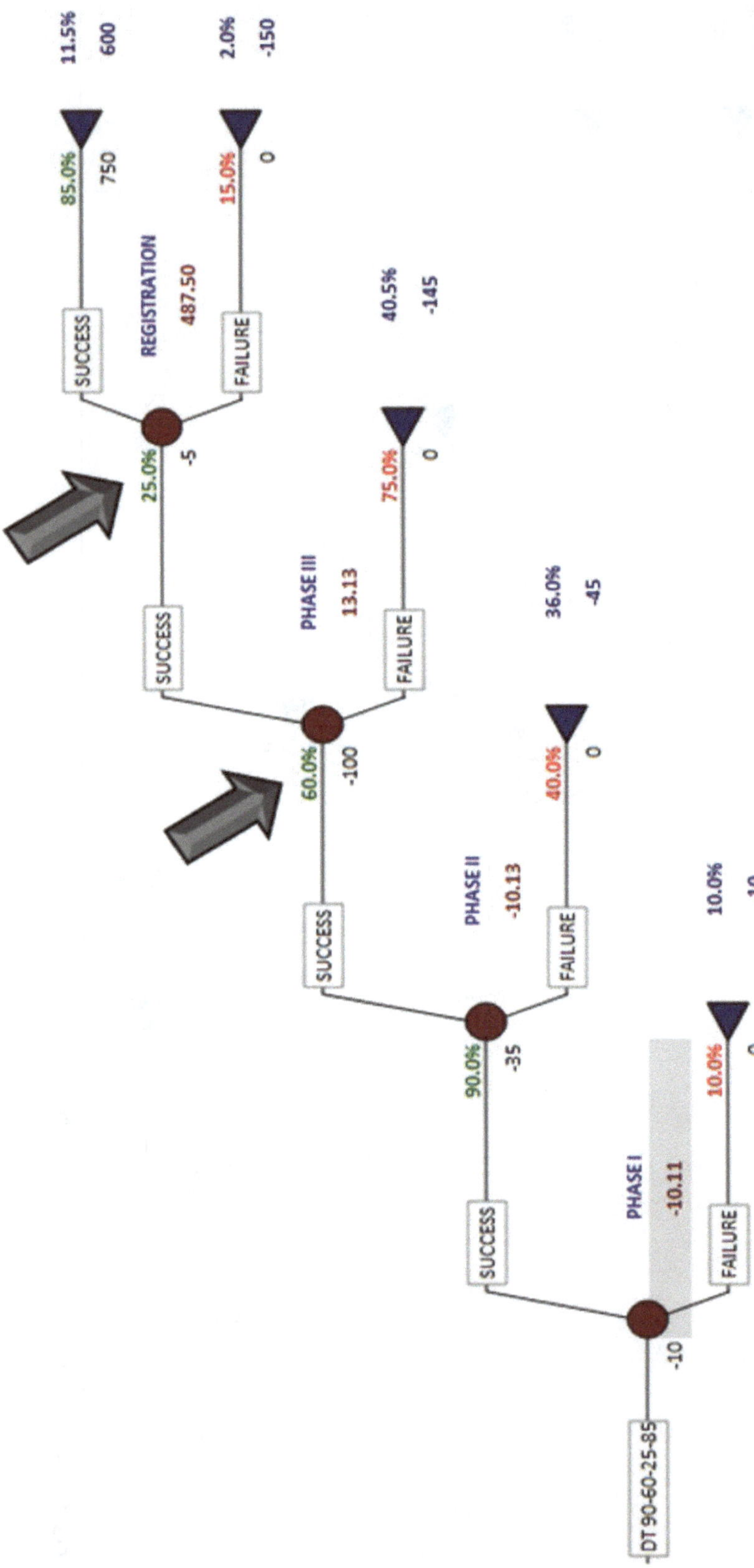

Figure 6.6 Decision tree calculation of eNPV (II).

6.3 FROM PROJECT EVALUATION TO PROGRAM EVALUATION

When sound financial principles are applied and decision trees are rigorously constructed, calculating a project's integrated value (eNPV) is relatively straightforward. Explicit modeling of strategic alternatives—such as develop, partner, defer, or terminate—often yields materially different eNPVs for the same asset, highlighting the importance of explicitly representing managerial choice.

Program-level valuation, however, is inherently more complex. It depends not only on individual project economics but also on how program success is defined and on the interdependencies among projects. Such interdependencies may arise from shared technical risk, overlapping development or manufacturing costs, regulatory linkages, or correlated commercial outcomes.

To illustrate, consider a program comprising two statistically independent projects targeting mild cognitive impairment (MCI) and Alzheimer's disease (AD). Each project has its own *p*(Success), CTL, and CVGS, as summarized in Table 6.3. Both projects are assumed to be entering Phase III, with regulatory approval conditional on technical success and therefore effectively certain.

Even under these simplifying assumptions, program-level eNPV cannot be inferred by simply summing project values without specifying the definition of program success and evaluating interactions between projects. Under a strategy in which MCI is pursued first and AD development follows sequentially regardless of MCI outcome, decision tree analysis yields a program-level eNPV of $825M (see Figure 6.7).

Table 6.3 Data for sequential pursuit of MCI and AD by independent projects.

	MCI	AD
p(Success)	0.30	0.27
p(Failure)	0.70	0.73
CTL ($M)	500	400
CVGS ($M)	3,500	2,500

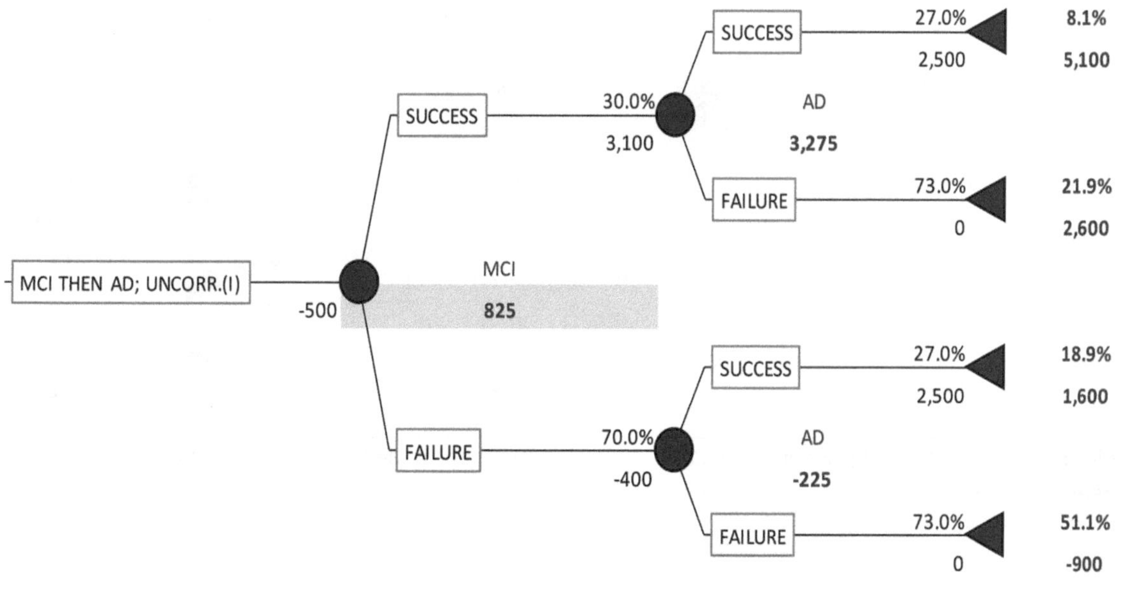

Figure 6.7 Sequential pursuit of MCI and AD by independent projects.

By construction, the two projects are technically and commercially independent. The cumulative probability of success for AD (27%) is unaffected by MCI outcomes, and the CVGS for AD ($2.5B) is assumed invariant regardless of whether MCI succeeds or fails. If AD is pursued only conditional on MCI success, the rolled-back eNPV declines to $633M (not shown). Both strategies may be defensible in practice. Sequential development reduces simultaneous capital exposure and allows early MCI revenues to partially de-risk subsequent AD investment.

Alternatively, if organizational capacity permits concurrent development of both projects, four mutually exclusive outcome states arise (success/failure combinations for MCI and AD). Under the same independence assumptions, concurrent development yields the same program-level eNPV of $825M (see Figure 6.8).

This equivalence arises solely because the CVGS for AD is assumed to be time-invariant. In reality, initiating AD development earlier would likely accelerate market entry and increase its NPV. Thus, even when technical risks are independent, commercial value is often time-dependent, and development sequencing can materially affect program value. Introducing commercial dependencies illustrates this effect. If AD CVGS increases to $4.0B when MCI succeeds and decreases to $2.0B when MCI fails (versus $2.5B with no dependency), the program eNPV increases to $852M relative to $825M under independence (not shown).

Technical dependence further complicates evaluation. As shown in Table 6.4, if MCI and AD are weakly positively correlated (r = 0.19) and the conditional probability of AD success given MCI success rises to 40%, the program-level eNPV becomes $818M, assuming no commercial dependency (see Figure 6.9).

Increasing the correlation strength (r = 0.71) and conditional success probability to 75% does not change the rolled-back eNPV under these assumptions, illustrating that correlation alone does not necessarily increase value absent changes in expected commercial outcomes. When both technical correlation and commercial dependency are present—such that AD CVGS is higher following MCI success and lower following failure—the eNPV of the program increases under weak positive correlation.

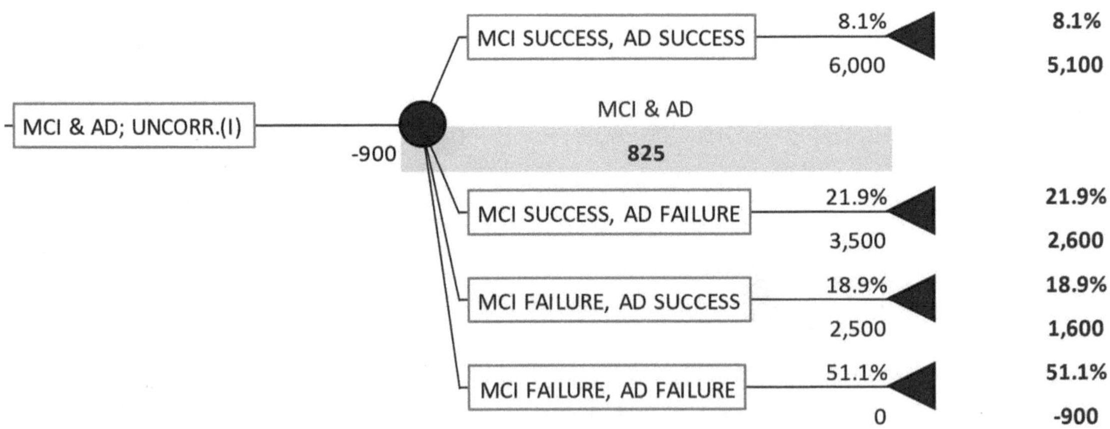

Figure 6.8 Concurrent pursuit of MCI and AD by independent projects with no commercial dependencies.

Table 6.4 Data for sequential pursuit of MCI and AD by technically dependent projects.

Probabilities and Correlation Coefficient			
Project	p(Success)	1 - p(Success)	
MCI	30%	1-30% = 70%	Elicited
AD	27%	1-27% = 73%	Calculated
$p(AD_S \mid MCI_S)$	40%	$p(MCI_S \mid AD_S)$	(40%*30%) / 27% = 44%
$p(AD_S \mid MCI_F)$	27% - (40%*30%) / (1-30%) = 21%	$p(MCI_S \mid AD_F)$	(1-40%)*30% / (1-27%) = 25%
		Corr(MCI,AD)	SQRT((40%-21%)*(44%-25%)) = 0.19

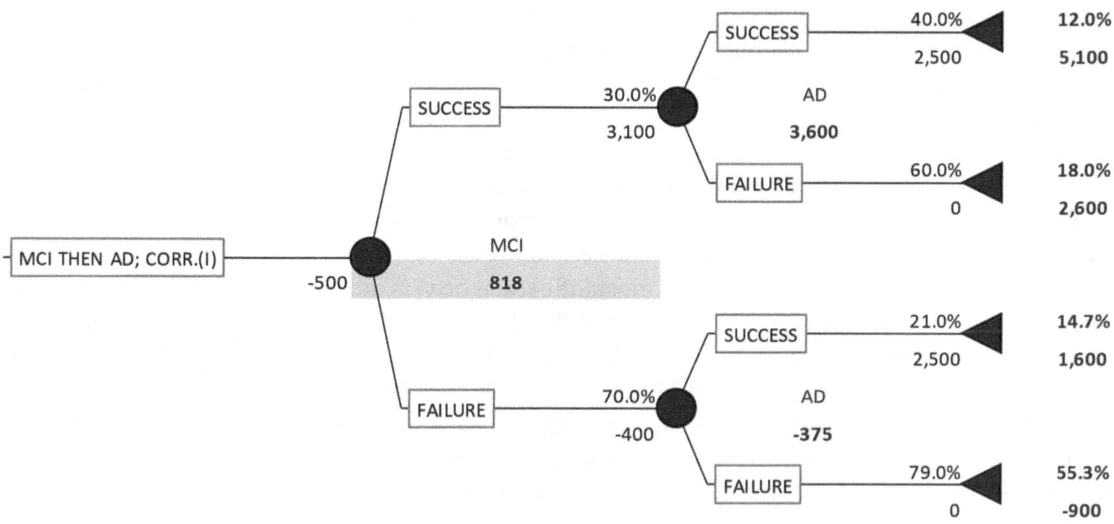

Figure 6.9 Sequential pursuit of MCI and AD by weakly positively correlated projects with no commercial dependencies.

In practice, biopharma organizations often defer initiation of a second indication until proof-of-concept (PoC) is achieved for the first, reducing capital exposure and enabling prioritization based on early clinical evidence. WAV Figure 6.10 illustrates this staged strategy, where the MCI PoC outcome serves as the pivotal decision node. If MCI PoC succeeds, management must choose between a sequential (slow-to-market) AD strategy that staggers investment and a concurrent (fast-to-market) strategy that accelerates AD entry at the cost of higher simultaneous capital deployment. If MCI PoC fails, the organization must reassess whether AD remains viable as a standalone opportunity. This framework demonstrates how program-level decision trees integrate technical risk, commercial timing, and strategic choice into a unified valuation approach.

6.4 SENSITIVITY ANALYSIS TO DECISION TREE PARAMETERS

With the decision tree structure established (see Figure 6.2), attention turns to the parameters that most strongly influence investment decisions: probabilities of success, development costs, and CVGS. Sensitivity analysis evaluates how variation in these inputs can change which strategic alternative dominates.

Consider a choice between two development opportunities (see Figure 6.11):

1. **Innovative drug—**a novel molecular entity with higher scientific risk
2. **Line extension—**a reformulation of an existing marketed product

The Innovative drug has a lower probability of technical success (PTS = 40%) than the line extension (PTS = 70%), while both assume a probability of regulatory success (PRS) of 100%. Development costs are higher for Innovative ($100M vs. $40M), but its commercial upside is greater (CVGS high/low of $500M/$250M vs. $175M/$125M). The probability of achieving a high commercial outcome is 80% for Innovative and 50% for line extension. For Innovative, the joint probabilities of R&D success with high and low commercial outcomes are 32% and 8%, respectively, while R&D failure (60%) results in a loss of –$100M. All calculated path values for both alternatives are summarized in Figure 6.11.

Rolled-back decision tree analysis yields eNPVs of $80M for Innovative and $65M for line extension, making Innovative the dominant alternative under risk-neutral assumptions.

Because all decision tree inputs are uncertain, sensitivity analysis examines how much an assumption must change before the dominant decision reverses. Global sensitivity analysis varies all key inputs simultaneously to identify the most influential drivers. For Innovative, varying R&D cost from –$200M to –$50M and PTS from 10% to 90% shows that PTS is the most influential variable: eNPV ranges from $65M to $305M (see WAV Figure 6.12). High CVGS is the second most influential, shifting eNPV from $65M to $240M across plausible bounds.

A detailed one-way sensitivity analysis (see Figure 6.13) varies Innovative's PTS around its base value of 40%. The eNPV of Innovative equals that of line extension ($65M) at a PTS of approximately 36.7%, identified using Excel Goal Seek. A symmetric analysis shows that line extension would require a PTS of 80% to achieve indifference.

Two-way sensitivity analysis evaluates joint variation in critical parameters. Figure 6.14 shows how Innovative's dominance depends on both PTS and the probability of high commercial success. When PTS is below ~35%, line extension dominates; when PTS is above ~45%, Innovative dominates regardless of commercial assumptions. Between these thresholds, dominance depends on the combination of technical and commercial probabilities. This two-way sensitivity analysis can be repeated with the use of PrecisionTree and Excel data tables to yield the same result which may lend itself to clearer communication with decision makers (see Table 6.5).

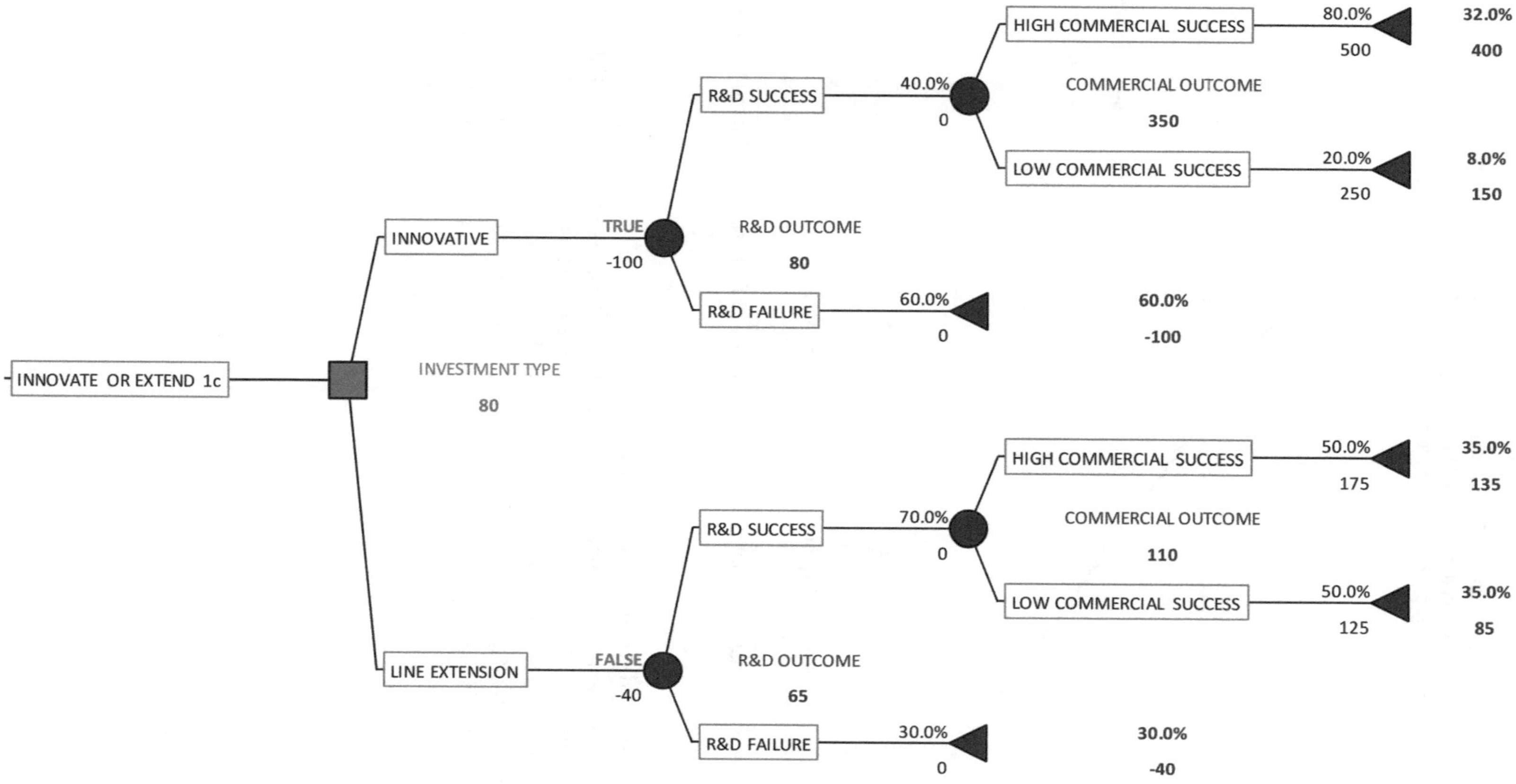

Figure 6.11 Invest in an Innovative or line extension project.

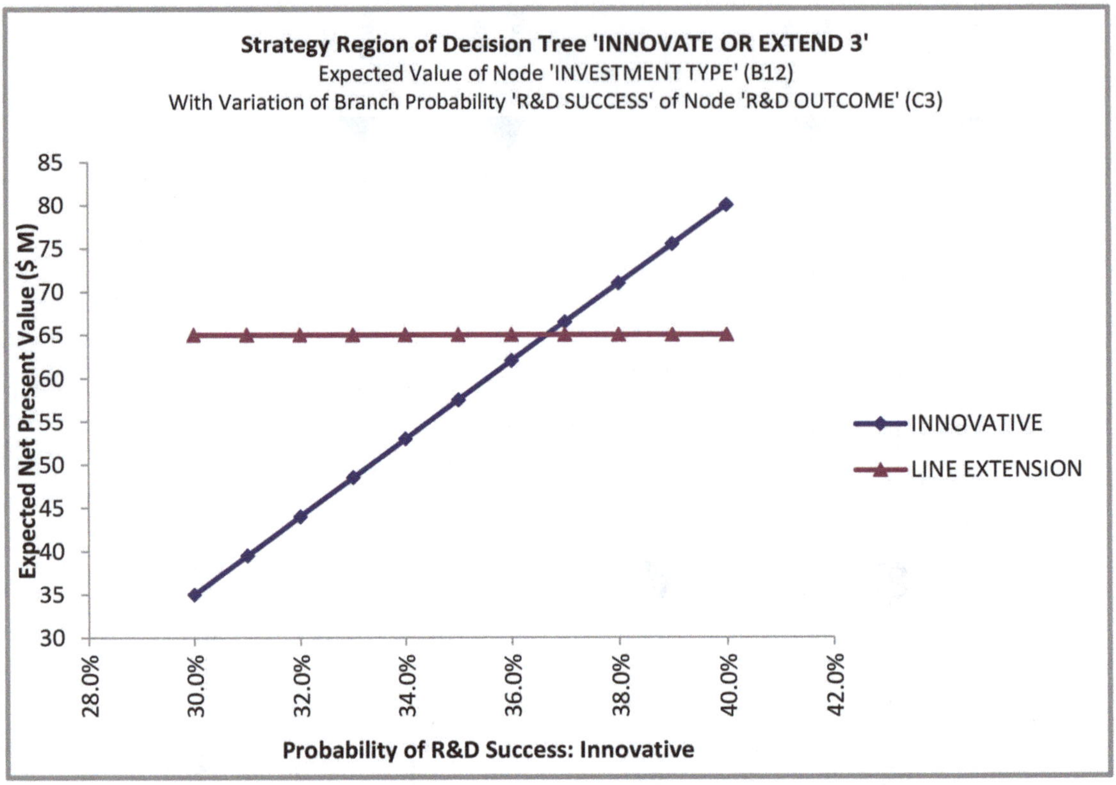

Figure 6.13 One-way sensitivity analysis of the probability of R&D success for the Innovative project.

The Impact of Uncertainty on Decision Making

Beyond deterministic sensitivity analysis, uncertainty in forecasts can be modeled explicitly using probability distributions. R&D costs and CVGS are represented as triangular distributions, capturing minimum, most-likely, and maximum values. For Innovative, R&D costs range from –$130M to –$80M (mode –$100M); and for line extension, from –$50M to –$20M (mode –$40M). Similar ranges are applied to CVGS. Linking these distributions converts the decision tree into a stochastic model (see Figure 6.15).

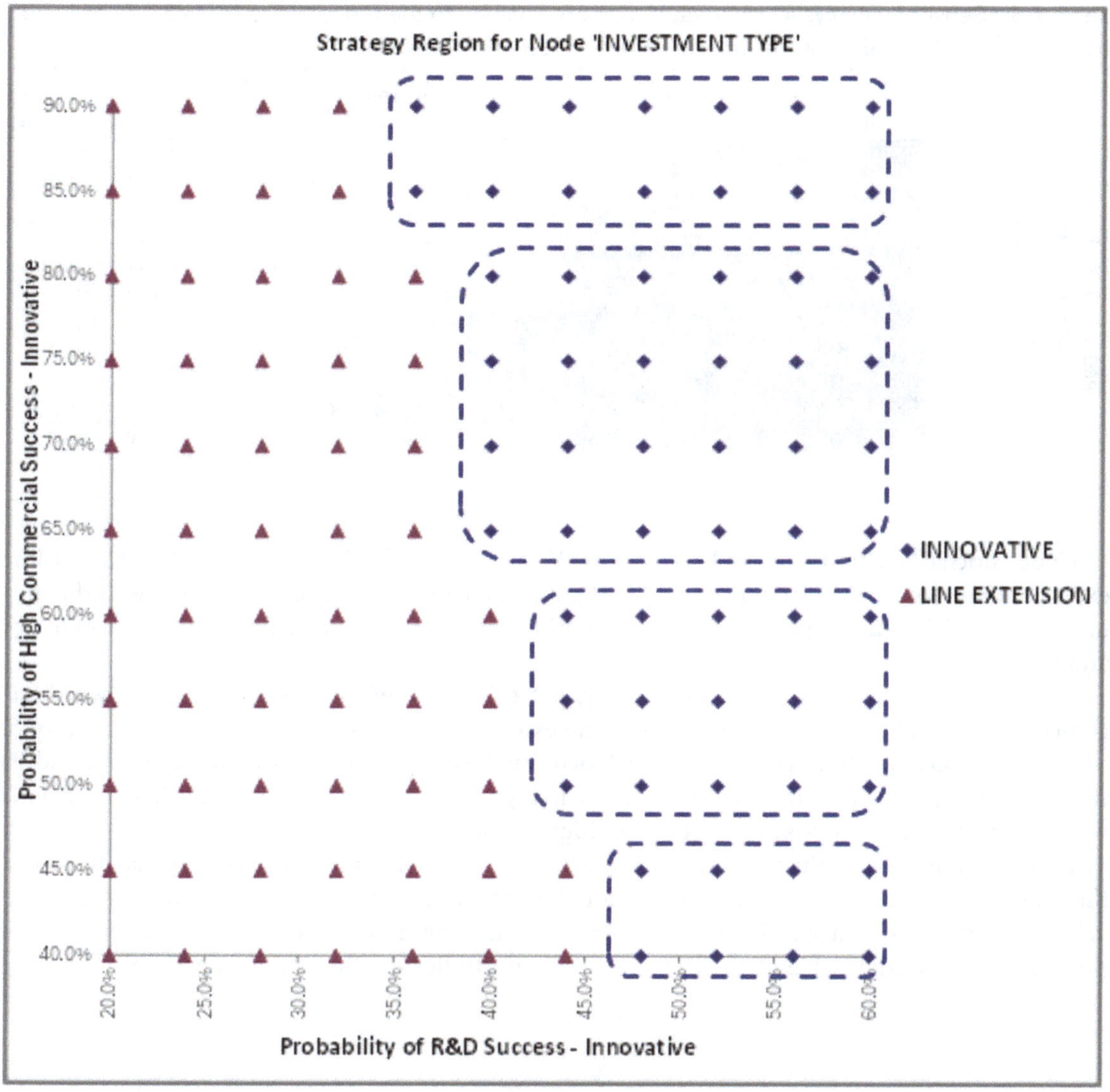

Figure 6.14 Strategy region—quantitative two-way sensitivity analysis of the probabilities of R&D success and high commercial success for the Innovative project using PrecisionTree.

Table 6.5 Sensitivity analysis—two-way sensitivity analysis to the probabilities of R&D success and high commercial success for the Innovative project using data tables.

		PROBABILITY OF R&D SUCCESS - INNOVATIVE									
	INNOV	0.20	0.25	0.30	0.35	0.40	0.45	0.50	0.55	0.60	0.90
PROBABILITY OF HIGH COMMERCIAL SUCCESS - INNOVATIVE	0.90	LIN EXT	LIN EXT	LIN EXT	INNOV	INNOV	INNOV	INNOV	INNOV	INNOV	INNOV
	0.85	LIN EXT	LIN EXT	LIN EXT	LIN EXT	INNOV	INNOV	INNOV	INNOV	INNOV	INNOV
	0.80	LIN EXT	LIN EXT	LIN EXT	LIN EXT	INNOV	INNOV	INNOV	INNOV	INNOV	INNOV
	0.75	LIN EXT	LIN EXT	LIN EXT	LIN EXT	INNOV	INNOV	INNOV	INNOV	INNOV	INNOV
	0.70	LIN EXT	LIN EXT	LIN EXT	LIN EXT	INNOV	INNOV	INNOV	INNOV	INNOV	INNOV
	0.65	LIN EXT	LIN EXT	LIN EXT	LIN EXT	INNOV	INNOV	INNOV	INNOV	INNOV	INNOV
	0.60	LIN EXT	LIN EXT	LIN EXT	LIN EXT	LIN EXT	INNOV	INNOV	INNOV	INNOV	INNOV
	0.55	LIN EXT	LIN EXT	LIN EXT	LIN EXT	LIN EXT	INNOV	INNOV	INNOV	INNOV	INNOV
	0.50	LIN EXT	LIN EXT	LIN EXT	LIN EXT	LIN EXT	INNOV	INNOV	INNOV	INNOV	INNOV
	0.45	LIN EXT	LIN EXT	LIN EXT	LIN EXT	LIN EXT	LIN EXT	INNOV	INNOV	INNOV	INNOV
	0.40	LIN EXT	LIN EXT	LIN EXT	LIN EXT	LIN EXT	LIN EXT	INNOV	INNOV	INNOV	INNOV

Under stochastic simulation, rolled-back eNPVs are $64.4M for Innovative and $63.1M for line extension, narrowing the difference and highlighting sensitivity to uncertainty. Monte Carlo simulation (10,000 iterations) generates full distributions of eNPV outcomes (see WAV Figures 6.16A and B).

For Innovative, there is a 52.2% probability that eNPV exceeds its mean of $64.4M, with outcomes ranging from −$10.8M to $129.4M. Line extension shows a symmetric 50% probability of exceeding its mean, with a narrower range. Stochastic tornado analyses (see Figures 6.17A and B) confirm that Innovative's eNPV is most sensitive to high CVGS and R&D cost, while line extension's eNPV is most sensitive to R&D cost and both high and low CVGS.

Finally, simulation results indicate that Innovative is the dominant choice approximately 54% of the time across all sampled uncertainties (not shown). Although Innovative maximizes expected value on average, the substantial overlap in outcome distributions underscores the importance of probabilistic, risk-adjusted analysis when informing portfolio-level investment decisions.

	INNOVATIVE	LINE EXTENSION
High CVGS	467	167
Low CVGS	230	118
CTL	-103	-37

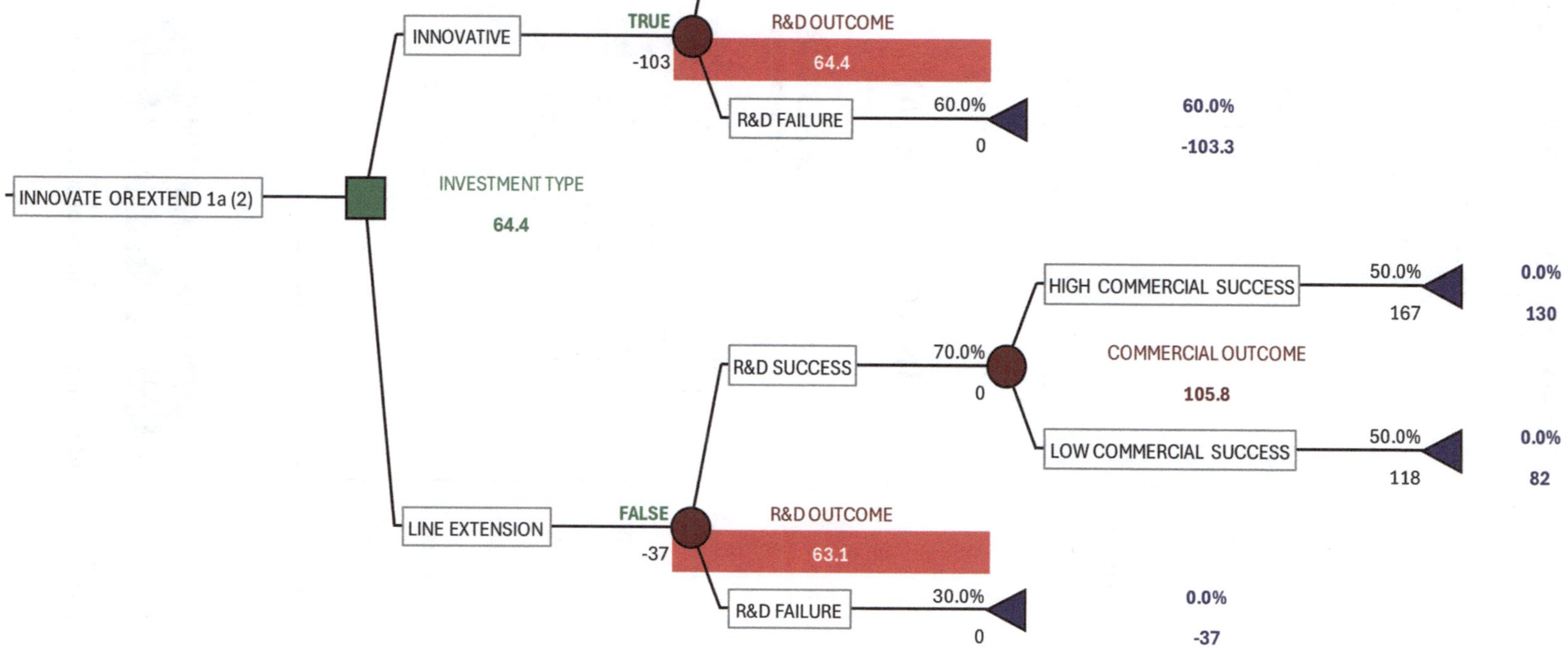

Figure 6.15 Stochastic decision tree for Innovative and line extension accounting for uncertainty in R&D costs and CVGS.

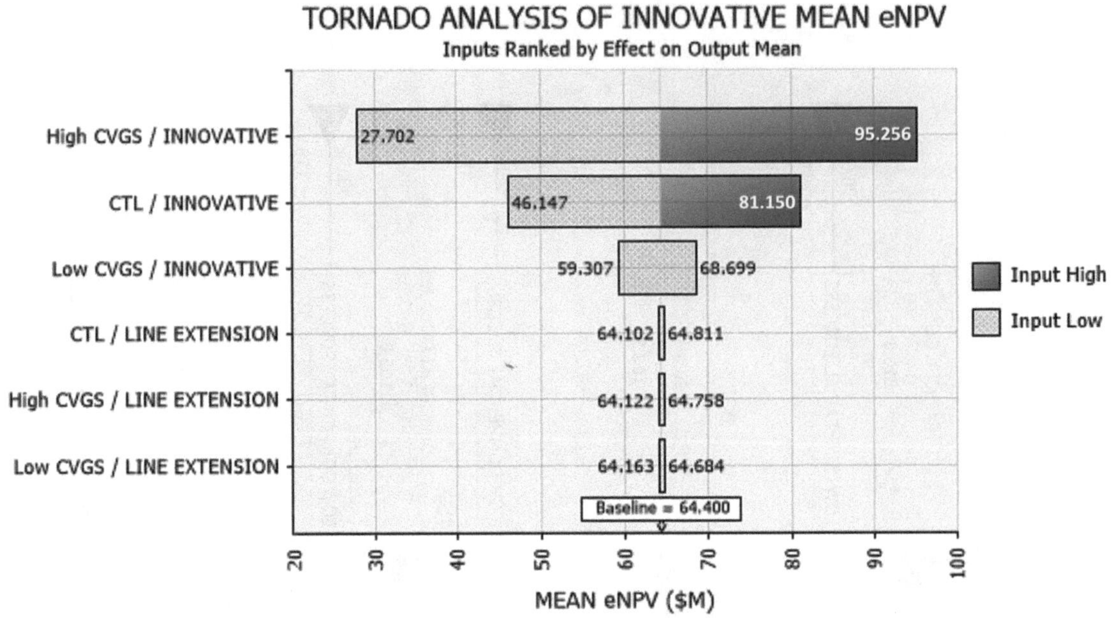

INNOVATIVE (A)

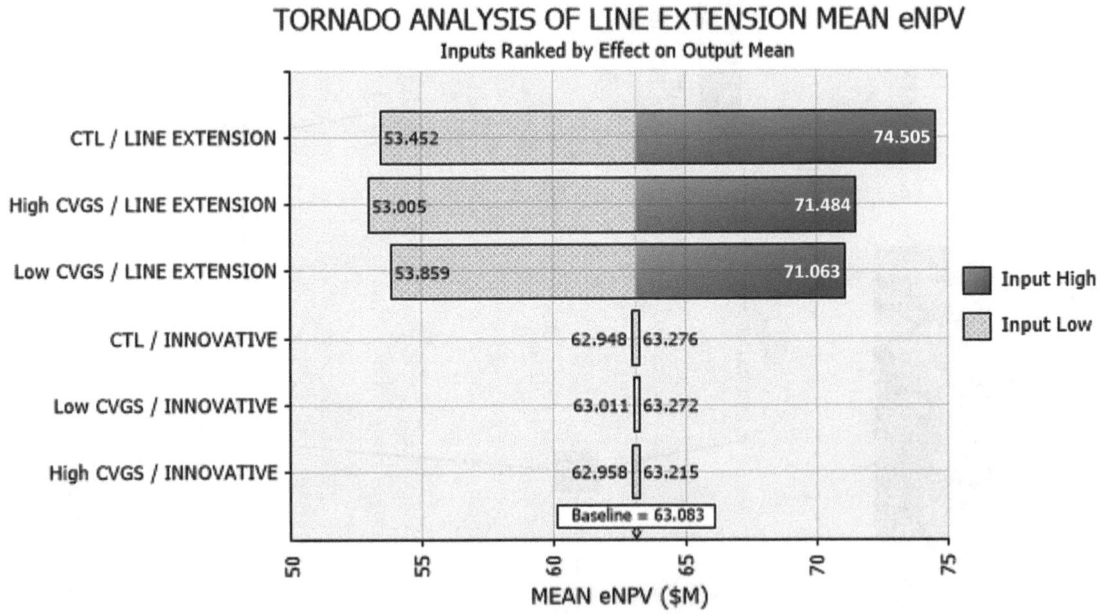

LINE EXTENSION (B)

Figures 6.17A and B Tornado sensitivity analysis of simulation of Innovative and line extension alternatives.

6.5 DISEASE SEQUENCING: MAXIMIZING RISK-ADJUSTED VALUE FOR A DRUG IN PURSUIT OF MULTIPLE DISEASES

Disease Sequencing: Maximizing Risk-Adjusted Value Across Multiple Indications

Once a novel molecule secures patent protection, the window for value capture is finite. In competitive therapeutic areas, sequential development across indications often underutilizes this window. Instead, once clinical PoC is achieved in at least one indication, firms are incentivized to accelerate development across multiple diseases to maximize risk-adjusted value before exclusivity expires. Strategic sequencing, therefore, requires explicit consideration of risk, capital intensity, and launch timing across indications.

In the United States, exclusivity rules define this constraint set:

- **New molecular entity (NME):** A 20-year patent term from filing; five years of exclusivity for the first new drug application (NDA)
- **Supplemental NDA (sNDA):** Three years of exclusivity for additional indications
- **New formulations:** Three years of exclusivity
- **First generic entry:** 180-day exclusivity post-abbreviated new drug application (ANDA)
- **Orphan drugs:** Seven years of exclusivity for populations <200,000

Program Context and Strategic Framing

Consider a novel molecule with a shared mechanism of action across three immune-mediated diseases: asthma, inflammatory bowel disease (IBD), and lupus erythematosus (LE). While preclinical evidence supports activity in all three, no initial clinical PoC exists. From a portfolio perspective, this program is structured into three indication-specific projects, with early PoC outcomes informing subsequent sequencing and investment decisions.

To structure these choices, a *strategy table* is constructed to define compelling, actionable, and differentiable (CAD) program-level alternatives, separating *what* decisions can be made from *how* they are operationalized (see Table 6.6).

Four decision domains span the life cycle:

1. **Early clinical development:** Asthma PoC, IBD PoC, or LE PoC first
2. **Late clinical development:** Parallel or sequential Phase III strategies across indications
3. **Registration:** U.S. alone; U.S. then European Union sequentially; U.S. and European Union concurrently
4. **Commercialization:** Leverage existing infrastructure, partner, or build new capabilities

Table 6.6 Strategy table for clinical development, registration, and commercialization.

STRATEGIC ALTERNATIVE	STRATEGIC DECISION 1: EARLY DEVELOPMENT	STRATEGIC DECISION 2: LATE DEVELOPMENT	STRATEGIC DECISION 3: REGISTRATION	STRATEGIC DECISION 4: COMMERCIALIZATION
ASTHMA POC FIRST	ASTHMA POC	ASTHMA, IBD, LE PARALLEL PHASE III	US ALONE	LEVERAGE
INFLAMMATORY BOWEL DISEASE POC FIRST	IBD POC	ASTHMA, IBD PARALLEL PHASE III; LE SEQUENTIAL PHASE III	US THEN EU SEQUENTIALLY	PARTNER
LUPUS ERYTHEMATOSUS POC FIRST	LE POC	LE, IBD PARALLEL PHASE III; ASTHMA SEQUENTIAL PHASE III	US & EU CONCURRENTLY	BUILD

Although 81 combinations (3^4) are theoretically possible, three CAD strategies are evaluated:

- **Asthma PoC first** (Asthma, IBD, LE parallel Phase III—U.S. and European Union concurrently—Partner)
- **IBD PoC first** (Asthma and IBD parallel Phase III; LE sequential—U.S. then European Union—build)
- **LE PoC first** (LE and IBD parallel Phase III; Asthma sequential—U.S. alone—leverage)

Phase-specific PTS and present values of after-tax cash flows are assigned. Two simplifying assumptions apply:

1. Indication-specific projects are probabilistically independent
2. U.S. and European Union regulatory approval is certain (100%)

Under these assumptions, Phase III success is equivalent to Phase III plus registration (see Table 6.7).

The abbreviated decision tree (see Figure 6.18) identifies IBD PoC first as dominant, with eNPV = \$485M, versus \$271M (Asthma PoC first) and \$319M (LE PoC first).

The full decision tree (WAV Figure 6.19) clarifies that, for IBD PoC first, LE Phase III is pursued only if both IBD and Asthma succeed in Phase III; partial success is insufficient. The same logic applies symmetrically to LE PoC first. A more realistic scenario introduces:

1. **Positive clinical correlation:** PoC success in one indication increases Phase III success probabilities in others
2. **Commercial timing effects:** Sequential launches reduce CVGS due to delayed market entry

Marginal probabilities are retained for comparison, while conditional probabilities and time-adjusted CVGS values are introduced (see Table 6.8). Key implications include:

- PoC success in one indication increases Phase III success probabilities in others (e.g., Asthma PoC success raises IBD and LE Phase III PTS)
- Delayed launches reduce CVGS (e.g., LE CVGS falls from \$350M to \$250M if launched later)
- Phase-specific costs remain unchanged

The abbreviated decision tree (see Figure 6.20) again identifies IBD PoC first as dominant, now with eNPV = \$552M. Asthma PoC first and LE PoC first converge at \$348M and \$346M, respectively, illustrating how interdependencies materially alter relative rankings.

In drug development, it is common for a primary molecule (Lead 1) to target a specific disease pathway to be developed alongside a backup molecule (Backup 1) that shares similar chemical or biological properties. Both are advanced in parallel within the same disease program to mitigate technical risk. Additionally, a second lead molecule (Lead 2)—which may modulate a different target implicated in the disease's pathogenesis—is often included in the program. Larger programs may

Table 6.7 Decision tree data assuming probabilistic independence between projects.

	MARGINAL POC POS	POC COST ($M)	MARGINAL PHASE III & REG POS	PHASE III COST ($M)	CVGS ($M)
ASTHMA	0.40	-40	0.60	-200	1,000
IBD	0.70	-20	0.75	-100	500
LE	0.60	-25	0.65	-125	350

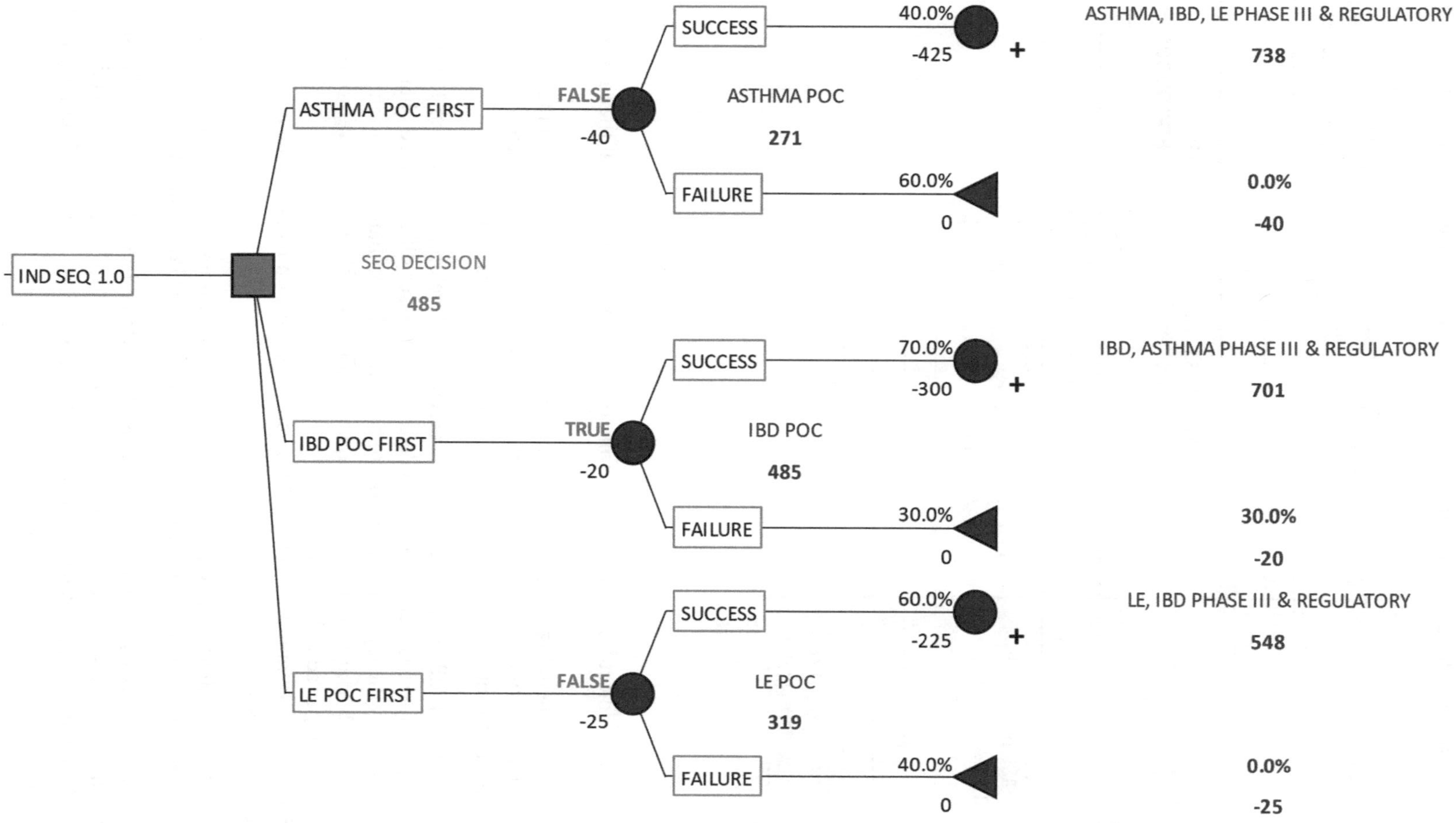

Figure 6.18 Abbreviated decision tree of strategic alternatives, assuming probabilistic independence.

Table 6.8 Decision tree data assuming probabilistic dependence between projects and commercial delays.

	MARGINAL POC POS	POC COST ($M)	MARGINAL PHASE III & REG POS	PHASE III COST ($M)	CVGS ($M)
ASTHMA	0.40	-40	0.60	-200	1,000
IBD	0.70	-20	0.75	-100	500
LE	0.60	-25	0.65	-125	350

	PHASE III & REG POS | ASTHMA POC SUCCESS	PHASE III & REG POS | IBD POC SUCCESS	PHASE III & REG POS| LE POC SUCCESS
ASTHMA	0.75	0.65	N/A
IBD	0.80	0.85	0.80
LE	0.70	N/A	0.80

	PHASE III & REG POS | IBD & ASTHMA PHASE III & REG SUCCESS	PHASE III & REG POS | LE & IBD PHASE III & REG SUCCESS
ASTHMA	N/A	0.70
IBD	N/A	N/A
LE	0.80	N/A

	CVGS | DELAYED LAUNCH ($M)
ASTHMA	650
IBD	N/A
LE	250

house multiple such molecules, creating a portfolio of potential assets aimed at the same therapeutic indication.

As with disease-sequencing programs, a central challenge for the development team is determining when sufficient information from Lead 1's preclinical or early clinical data is available to inform decisions about the downstream fate of Backup 1 and Lead 2. For instance:

- If Lead 1 successfully progresses through preclinical development, Backup 1 may be discontinued unless there is a credible rationale that it possesses differentiated characteristics that could enhance efficacy, safety, or mitigate commercial cannibalization upon launch
- Conversely, if Backup 1 demonstrates superior safety or efficacy, the development trajectory for Lead 1 may need to be reevaluated
- Similarly, the development decision for Lead 2 depends on its potential value in the context of Lead 1 and Backup 1

WAV Figure 6.21 illustrates this integrated decision-making framework. Here, an initial decision is made to pursue PoC for the primary disease target. Based on the PoC outcome:

- **Success:** In this case, the team may choose to advance Lead 1 alone into Phase III, or to simultaneously pursue Phase III for Lead 1 while establishing PoC for Lead 2. Given the redundancy of Backup 1 relative to Lead 1, additional PoC trials for Backup 1 are typically unnecessary.

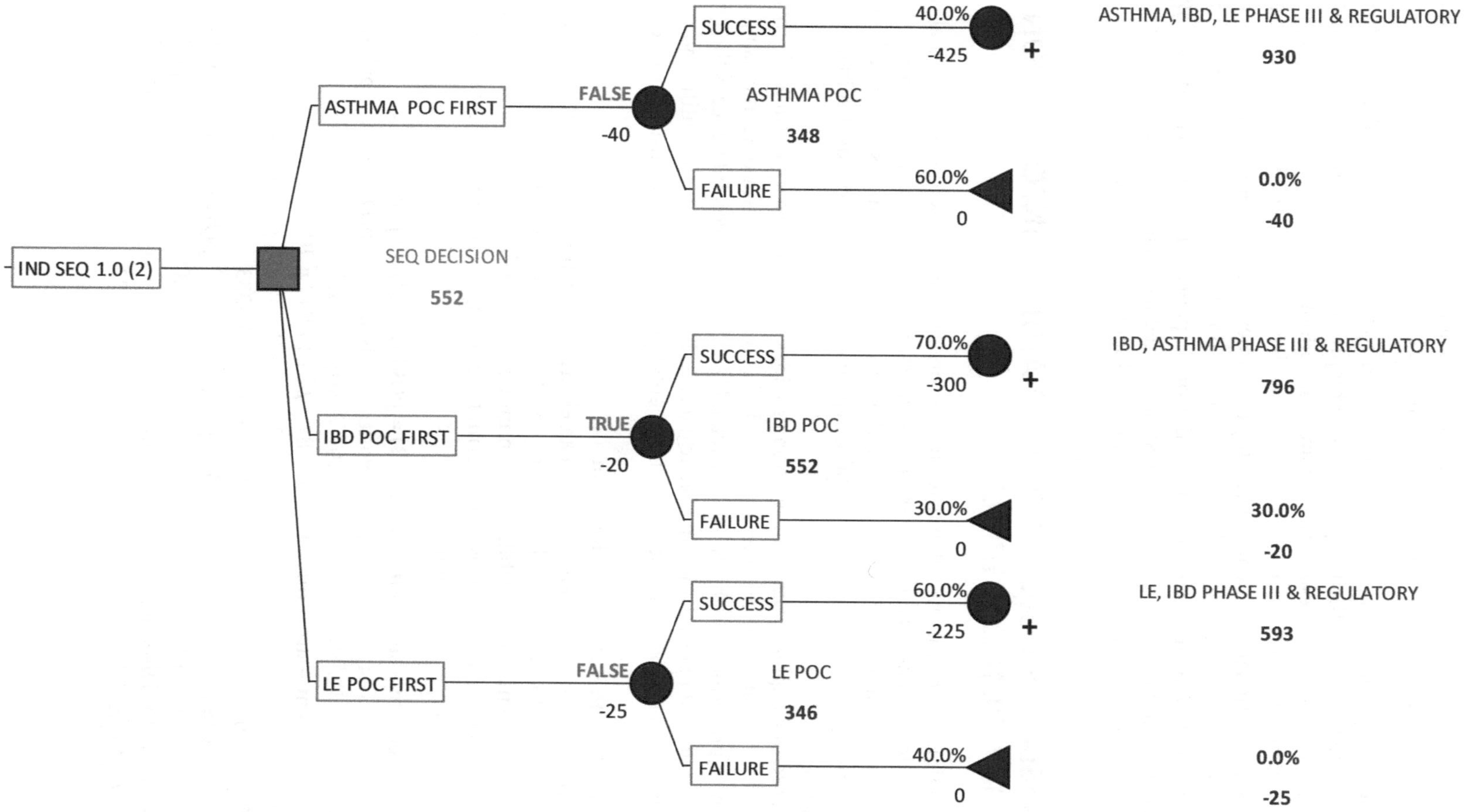

Figure 6.20 Abbreviated decision tree for first indication PoC, assuming positively correlated projects.

- **Failure:** In this instance, the team may opt to establish PoC for Lead 2 alone or for both Backup 1 and Lead 2, as shown in WAV Figure 6.21.

While decision trees provide a rigorous framework for mapping the sequence of events that inform strategic development choices, their application can be prohibitively resource-intensive for smaller biopharma organizations. As a result, many companies adopt a staged, risk-mitigated approach to early development decisions. In practice, this often involves conducting an elaborated Phase I trial that goes beyond traditional safety and tolerability assessments to incorporate early efficacy signals. This trial serves as a cost-effective, preliminary decision point, enabling the organization to gauge whether advancing to a formal, larger-scale, and more expensive PoC trial is justified. By front-loading information in this manner, the organization can reduce financial exposure while maintaining a defensible, data-driven basis for downstream development decisions.

6.6 BAYESIAN REVISION AND THE VALUE OF INFORMATION FROM EARLY CLINICAL TRIALS

Before committing to large, irreversible investments, organizations often seek additional information to improve decision quality. Examples include market research, geological surveys, diagnostic testing, and early clinical PoC trials. In all cases, the information obtained is imperfect and characterized by false positives (Type I errors) and false negatives (Type II errors), determined by sensitivity and specificity.

As introduced in Chapter 3, Bayesian revision provides the formal mechanism for incorporating such imperfect information. It updates a prior belief about an uncertain outcome using new evidence, yielding a posterior probability. As articulated by Thomas Bayes, inference is the revision of prior opinion in light of relevant data. In practical terms, Bayesian revision quantifies how information from diagnostics or PoC trials alters the judgmental probability of downstream success.

Diagnostic Illustration of Bayesian Revision

Consider Disease X with a prevalence of 5% in an elderly population. A diagnostic test has: (a) *sensitivity*: 95% (true positive rate) and (b) *specificity*: 90% (true negative rate).

Using Bayes' theorem, the probability that an individual actually has Disease X given a positive test result can be derived using Bayes' short form, long form, or joint probability formulations, all yielding the same posterior probability.

1. **Bayes' theorem—short form I:** $p(A \mid B) = p(B \mid A) * p(A) / p(B)$
   p(Disease X | Test Positive) = p(Test Positive | Disease X) * p(Disease X) / p(Test Positive)
   = (0.95) * (0.05) / [(0.95 * 0.05) + (0.10 * 0.95)] = 0.3333
2. **Bayes' theorem—short form II:** $p(A \mid B) = p(A \cap B) / p(B)$
   p(Disease X | Test Positive) = p(Disease X * Test Positive) / p(Test Positive)
   = (0.05) * (0.95) / [(0.95 * 0.05) + (0.10 * 0.95)] = 0.3333
3. **Bayes' theorem—long form:** $p(A \mid B) = p(B \mid A) * p(A) / [p(B \mid A) * p(A) + p(B \mid \bar{A}) * p(\bar{A})]$
   p(Disease X | Test Positive) = p(Test Positive | Disease X) * p(Disease X) / [p(Test Positive | Disease X) * p(Disease X) + p(Test Positive | No Disease X) * p(No Disease X)]
   = (0.95) * (0.05) / [(0.95 * 0.05) + (0.10 * 0.95)] = 0.3333

While Bayesian calculations can be performed manually in Excel, PrecisionTree provides an integrated environment to execute them seamlessly. To illustrate, consider a simple decision tree modeling the probabilistic analysis of Disease X (see Figure 6.22).

- The prior probability that an elderly patient has Disease X is equal to the known disease prevalence, 5% (0.05). This represents the decision maker's initial belief before any diagnostic evidence is obtained.
- Empirical data from the diagnostic test provide conditional probabilities:
    - Sensitivity: If the patient has Disease X, the test yields a true positive 95% of the time and a false negative 5% of the time
    - Specificity: If the patient does not have Disease X, the test yields a true negative 90% of the time and a false positive 10% of the time

The key question is: Given a positive test result, what is the probability that the patient actually has Disease X? Formally, we are interested in $p(\text{Disease X} \mid \text{Test Positive})$, the posterior probability. By *flipping the tree*—reordering uncertain nodes so the diagnostic test precedes disease status—posterior probabilities are computed directly (see Figure 6.23).

Despite high sensitivity, the posterior probability remains modest due to low disease prevalence. Parenthetically, if prevalence increased to 10%, the posterior probability following a positive test would rise to 0.5135, illustrating the interaction between base rates and test performance.

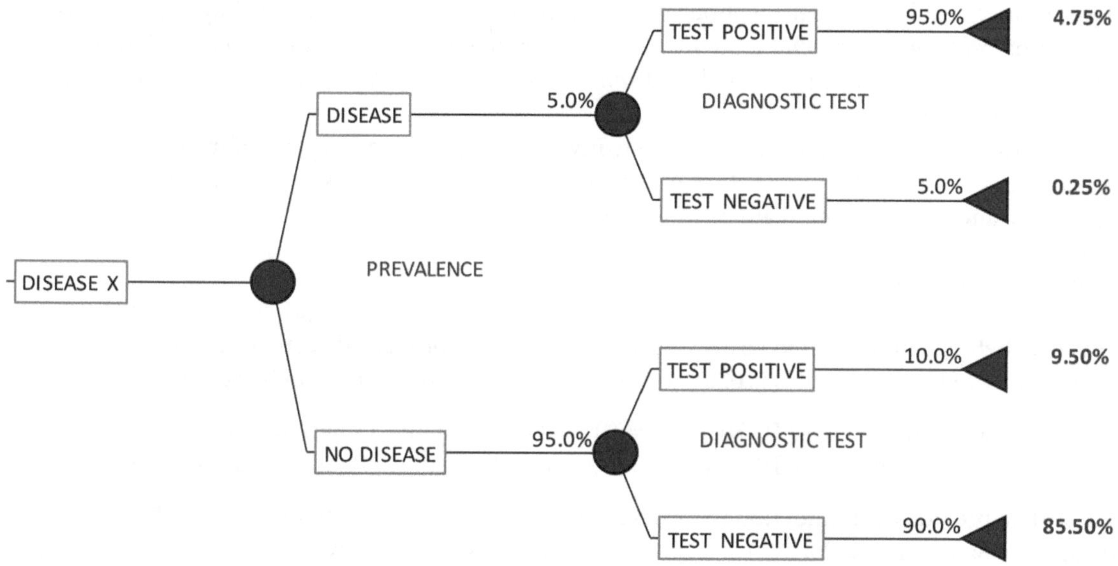

Figure 6.22 Prior belief in patient disease before considering conditional diagnostic evidence.

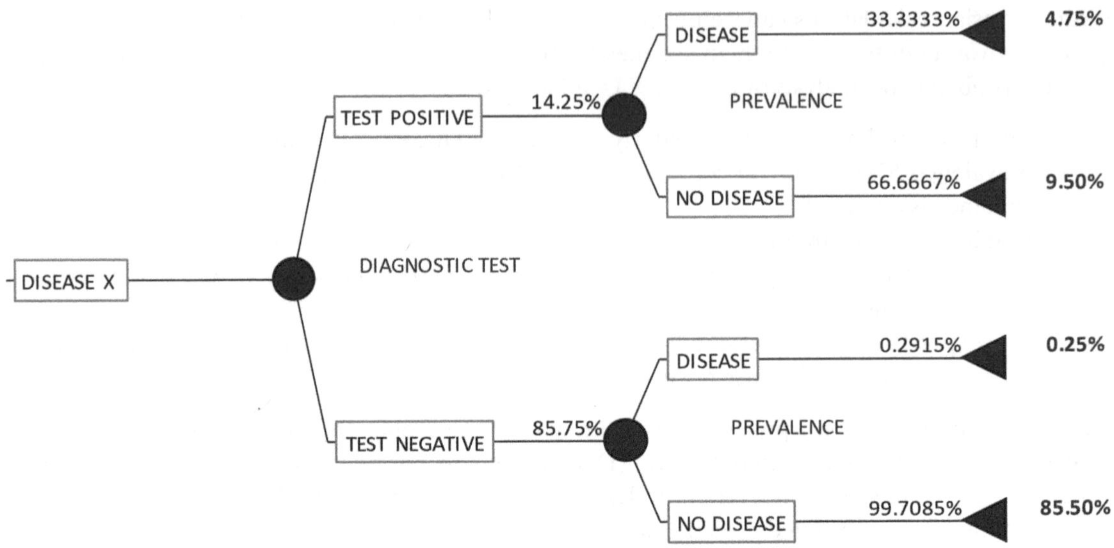

Figure 6.23 Revised belief in patient disease after considering conditional diagnostic evidence.

Extension to Drug Development Decisions

This logic extends naturally to drug development. Outcomes of early trials (e.g., PoC) provide imperfect but valuable information about the likelihood of later-phase success. Bayesian revision allows prior probabilities of Phase III success to be updated based on PoC outcomes, forming the basis for calculating the *expected value of sample information* (EVSI), also referred to as value of information (VOI).

PoC trials are designed with explicit sensitivity and specificity targets informed by historical benchmarks. While PoC success does not guarantee Phase III success, it meaningfully revises expectations and informs capital allocation decisions.

Illustrative PoC–Phase III Example

Consider the case of Therapies 'R' Us, which must decide whether to (a) conduct an inexpensive PoC trial before committing to costly Phase III trials or (b) proceed directly to Phase III. The prior probability of Phase III success is 0.20, while empirical conditional probabilities are:

- *p*(PoC Positive | Phase III Success) = 70%; *p*(PoC Negative | Phase III Success) = 30%
- *p*(PoC Positive | Phase III Failure) = 15%; *p*(PoC Negative | Phase III Failure) = 85%

These data provide the foundation for applying Bayesian revision to update the probability of Phase III success based on the PoC outcome, and, subsequently, to quantify the value of obtaining early information through a PoC trial (see Table 6.9).

The PoC trial in this example exhibits relatively high sensitivity and specificity, with a 15% probability of a Type I error (false positive) and a 30% probability of a Type II error (false negative). The key Bayesian question is: "*Given the empirical, conditional evidence provided by the PoC trial, what is the probability that Phase III will succeed if the PoC trial outcome is positive?*" To answer this, a decision tree is constructed containing only the probability data (see Figure 6.24), which shows the following:

Table 6.9 Prior and conditional probabilities for Phase III and PoC clinical trials.

Prior probability of Phase III	Success 0.20	Failure 0.80
Observed (conditional) probability of positive PoC	When Phase III succeeds 0.70	When Phase III fails 0.15
Observed (conditional) probability of negative PoC	When Phase III succeeds 0.30	When Phase III fails 0.85
Revised probability of Phase III	Success if PoC positive?	Failure if PoC negative?

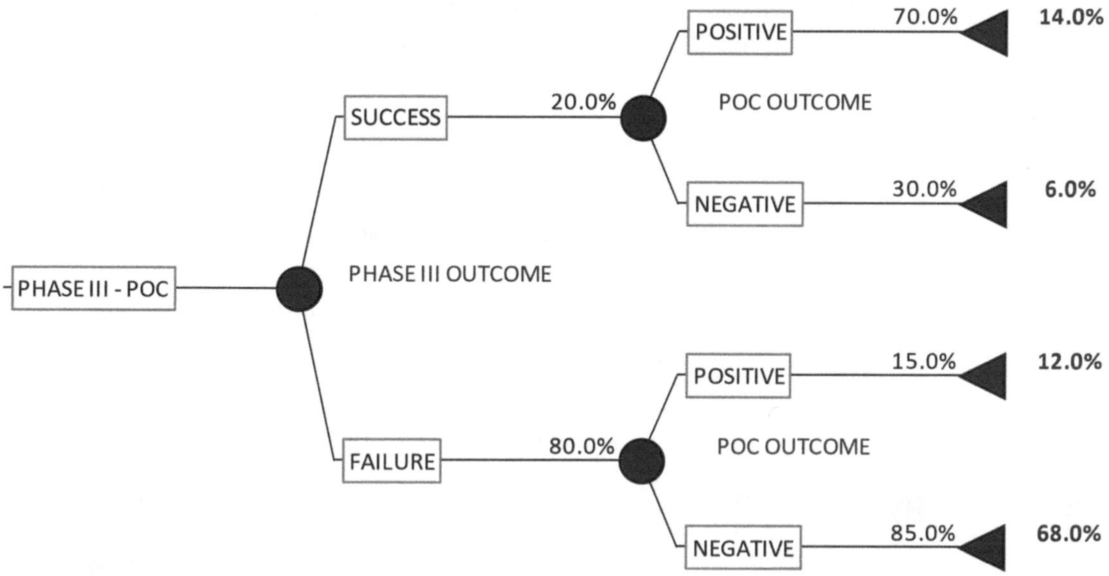

Figure 6.24 Decision tree for prior and conditional probabilities of Phase III and PoC trial outcomes.

1. The prior probability of Phase III success (p(Phase III Success)) = 0.20
2. The conditional probabilities associated with the PoC outcome:
   - p(PoC Positive | Phase III Success) = 0.70
   - p(PoC Positive | Phase III Failure) = 0.15
3. The complementary probabilities:
   - p(PoC Negative | Phase III Success) = 0.30
   - p(PoC Negative | Phase III Failure) = 0.85

Using Bayesian revision, the prior belief about Phase III success is updated based on the PoC outcome. Specifically, the posterior probability of Phase III success given a positive PoC result is calculated as:

$$p(\text{Phase III Success} \mid \text{PoC Positive}) = \frac{p(\text{PoC Positive} \mid \text{Phase III Success}) \cdot p(\text{Phase III Success})}{p(\text{PoC Positive})}$$

where

$$p(\text{PoC Positive}) = p(\text{PoC Positive} \mid \text{Phase III Success}) \cdot p(\text{Phase III Success}) + p(\text{PoC Positive} \mid \text{Phase III Failure}) \cdot p(\text{Phase III Failure})$$

By organizing the problem in a decision tree (see Figure 6.25), the Bayesian calculation becomes transparent, systematic, and visually intuitive, allowing decision makers to clearly see how early information from a PoC trial updates the expected likelihood of Phase III success.

We can see that PoC is positive 26% of the time, as determined by the following calculation:

$$p(\text{PoC Positive} \mid \text{Phase III Success}) \cap p(\text{PoC Positive} \mid \text{Phase III Failure}) = (0.20 * 0.70) + (0.80 * 0.15) = 0.26$$

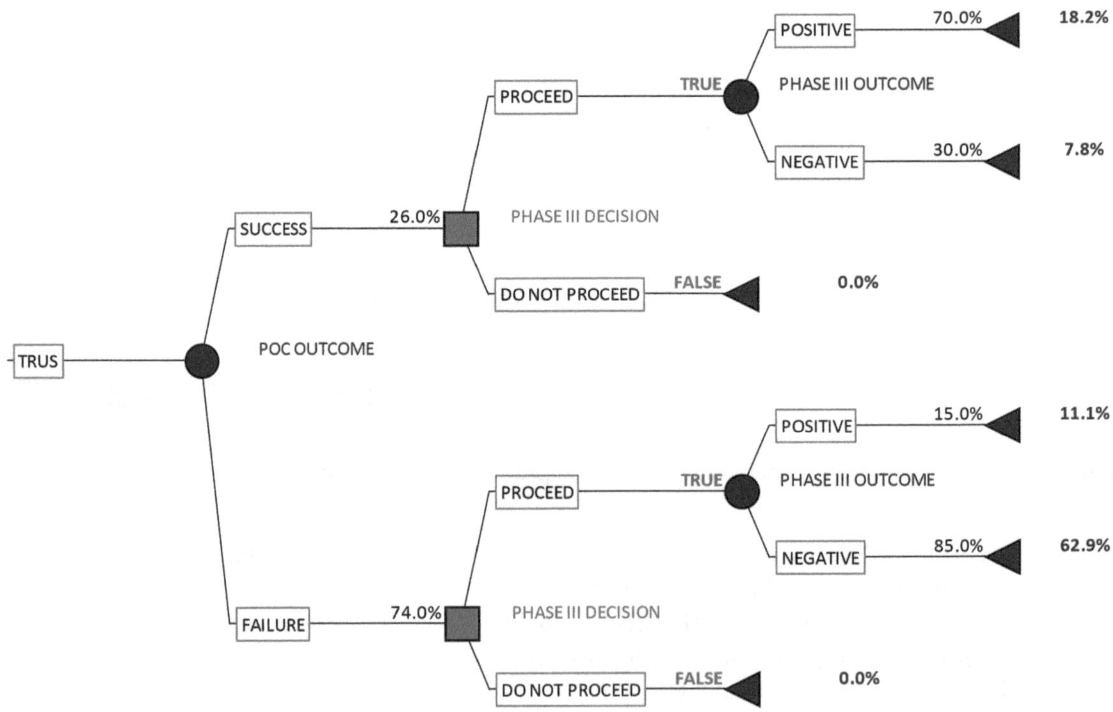

Figure 6.25 Decision tree for Phase III based on the outcomes of PoC trial.

This calculated data is now reflected in Figure 6.26, where the Phase III decision needs to be made after the PoC outcome is known. Posterior probabilities (after *flipping the tree*):

- If PoC is positive:
  p(Phase III Success | Positive PoC) = *p*(Phase III Success ∩ Positive PoC) / *p*(Positive PoC) = 0.14 / 0.26 = 0.5385
- If PoC is negative:
  p(Phase III Success | Negative PoC) = *p*(Phase III Success ∩ Negative PoC) / *p*(Negative PoC) = 0.06 / 0.74 = 0.0811

These posterior probabilities (see Figure 6.26) reflect how the PoC trial updates the prior belief (*p*(Phase III Success) = 20%) based on empirical evidence about sensitivity and specificity. With these revised probabilities, Therapies 'R' Us can now calculate the maximum amount it would rationally pay for the PoC trial:

1. Compute the expected NPV of Phase III without PoC (based on the prior probability of 20%)
2. Compute the expected NPV of Phase III with PoC by weighting the Phase III outcomes with the posterior probabilities (~53.85% if positive, ~8.11% if negative)
3. The difference between these two expected NPVs is the theoretical maximum price to pay for the PoC trial—the EVSI

Finally, we can turn our attention to using these marginal, conditional, and posterior probabilities in a decision tree to quantify the difference in eNPV between (a) conducting an inexpensive PoC

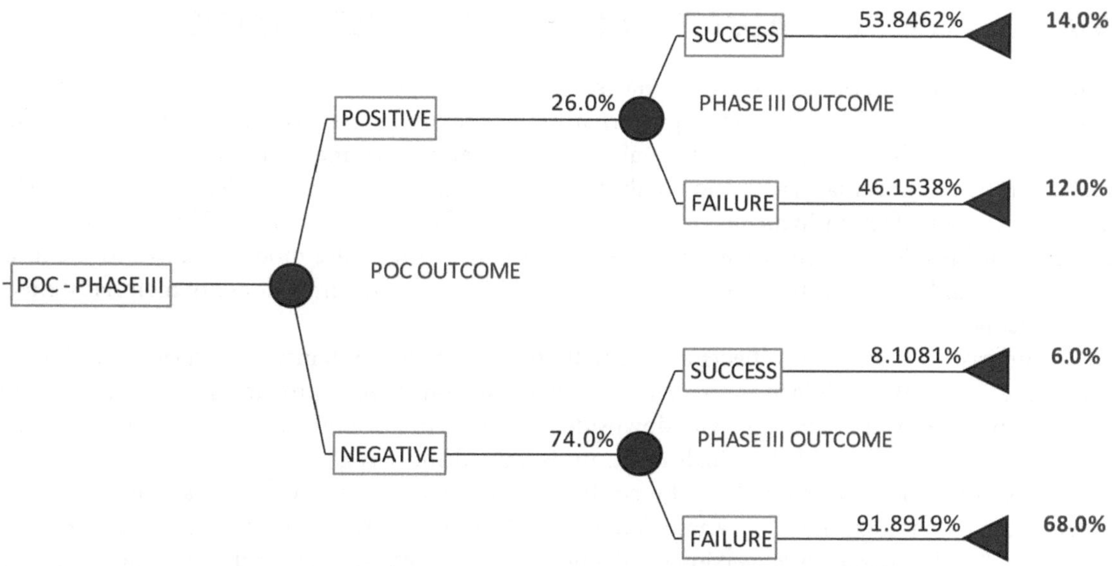

Figure 6.26 Decision tree for posterior probabilities of Phase III given PoC trial outcomes.

trial before committing to expensive Phase III trials and (b) proceeding directly to Phase III trials without a PoC trial. The following assumptions are used in the decision tree (see WAV Figure 6.27):

- The cost of Phase III is \$600M and, if the drug is successful in Phase III, it is expected to yield a CVGS = \$5,500M
- If a PoC trial is conducted before proceeding to Phase III trials, launch of the drug will be delayed, resulting in a lower CVGS = \$5,000M

On an eNPV basis, the strategic alternative of proceeding directly to Phase III trials is valued at \$500M, whereas if a PoC trial is conducted before deciding to proceed to Phase III, this alternative is valued at \$544M. From this, one can deduce that the maximum to pay for the PoC trial (or the VOI gained from this trial) is: VOI or max PoC cost = eNPV(PoC first) – eNPV(Direct Phase III) = \$544M – \$500M = \$44M. Note that, in the decision tree, no cost is attributed to conducting the PoC trial since the purpose of the exercise is to determine its maximum cost. In addition, the calculated posterior probabilities (e.g., p(Phase III Success | Positive PoC) = 53.8462%) above are used.

Since an imperfect PoC trial provides a VOI of \$44M, one would expect that a perfect PoC trial (if it did exist) would provide even greater value but would also cost much more. While there is an approximately 46% likelihood of failing in Phase III even if the imperfect PoC trial is positive, there would be 0% likelihood of failing in Phase III (i.e., 100% success) if the PoC trial were perfectly predictive. In this case, the expected value of perfect information is now \$644M and represents the theoretical maximum one would pay for *perfect information* due to the PoC trial (see WAV Figure 6.28). Of course, the robustness of such an analysis could be supported by several sensitivity analyses of various parameters, including marginal probabilities, the cost of Phase III trials, and the impact on CVGS of a delayed market launch.

6.7 INTEGRATED VALUATION BASED ON UTILITY VALUE

Unless stated otherwise, decision tree analyses conventionally assume that decision makers are risk-neutral and therefore select among alternatives by maximizing expected economic value. Under risk neutrality, gains and losses of equal magnitude carry a symmetric impact on the decision-maker's utility. Conceptually, a decision maker who derives the same level of satisfaction from a \$100 gain as the dissatisfaction incurred from a \$100 loss is defined as risk neutral. This assumption is generally defensible when the financial stakes under consideration are modest relative to the total wealth of the individual or the enterprise, such that variability in outcomes does not materially affect overall economic position.

By contrast, risk aversion reflects a systematic preference for certainty over variability. A risk-averse decision maker will favor a smaller, guaranteed outcome over a probabilistic alternative with a higher expected value but nontrivial downside exposure. Importantly, risk aversion represents a rational decision framework in which certainty is valued more highly than probability-weighted outcomes. This construct should not be conflated with loss aversion, which is a behavioral bias wherein the psychological disutility associated with a loss exceeds the utility derived from an equivalent gain. In other words, risk aversion is a deliberate and analytically coherent approach to limiting exposure to uncertainty, whereas loss aversion is an emotional response to potential losses that can induce both excessively conservative and, paradoxically, risk-seeking behaviors.

At the opposite end of the spectrum, risk-seeking behavior is characterized by a preference for high-variance outcomes, where decision makers accept elevated uncertainty in exchange for the possibility of outsized gains, even when a more certain alternative with a lower expected value is available. Collectively, these behavioral orientations—risk aversion, risk neutrality, and risk seeking—define an individual's or an organization's risk tolerance or risk appetite, which materially influences strategic decision making in R&D-intensive environments such as biopharma R&D.

To formally incorporate risk preferences into decision tree analysis, utility theory can be employed. A utility function transforms monetary outcomes, expressed, for example, in dollars, into utility values measured in *utils*. Once this transformation is applied, all calculations within the decision tree are performed using utility and expected utility rather than NPV and eNPV. Consequently, the preferred decision under a utility-based framework maximizes expected utility, which reflects the decision maker's subjective satisfaction with outcomes and may diverge from the alternative that yields the highest expected financial value.

As illustrated in Figure 6.29, a risk-averse utility function is concave, indicating diminishing marginal utility as payoffs increase and as uncertainty grows. In this case, incremental gains deliver progressively smaller increases in perceived value. In contrast, a risk-seeking utility function is convex, reflecting increasing marginal utility as potential payoffs rise and uncertainty intensifies, thereby reinforcing a preference for high-variance, high-upside outcomes.

Two of the most popular utility functions are exponential and logarithmic, both of which are implemented in Lumivero's PrecisionTree. And utility functions typically have a *risk-tolerance* parameter, R, associated with them.

- For exponential utility, where $U(x) = 1 - e^{-x/R}$, R determines how sensitive the decision maker is to outcome variability, while x is the financial payoff at a terminal node. A small R is indicative of a more risk-averse predisposition while a large R approximates risk neutrality.

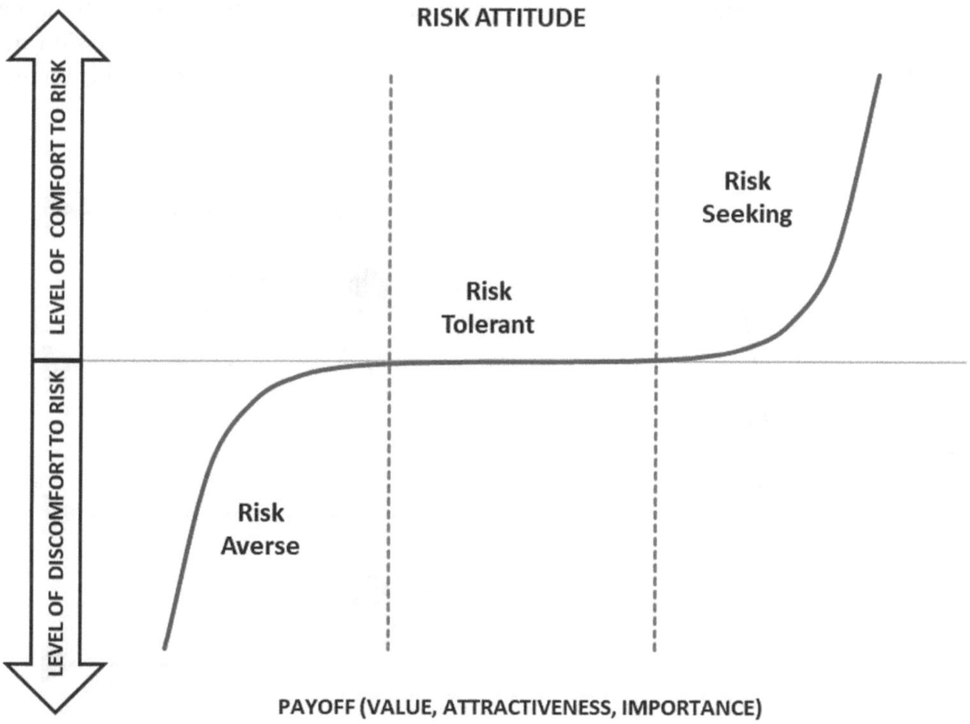

Figure 6.29 Risk-averse and risk-seeking utility functions.

- For logarithmic utility, where $U(x) = \ln(x + C)$, there is diminishing marginal utility for increasing payoff. C is a positive constant introduced to ensure the argument of the logarithm remains strictly positive across all feasible outcomes.

Let's consider a simple decision tree with a decision to be made between two strategic alternatives, each decision having binary outcomes with the same probabilities of success and failure, as shown in Figure 6.30. The difference between strategic alternatives 1 (SA1) and 2 (SA2) is the much larger financial payoff and expected financial value in SA1. Consequently, the decision tree recommends SA1 on the basis of a higher or dominant expected financial value and risk neutrality.

Drug development decisions often involve a high upside and high risk (e.g., an innovative drug vs. a line extension). Using risk-adjusted utility functions allows an organization to formally incorporate its risk appetite into the project selection process. Expected NPV alone may recommend the big bet, but expected utility may favor a safer, smaller bet, reflecting management's risk tolerance.

A decision maker may want to consider that the potential loss is 40 times greater with SA1 than with SA2. Consequently, we need to account for this consideration that may reflect a risk-averse disposition. The exponential utility function has the following mathematical form:

$$Utility = 1 - \exp\left(-\frac{value}{R}\right)$$

If we use R=100 to reflect the decision maker's risk tolerance (in this case, moderate risk aversion), the decision tree recommends SA2 on the basis of its higher or dominant expected utility value (see Figure 6.31). We can see that high variability in SA1's payoff reduces its utility relative to its EV, while for SA2, with smaller but more certain payoffs, the alternative may now become preferred, even though its eNPV is lower. Stated otherwise, the exponential utility captures diminishing marginal utility where a gain of $500M is not 10× as *useful* as a gain of $50M when the decision maker is risk-averse, but the potential $400M loss from SA1 has a disproportionate negative impact on utility.

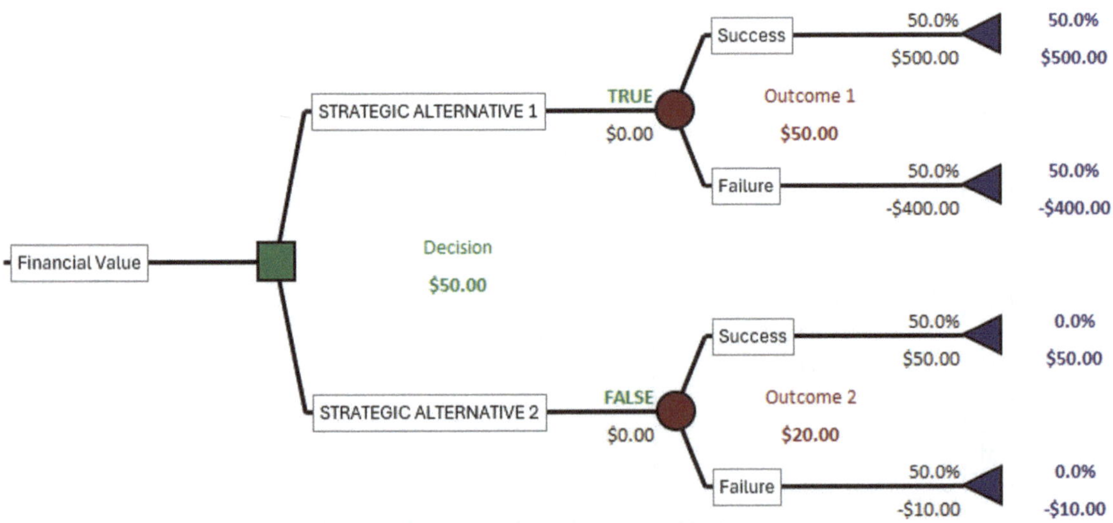

Figure 6.30 Decision tree with rolled back expected financial values.

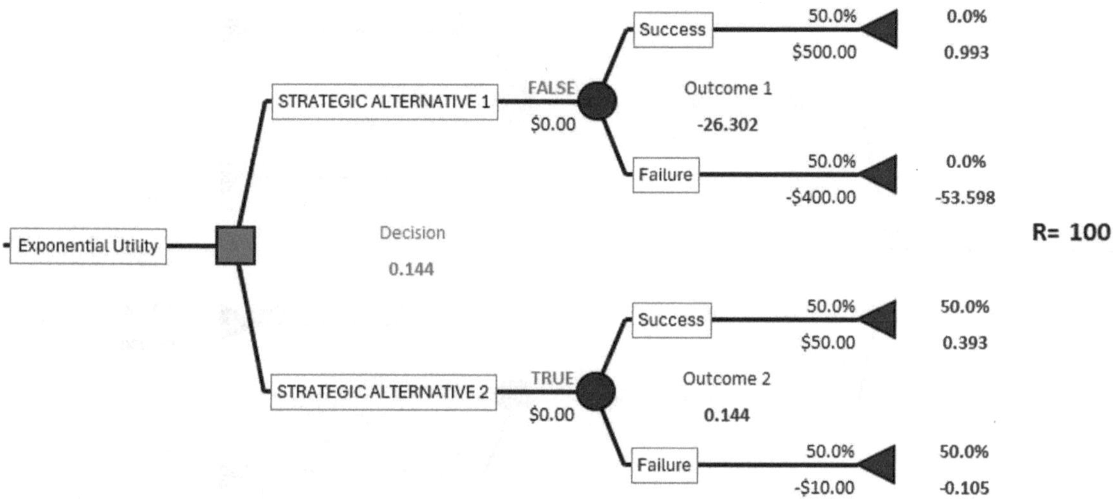

Figure 6.31 Decision tree with rolled back expected utility values.

Applying the utility function $U(\text{value}) = 1 - e^{-\text{value}/R}$

- SA1 success branch: \$500M

$$U = 1 - e^{-500/100} = 1 - e^{-5} \approx 0.993 \text{ utils}$$

- SA1 low outcome: −\$400M

$$U = 1 - e^{400/100} \approx -53.598 \text{ utils}$$

The expected utility at a decision node is the probability-weighted average of utilities at the terminal branches:

$$EU = \sum_i p_i \cdot U(\text{value}_i)$$

This allows a risk-averse decision maker to account for downside risk, not just expected financial value.

The expected utility value can be displayed as the *certainty equivalent* (CE) value. It is calculated using the inverse of the aforementioned utility function and the expected utility value at a node. The CE is the monetary value an individual would pay or accept to avoid having to make a decision. The decision is the same as when the expected utility is displayed, but the units and values may be easier to understand when displayed in the decision tree containing the CE (see Figure 6.32).

Stated otherwise, the CE is the monetary value that yields the same utility as the expected utility (EU):

$$CE = -R \cdot \ln(1 - EU)$$

CE converts utils back into dollars, making the result easier to interpret. If CE < eNPV, the decision maker is risk averse. On the other hand, if CE ≈ eNPV, the decision maker is risk neutral.

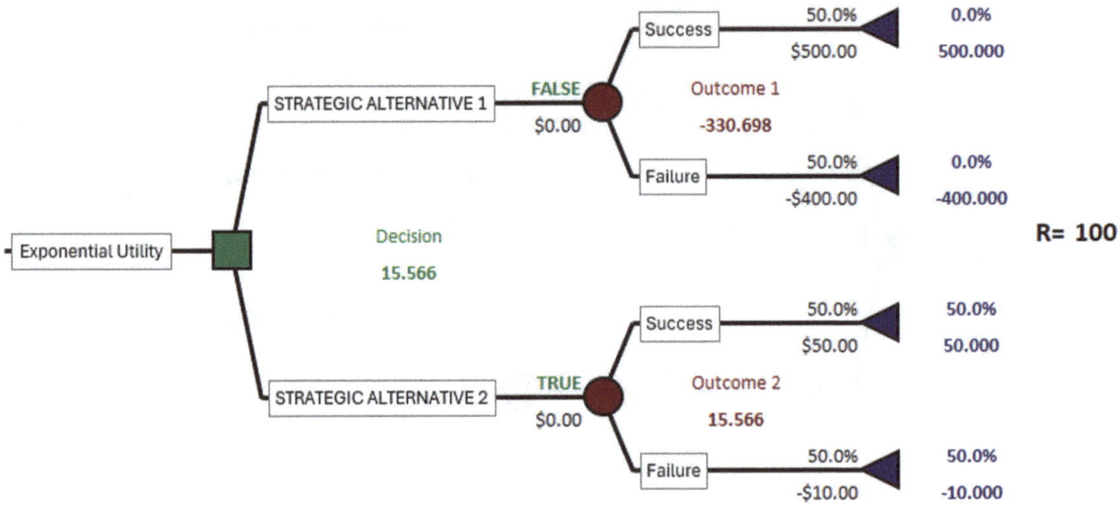

Figure 6.32 Decision tree displaying certainty equivalence.

In our example, since the EU for SA1 = −26.3 utils:

$$CE = -100 \cdot \ln(1 - -26.3) \approx -330.7\text{M}$$

Even though the eNPV might be $250M, the risk-averse CE is −$330.7M—the decision maker is pricing in the risk.

Likewise, the CE value for SA2 is calculated to be $15.57, so a decision maker with a risk tolerance of 100 would accept $15.57 to avoid having to make a decision altogether.

Let us now revisit the Asthma-IBD-LE inflammation indication sequencing case study discussed earlier in this chapter, in which, of the three strategic alternatives listed in Table 6.6, and assuming independent projects, the dominant strategy is to conduct the IBD PoC first.

As described in Figure 6.18 and WAV Figure 6.19, the expected value of the dominant decision is $485M; however, this strategic alternative requires an investment exposure of $320M (to conduct PoC and Phase III studies) before success can potentially materialize. On the other hand, the next-best strategic alternative (LE PoC first) has an expected value of $319M with an investment exposure of $250M (to conduct PoC and Phase III studies).

On the assumption that the organization is risk averse and has doubts about whether the additional investment exposure in IBD PoC first is offset by the higher expected value of the strategy, it has estimated its risk tolerance (R value) to be 140. That is, it is willing to risk $70 to win $140 on a 50/50 coin toss. A summary of financial outcomes and their associated probabilities is shown in WAV Table 6.10 for each of the three mutually exclusive strategic alternatives.

Changing the risk-adjusted financial evaluation of the decision tree to a risk-adjusted utility evaluation via the use of an exponential utility function results in a switch of the dominant strategic alternative from IBD PoC first to the LE PoC first strategy, as shown in Figure 6.33.

A complete list of outcomes based on the utility function where R = 140 (see WAV Table 6.11) can be used to understand why there is a change in the dominant strategy from IBD PoC first to LE PoC first. While IBD PoC first has the highest expected value ($485M) among the three strategic alternatives, it also requires the largest upfront investment exposure ($320M) and carries the possibility of substantial negative outcomes if the PoC or Phase III trials fail. LE PoC first, in contrast,

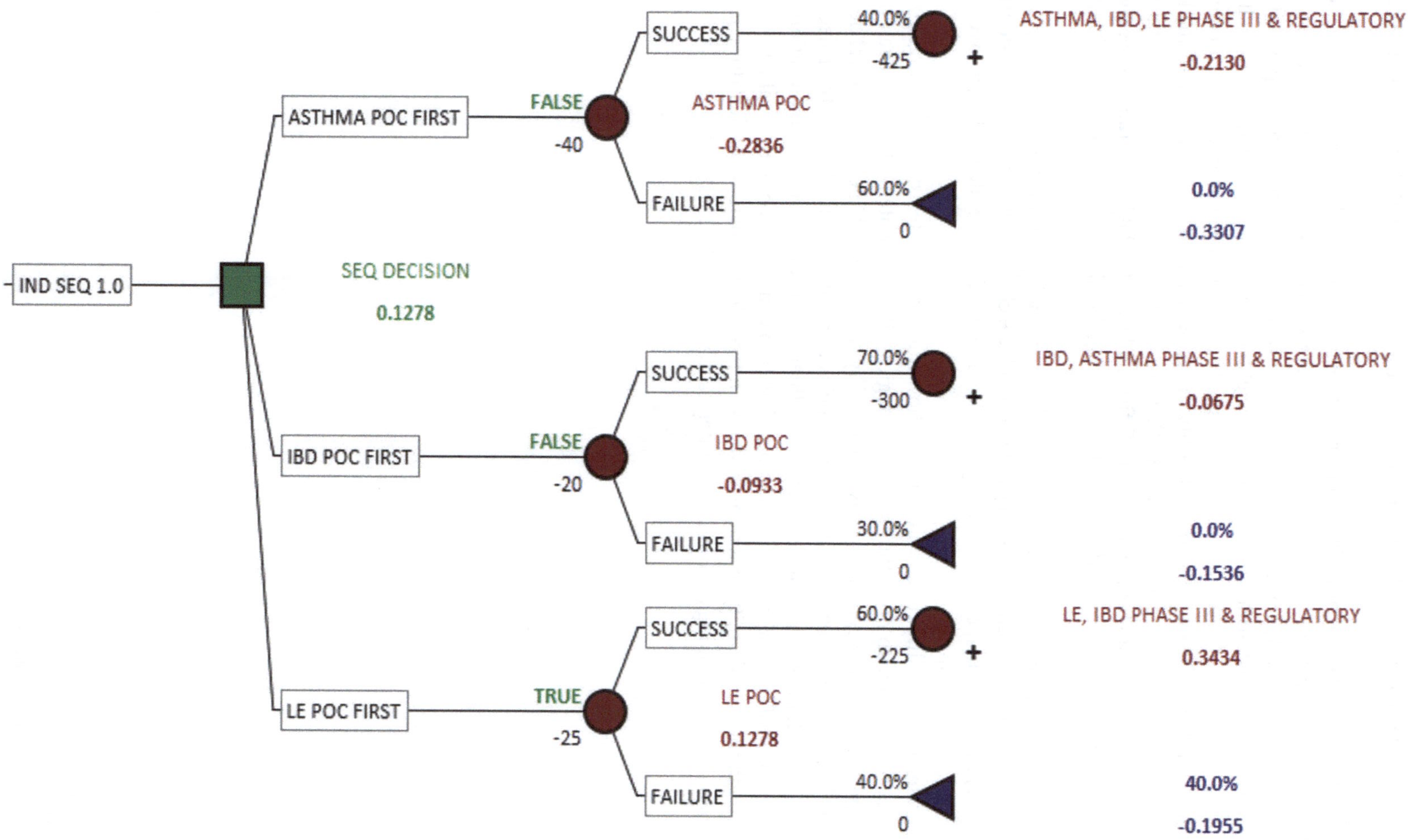

Figure 6.33 Summarized decision tree using transformed utility values and displaying the expected utilities of each strategic alternative.

has a lower expected value ($319M) but requires a smaller investment ($250M) and therefore carries lower downside risk.

The utility of the positive outcomes in the decision tree is similar for all success cases and is close to 1.0. Due to the organization's risk tolerance, as reflected in the value of R used in the analysis, the utility value is significantly lower at higher levels of investment exposure with a possibility of negative outcomes. Expected utility is a probability-weighted average of utility values across all possible outcomes. For IBD PoC first, high upside plus significant negative outcomes result in a utility penalty large enough that the overall expected utility drops. By contrast, for LE PoC first, the lower upside but less severe downside exposure ensures that expected utility remains relatively higher. In short, the negative utility values for the Asthma PoC first and the IBD PoC first strategies are significant enough to result in lower expected utility values, making the LE PoC first strategy the dominant strategic alternative.

It is important to note that if the risk tolerance of the organization was higher—for example, with an R parameter of 210—the dominant strategy would not have changed from the original dominant strategy, that is, IBD PoC first, because the organization is willing to tolerate the larger investment risk in pursuit of a higher upside.

The R parameter in a utility function quantifies how much downside exposure an organization is willing to accept relative to potential upside. Strategies with high potential losses (even if they have high expected value) are penalized in expected utility calculations. Unless risk neutral decision makers are willing to employ expected financial value as a major consideration in their decisions, it is clear that the risk tolerance of an organization influences the decisions the organization eventually makes. This profile, once quantified, can be taken into consideration in the evaluation of any decision through the use of utility functions in decision trees. This provides another dimension of analysis to organizations making decisions on strategies that reflect their risk tolerance.

7

PORTFOLIO OPTIMIZATION

Some of the figures and tables in this chapter are too large or complex to properly format in printed form. Figures and tables marked with a "WAV" label can be downloaded from the Web Added Value Download Resource Center at www.jrosspub.com/wav for proper viewing.

7.1 KEY CONCEPTS

Optimization refers to a disciplined analytical process aimed at configuring a system to achieve the highest attainable level of performance, effectiveness, or efficiency relative to defined objectives. In a formal economic and decision science context, optimization can be interpreted as: (a) a structured mechanism for deploying scarce economic resources in their most productive configuration or (b) a mathematical and managerial approach for maximizing or minimizing one or more performance criteria subject to explicit and implicit conditions.

Within biopharma portfolio management, optimization methodologies are applied to identify the most advantageous combination of investments—typically research and development (R&D) projects and programs, and life-cycle initiatives that collectively advance enterprise objectives. These objectives commonly include maximizing portfolio value (e.g., long-term revenue growth or risk-adjusted net present value), achieving an acceptable risk profile, or maintaining strategic balance across therapeutic areas and development stages. Optimization is performed under a set of constraints that reflect organizational realities, such as capital availability, human resource capacity, manufacturing limitations, and development throughput bandwidth. In addition to strictly binding (*hard*) constraints, portfolios often incorporate aspirational or policy-driven targets or requirements—such as launching approximately five new products per year—that function as *soft* constraints and may be relaxed or traded off when necessary to improve overall portfolio performance.

Portfolio optimization is a mature, analytically rigorous methodology derived from the disciplines of management science and operations research, where it is more broadly categorized under the umbrella of *mathematical programming* (MP). This class of methods encompasses a wide range of optimization techniques, including linear and nonlinear programming (LP, NLP), integer programming (IP), goal programming, single- and multiple-objective linear programming (SOLP, MOLP), and stochastic programming.

A *decision*, in the context of portfolio selection and optimization, represents the allocation of resources—typically financial, human, or infrastructural—to an asset, most often in an irrevocable manner until the next formal decision gate or phase transition. While such decisions commit resources over a defined time horizon, they do not preclude subsequent reallocation as additional information becomes available through non-clinical, clinical, regulatory, or commercial learning.

Nor does approval of a project for a given phase imply automatic continuation into all future phases; each development stage is typically reassessed independently.

From a modeling perspective, prior to a formal commitment, each project or program is treated as a *decision variable*. The optimization process determines whether, when, and to what extent these variables are operationalized, consistent with portfolio objectives and constraints. An *objective* defines a specific business outcome or performance target that the organization seeks to achieve through its portfolio. Common examples include maximizing total portfolio value (e.g., cumulative sales over a planning horizon), accelerating revenue growth, or minimizing expenditures within a defined planning period (e.g., ensuring that total investment in year X does not exceed $\$Y$ million).

When optimization is structured around a single objective, the resulting portfolio solution is designed to achieve the best possible value of that objective—such as the highest achievable sales forecast or the lowest feasible cost above a predefined threshold. In contrast, when multiple objectives are pursued simultaneously, the optimization problem generally yields a set of alternative portfolio solutions. In such cases, improving performance along one objective dimension often necessitates making concessions along another, and no single solution may be optimal across all objectives simultaneously.

Constraints represent limitations on resources or conditions that restrict the feasible set of portfolio solutions. These may be *fungible*, such as financial budgets that can be reallocated across projects, or *nonfungible*, such as specialized human resources, unique technical capabilities, or fixed manufacturing assets.

Constraints are commonly classified as either hard or soft. Hard (binding) constraints—such as maximum manufacturing capacity or regulatory-imposed limits—cannot be violated without invalidating the solution. Soft constraints, more appropriately described as requirements, express desired conditions (e.g., a minimum number of annual product launches) that guide portfolio construction but may be relaxed if doing so yields superior overall outcomes. Notably, some requirements, such as the mandatory inclusion of a strategically critical or externally obligated project, effectively become hard constraints within the optimization framework.

SOLP

SOLP is a mathematical optimization technique used to solve problems characterized by a single objective function and linear relationships among decision variables and constraints. In SOLP, both the objective function (e.g., maximize portfolio value) and the constraints (e.g., budget ceilings, headcount limits) are expressed as linear equations or inequalities. This approach is particularly useful when the decision environment can be reasonably approximated as linear and when management seeks a single, clearly defined performance outcome.

MOLP

MOLP extends the SOLP framework by allowing two or more objectives to be optimized concurrently, while remaining subject to a common set of linear constraints. Because portfolio objectives in biopharma R&D—such as value maximization, risk diversification, and strategic alignment—are often inherently conflicting, MOLP explicitly accommodates trade-offs among objectives. Rather than producing a single *optimal* solution, MOLP typically generates a set of efficient, or Pareto-optimal, portfolios from which decision makers can select based on their strategic preferences and risk tolerance.

Deterministic Optimization

Deterministic optimization applies to decision problems in which all relevant inputs—parameters, coefficients, and constraints—are assumed to be known with certainty. Under this assumption, the resulting portfolio solution is fully determined by the input data, and outcomes are predictable and reproducible. Deterministic models are analytically tractable and useful for baseline planning, sensitivity analyses, and scenarios where uncertainty is either negligible or intentionally abstracted.

Stochastic Optimization

Stochastic optimization addresses decision problems characterized by uncertainty in one or more parameters, such as clinical success probabilities, development timelines, market uptake, or pricing assumptions. In these models, uncertain parameters are represented by probability distributions rather than fixed values. The optimization process, therefore, seeks solutions that perform optimally in expectation or that balance performance and risk across a range of possible future states. As a result, stochastic optimization explicitly incorporates randomness and uncertainty, making it particularly well-suited for biopharma R&D portfolios, where technical and commercial outcomes are inherently uncertain.

Over several decades, portfolio optimization has been extensively applied across diverse functional domains, including financial asset allocation, supply chain and logistics planning, sales force deployment, manufacturing capacity utilization, and R&D project and program selection. Its enduring relevance is attributable to its ability to formally integrate risk and uncertainty into decision-making processes—an essential capability in biopharma R&D environments.

Despite the diversity of mathematical formulations, all portfolio optimization problems share three foundational components:

1. **Objective function(s)**—Objectives articulate what the organization seeks to optimize. Typical objectives include:
   a. Maximization of growth or value, commonly quantified using metrics such as forecasted sales revenue, net present value (NPV), or expected NPV (eNPV).
   b. Minimization of risk, which may be expressed quantitatively using measures such as the standard deviation of expected return on investment or other portfolio-level risk metrics.

   Optimization problems may be structured around a single objective or multiple, potentially competing objectives.
2. **Decision variables**—Decision variables represent the selectable elements of the portfolio, such as R&D projects, programs, life-cycle initiatives, or financial instruments. In binary formulations typical of project selection problems, each decision variable assumes a value of 0 or 1, indicating whether a given project is excluded from or included in the portfolio.
3. **Constraints and requirements**—Constraints define the limits within which the portfolio must be constructed. These commonly include financial budgets, headcount availability, and other scarce resources. In addition, organizations often impose portfolio-level requirements—such as achieving a target of five product launches per year—that function operationally as constraints. Certain obligatory business needs, including mandatory regulatory or compliance projects, effectively become binding constraints because they must be included in any feasible portfolio configuration.

The central aim of portfolio optimization is to identify the specific combination of decision variables that maximizes or minimizes the defined objective function(s), subject to the stated constraints and requirements. In doing so, optimization yields a portfolio that uses constrained resources in the most efficient and value-generating manner possible.

When compared with more heuristic portfolio management techniques, such as simple project scoring or rank-based prioritization, optimization offers several distinct advantages (Ragsdale 2017):

- **Exhaustive evaluation of portfolio combinations:** Optimization systematically evaluates a collectively exhaustive set of feasible asset combinations to identify the portfolio that best satisfies organizational objectives while respecting all constraints and requirements.
- **Quantification of opportunity cost:** This methodology enables explicit calculation of the value *left on the table* when selecting any portfolio other than the optimal one. By definition, such alternatives represent suboptimal portfolios in which constrained resources are deployed inefficiently.
- **Explicit handling of project interdependencies:** Optimization models can formally incorporate dependencies among projects, such as situations where the success or viability of one program is contingent on investment in another.
- **Implementation of logical selection rules:** Project selection logic—such as mutually exclusive choices or the requirement to select exactly one project from a defined subgroup—can be encoded directly within the optimization framework.
- **Real-time scenario and sensitivity analysis:** Once established, optimization models support dynamic scenario analysis during portfolio governance and review meetings, enabling decision makers to explore trade-offs and stress-test assumptions in real time.

7.2 THE EFFICIENT FRONTIER IN PORTFOLIO SELECTION

A central concept underpinning the value of portfolio optimization is the *efficient frontier*. The efficient frontier can be conceptualized as a boundary of optimality that delineates portfolios that maximize return or value for a given level of risk or investment, or equivalently, minimize risk or investment for a given level of value. Importantly, portfolios that lie on the efficient frontier represent optimal combinations of projects in aggregate; they do not imply any intrinsic rank ordering or prioritization of individual projects within those portfolios. Each point on the frontier corresponds to a distinct, yet equally efficient, portfolio configuration under a specific set of constraints.

As illustrated in Figure 7.1, for a fixed level of investment—commonly represented in biopharma contexts by an annual R&D budget—portfolio B dominates portfolio A by delivering greater value. Conversely, portfolio A is dominated by portfolio C, which achieves the same level of value with a lower investment requirement. All portfolios residing along this frontier are classified as feasible and efficient (or optimal), whereas portfolios lying below the frontier are feasible but inefficient, and therefore suboptimal.

In the context of R&D project selection, the fundamental challenge addressed by optimization methodologies is to identify, for any given limiting factor—such as budget, aggregate risk exposure, or constrained human resources—the specific combination of projects that defines a point on the efficient frontier.

To illustrate the mechanics underlying the construction of an efficient frontier (adapted from Ragsdale 2017), consider the case of *Contract Research Leads R Us*, a hypothetical contract research organization operating under multiple binding operational constraints. The firm faces three critical and scarce resources: (a) formulation capacity: 8,000 labor-hours, (b) scale-up capacity: 10,000 labor-hours, and (c) stability testing capacity: 9,000 labor-hours. Within these limits, the organization seeks to maximize total profit over the next calendar year by producing chemical and biological lead molecules for external clients.

As summarized in Table 7.1, each chemical lead and biologic lead consumes different amounts of formulation, scale-up, and stability testing resources. Each successfully delivered chemical lead generates a profit of $5.0M, while each biologic lead generates a higher profit of $7.5M. The managerial challenge is therefore to determine the optimal mix of chemical and biologic leads that maximizes

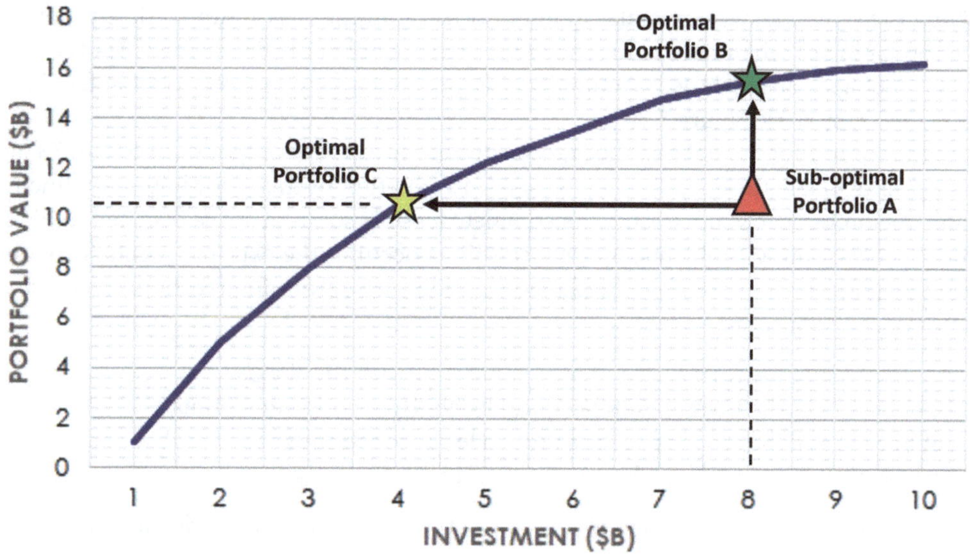

Figure 7.1 The efficient frontier.

Table 7.1 *Contract Research Leads R Us*—resource requirements and profits for chemical and biologic leads.

LEAD TYPE	FORMULATION HOURS REQUIRED PER LEAD	SCALE-UP HOURS REQUIRED PER LEAD	STABILITY TESTING HOURS REQUIRED PER LEAD	PROFIT PER LEAD ($M)
CHEMICAL (X_1)	125	220	180	5.0
BIOLOGIC (X_2)	225	200	235	7.5
TOTAL	7,925	8,460	8,920	275.0
CONSTRAINT (HOURS)	8,000	10,000	9,000	

total profit while remaining feasible with respect to all three resource constraints and while utilizing constrained financial capacity as efficiently as possible.

Given the higher unit profit associated with biologics, an intuitive heuristic approach would suggest allocating constrained resources preferentially toward biologic leads, then using any residual capacity to produce chemical leads. While this reasoning is directionally sensible, it does not guarantee that the resulting portfolio is optimal. To rigorously assess optimality, the problem must be evaluated within the full feasible solution space.

The feasible region represents the set of all portfolio combinations that satisfy the stated constraints simultaneously. This region can be visualized graphically by plotting the number of chemical leads (X_1) on the X-axis and the number of biologic leads (X_2) on the Y-axis (see WAV Figure 7.2).

Beginning with the formulation hours constraint, if the company allocates all formulation capacity exclusively to biologics ($X_1 = 0$), the maximum feasible output is 35 biologic leads. Conversely, if no biologics are produced ($X_2 = 0$), the same constraint allows for a maximum of 64 chemical leads. These two intercepts define a boundary line that constrains the feasible region and represents all portfolio combinations that fully utilize formulation capacity. This boundary is illustrated in WAV Figure 7.2.

Applying the same logic to the scale-up hours constraint, the maximum feasible production increases to 50 biologics when no chemicals are produced, while only 45 chemical leads can be produced if no biologics are included. Adding this second boundary line reduces the feasible region further, as shown in WAV Figure 7.3.

Finally, incorporating the stability-testing constraint, the feasible output is limited to 38 biologics when $X_1 = 0$ and 50 chemical leads when $X_2 = 0$. Introduction of this third boundary line narrows the feasible region further still, yielding the complete feasible solution space depicted in Figure 7.4.

In linear programming problems, optimal solutions occur at the extreme points (or vertices) of the feasible region. In this case, four candidate solutions arise from intersections of the constraint boundaries and the coordinate axes:

1. Intersection of the formulation hours constraint with the Y-axis
2. Intersection of the formulation hours and stability-testing constraints
3. Intersection of the stability-testing and scale-up constraints
4. Intersection of the scale-up hours constraint with the X-axis

Solving the corresponding simultaneous equations at each intersection yields the following feasible portfolios:

1. $X_1 = 0, X_2 = 35 \rightarrow$ Total profit = \$262.5M
2. $X_1 = 13.1, X_2 = 28.3 \rightarrow$ Total profit = \$277.8M
3. $X_1 = 35, X_2 = 11.5 \rightarrow$ Total profit = \$261.3M
4. $X_1 = 45, X_2 = 0 \rightarrow$ Total profit = \$225.0M

Among these continuous solutions, the second intersection clearly dominates the others in terms of total profit. However, before confirming this result, the problem can be formally reframed as a mathematical optimization model.

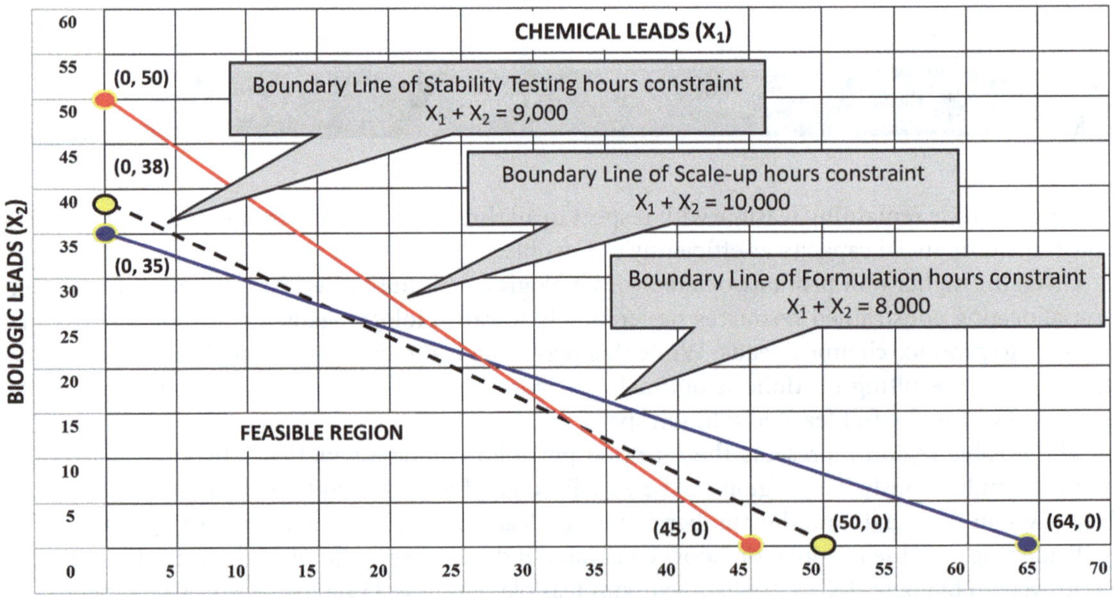

Figure 7.4 Identifying the feasible region of all possible portfolios with the formulation hours, scale-up hours, and stability-testing hours constraints.

1. **Decision variables**
   - X_1 = number of chemical leads produced
   - X_2 = number of biologic leads produced
2. **Objective function (profit maximization)**

$$\text{Maximize } Z = 5.0X_1 + 7.5X_2$$

3. **Resource constraints**
   - Formulation hours:

$$125X_1 + 225X_2 \leq 8{,}000$$

   - Scale-up hours:

$$220X_1 + 200X_2 \leq 10{,}000$$

   - Stability-testing hours:

$$180X_1 + 235X_2 \leq 9{,}000$$

Because both the objective function and the constraints are linear weighted sums of the decision variables, the problem is initially classified as an LP problem.

4. **Bounds and integrality conditions**
   - Non-negativity:

$$X_1, X_2 \geq 0$$

   - Integrality:

$$X_1, X_2 \in \mathbb{Z}$$

The requirement that the number of leads produced must be whole numbers transforms the problem from LP to an integer linear programming (ILP) formulation.

Applying the heuristic strategy of producing biologics exclusively (i.e., setting $X_1 = 0$), the constraints yield the following limits:

- Formulation:

$$225X_2 \leq 8{,}000 \Rightarrow X_2 = 35.5$$

- Scale-up:

$$200X_2 \leq 10{,}000 \Rightarrow X_2 = 50$$

- Stability testing:

$$235X_2 \leq 9{,}000 \Rightarrow X_2 = 38.2$$

The binding constraint is formulation capacity, resulting in $X_2 = 35$ biologic leads (after rounding down). This portfolio yields a total profit of $262.5M, which is clearly feasible. However, feasibility alone does not imply optimality. The critical question is whether this solution lies on the efficient frontier and therefore represents the best possible use of constrained resources.

Solving the ILP using Lumivero's Evolver® optimization software identifies the optimal integer solution as $X_1 = 13$ chemical leads; $X_2 = 28$ biologic leads.

This portfolio yields a total profit of $275.0M, exceeding the heuristic solution by a meaningful margin and demonstrating that exclusive prioritization of biologics does not lead to optimal resource utilization.

7.3 APPLICATION OF DETERMINISTIC SINGLE OBJECTIVE OPTIMIZATION FOR R&D PORTFOLIO SELECTION

Across much of biopharma, portfolio selection is conducted either immediately prior to, or in parallel with, finalization of the annual business plan and the associated budget rationalization process. When portfolio optimization is employed as the analytical mechanism supporting this exercise, the selection problem can be formally expressed as follows:

> *Identify the combination of candidate projects and programs (decision variables) that maximizes portfolio value—or equivalently minimizes risk, investment, or time—subject to the organization's binding resource constraints, such as financial budget, specialized human resources, and manufacturing capacity.*

The following case study illustrates how this problem can be framed and solved rigorously using optimization techniques in a realistic biopharma R&D context.

Innovative Therapies R Us manages a growing clinical development portfolio consisting of 10 projects, labeled A through J. Each project has defined requirements for financial resources and functional headcount in the upcoming calendar year (year 1), as summarized in Table 7.2.

To fund all 10 projects concurrently, and thereby maximize portfolio value as measured by eNPV, the organization would require discretionary incremental resources totaling:

(a) $2.0B in year 1 budget
(b) 11.8 full-time equivalents (FTEs) from function A (e.g., pharmacokinetics/pharmacodynamics)

Table 7.2 *Innovative Therapies R Us*—resource requirements for a portfolio of 10 projects in year 1.

PROJECT	DECISION VARIABLES	FUNCTION A FTEs REQUIRED YEAR 1	FUNCTION B FTEs REQUIRED YEAR 1	BUDGET REQUIRED ($M) YEAR 1	RISK-ADJUSTED NPV (eNPV) ($M)
A	1	0.70	2.40	200	650
B	1	1.60	2.50	175	550
C	1	0.90	1.60	250	600
D	1	0.80	1.80	275	895
E	1	0.90	0.70	300	900
F	1	0.60	1.60	150	550
G	1	1.50	0.50	125	450
H	1	1.60	2.00	145	375
I	1	2.40	1.80	160	525
J	1	0.80	1.90	220	750
TOTAL		11.80	16.80	2,000	6,245
CONSTRAINT		7.00	10.00	1,500	

(c) 16.8 FTEs from function B (e.g., safety, toxicology)

In practice, however, only the following resources are available in year 1:

(a) $1.50B in budget
(b) 7.0 function A FTEs
(c) 10.0 function B FTEs

If unconstrained, the full 10-project portfolio would generate an aggregate eNPV of $6.245B. The strategic challenge confronting *Innovative Therapies R Us* is, therefore, to determine which subset of projects maximizes value while operating within these binding financial and human resource constraints. In effect, the firm must identify its optimal portfolio from among the 10 candidates.

Although it is theoretically possible to enumerate and evaluate every feasible project combination manually using spreadsheet logic, the combinatorial complexity makes this approach impractical. Instead, optimization engines such as Lumivero's Evolver®, implemented as an Excel add-in, enable rapid identification of optimal solutions.

To illustrate a common industry practice—and a common analytical pitfall—the optimization is first performed using only the year 1 budget constraint, ignoring human resource limitations. The resulting optimal solution is presented in Table 7.3.

Under this single-constraint formulation, the optimal portfolio consists of seven of the 10 projects and achieves a maximum eNPV of $4.82B, while utilizing $1.48B of the available budget. However, this solution requires 7.8 function A FTEs and 12.7 function B FTEs, thereby violating both human resource constraints.

This result raises an important question: *Why optimize on the budget constraint alone if the resulting solution is infeasible from a staffing perspective?* The reason for this step is illustrative. In many biopharma organizations, the annual budget is treated—implicitly or explicitly—as the only binding constraint worthy of formal analysis. Extensive experience across biopharma, medical devices and diagnostics, consumer health, and agriculture consistently demonstrates that this assumption is

Table 7.3 Portfolio optimization subject to the year 1 budget constraint only using ILP.

PROJECT	DECISION VARIABLES	FUNCTION A FTEs REQUIRED YEAR 1	FUNCTION B FTEs REQUIRED YEAR 1	BUDGET REQUIRED ($M) YEAR 1	RISK-ADJUSTED NPV (eNPV) ($M)
A	1	0.70	2.40	200	650
B	1	1.60	2.50	175	550
C	0	0.90	1.60	250	600
D	1	0.80	1.80	275	895
E	1	0.90	0.70	300	900
F	1	0.60	1.60	150	550
G	0	1.50	0.50	125	450
H	0	1.60	2.00	145	375
I	1	2.40	1.80	160	525
J	1	0.80	1.90	220	750
TOTAL		7.80	12.70	1,480	4,820
CONSTRAINT		7.00	10.00	1,500	

flawed. In practice, scarce and specialized human resources frequently represent the true bottleneck in R&D execution, rather than capital availability.

In the budget-only optimal solution, projects C, G, and H are excluded. At first glance, this may appear counterintuitive, particularly because project C has a higher standalone eNPV than several projects included in the portfolio. The explanation lies in resource efficiency. For any feasible combination that includes project C, there exists an alternative combination—excluding project C—that delivers higher aggregate eNPV for the same or lower resource consumption. Consequently, the solution excluding project C cannot be improved upon without violating constraints and is therefore optimal under the stated conditions.

Importantly, optimization does not prescribe what management *must* do. *Innovative Therapies R Us* may elect to pursue any feasible project combination. However, any portfolio other than the optimal one is, by definition, suboptimal, meaning it yields a lower total eNPV than the maximum achievable value of $4.82B.

Suppose now that project C, with an eNPV of $600M, is deemed strategically essential and must be included in the portfolio. This requirement is implemented by constraining the project's binary decision variable to equal 1, effectively adding a new hard constraint to the optimization.

Re-figuring the problem under this condition yields the results shown in WAV Table 7.4. With project C mandated, the new optimal portfolio consists of eight projects and delivers a maximum eNPV of $4.695B, representing a $125M reduction relative to the unconstrained optimal solution. Notably, project E, the single highest value-adding project (eNPV = $900M), is now excluded, along with project A.

This outcome underscores a fundamental principle of optimization: *mandates and requirements carry opportunity costs*. If project C is indispensable for strategic or existential reasons, then—ceteris paribus—the best outcome the organization can achieve under the year 1 budget constraint is $4.695B.

An iterative optimization process is strongly recommended and can be conceptualized as the following sequence:

- Define portfolio objectives and identify all hard constraints (budgetary and human)
- Segregate must-do (nondiscretionary) projects from may-do projects, enforcing mandatory inclusion via binary constraints
- Optimize the portfolio and document the maximum value achieved, along with selected (list A) and excluded (list B) projects
- Sequentially force projects from list B into subsequent optimization runs to quantify the value foregone as a result of each mandate
- Alternatively, reserve resources for must-do projects up front and optimize only across discretionary projects using the remaining resource pool

Real-world portfolios frequently include technical or commercial dependencies. Returning to Table 7.2, assume (a) project F requires project H to be commercially successful (but not vice versa) and (b) projects B and J compete in the same market segment, and only one should be selected to avoid cannibalization.

These dependencies are encoded using binary integer logic as follows:

- Project F – Project H ≤ 0
- Project B + Project J = 1

Re-optimizing with these additional requirements yields the solution shown in WAV Table 7.5. The resulting optimal portfolio is valued at $4.595B, reflecting a $225M reduction relative to the prior solution in Table 7.3. As expected, layering additional requirements progressively erodes the maximum achievable value.

To further illuminate the relationship between investment level and portfolio value, an efficient frontier is constructed by varying the year 1 budget constraint in $5M increments from $1.40B to $1.50B, with all project requirements removed (see Figure 7.5).

From this analysis, key insights include the following:

1. At a budget of $1.405B, the optimal portfolio achieves an eNPV of $4.620B
2. No improvement in portfolio value is achieved until the budget exceeds $1.427B
3. Incremental value gains occur between $1.427B and $1.432B, reaching approximately $4.720B
4. The maximum eNPV of $4.820B is achieved at $1.484B, beyond which additional investment yields no further value

Accordingly, *Innovative Therapies R Us* should not invest more than $1.484B in year 1 if value maximization is the sole objective.

Given that clinical development investments span multiple years, it is analytically inconsistent to perform portfolio selection solely on the basis of the next year's budget. A more robust approach, in our experience, considers a minimum three-year planning horizon. The multi-year resource requirements for *Innovative Therapies R Us* are shown in WAV Table 7.6.

When optimization is performed across years 1–3 budget constraints, the optimal portfolio value is $4.620B, as shown in WAV Table 7.7. Projects A, C, and H are excluded, and efficient frontier analysis indicates that year 1 investment should not exceed $1.405B to avoid downstream budget violations (see WAV Figure 7.6).

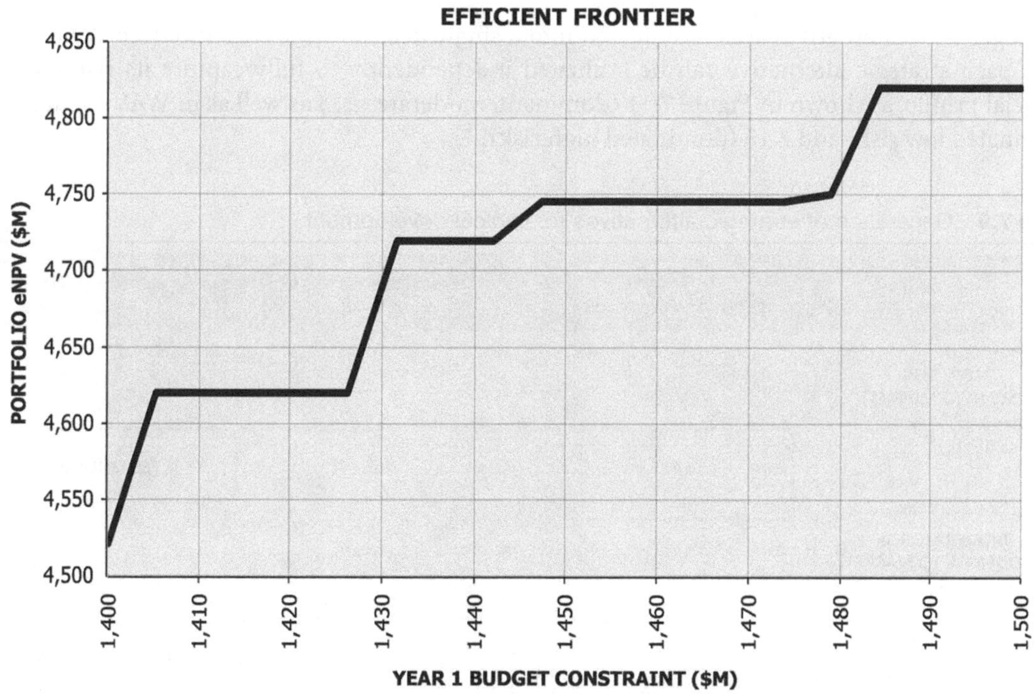

Figure 7.5 Efficient frontier analysis of the year 1 budget constraint.

Finally, applying all three year 1 constraints—budget, function A FTEs, and function B FTEs—yields the solution shown in WAV Table 7.8. As anticipated, the optimal portfolio value declines from $4.820B (see Table 7.2) to $4.345B, with projects B, G, H, and I excluded.

Efficient frontier analysis indicates that no more than $1.395B should be invested in year 1 (see WAV Figure 7.7). Analogous efficient frontier analyses for human resources suggest that year 1 staffing should be capped at approximately 4.8 function A FTEs and 10.2 function B FTEs, as shown in WAV Figures 7.8 and 7.9, respectively.

While the *Innovations R Us* case study demonstrates the utility and flexibility of the deterministic optimization technique for portfolio selection and value maximization, it presupposes that all executive decisions on project investments are binary (i.e., invest or do not invest). In reality, since there are multiple pathways to develop, register, and commercialize an asset, any organization would benefit from having multiple strategic alternatives to choose from. Recognizing this complexity, the *Diverse Therapies R Us* case study extends deterministic optimization to a more realistic decision environment in which strategic project alternatives serve as the fundamental decision variables.

Diverse Therapies R Us has a portfolio of 10 clinical programs that are approaching completion of Phase I trials. For each program, management has defined three mutually exclusive strategic development pathways, derived from a structured strategy table (see Table 7.9). These alternatives represent differentiated risk–reward profiles:

1. **Low-risk strategy** (slow to market; small line dashes): Phase 2a proof-of-concept (PoC)—Phase 2b dose finding—extended-release (ER) formulation for Phase 3 registrational trials
2. **Moderate-risk strategy** (moderate to market; large line dashes): Phase 2a PoC—ER formulation for Phase 3 registrational trials
3. **High-risk strategy** (fast to market; no line dashes): immediate release (IR) formulation for Phase 3 registrational trials

Although these alternatives are consolidated into a single decision-tree structure (see WAV Figure 7.10), each strategic alternative can be evaluated independently to fully capture its risk-adjusted financial profile, as shown in Figure 7.11 (dominant, moderate risk) as well as in WAV Figures 7.12 (dominated low risk) and 7.13 (dominated high risk).

Table 7.9 Generation of strategic alternatives for clinical development.

STRATEGIC ALTERNATIVE	STRATEGIC DECISION 1: Phase IIa Proof-Of-Concept	STRATEGIC DECISION 2: Phase IIb Dose Finding	STRATEGIC DECISION 3: Phase III Registrational Trials
LOW RISK (SLOW TO MARKET)			
	YES	YES	IR FORMULATION
MODERATE RISK (MODERATE TO MARKET)			
	NO	NO	ER FORMULATION
HIGH RISK (FAST TO MARKET)			

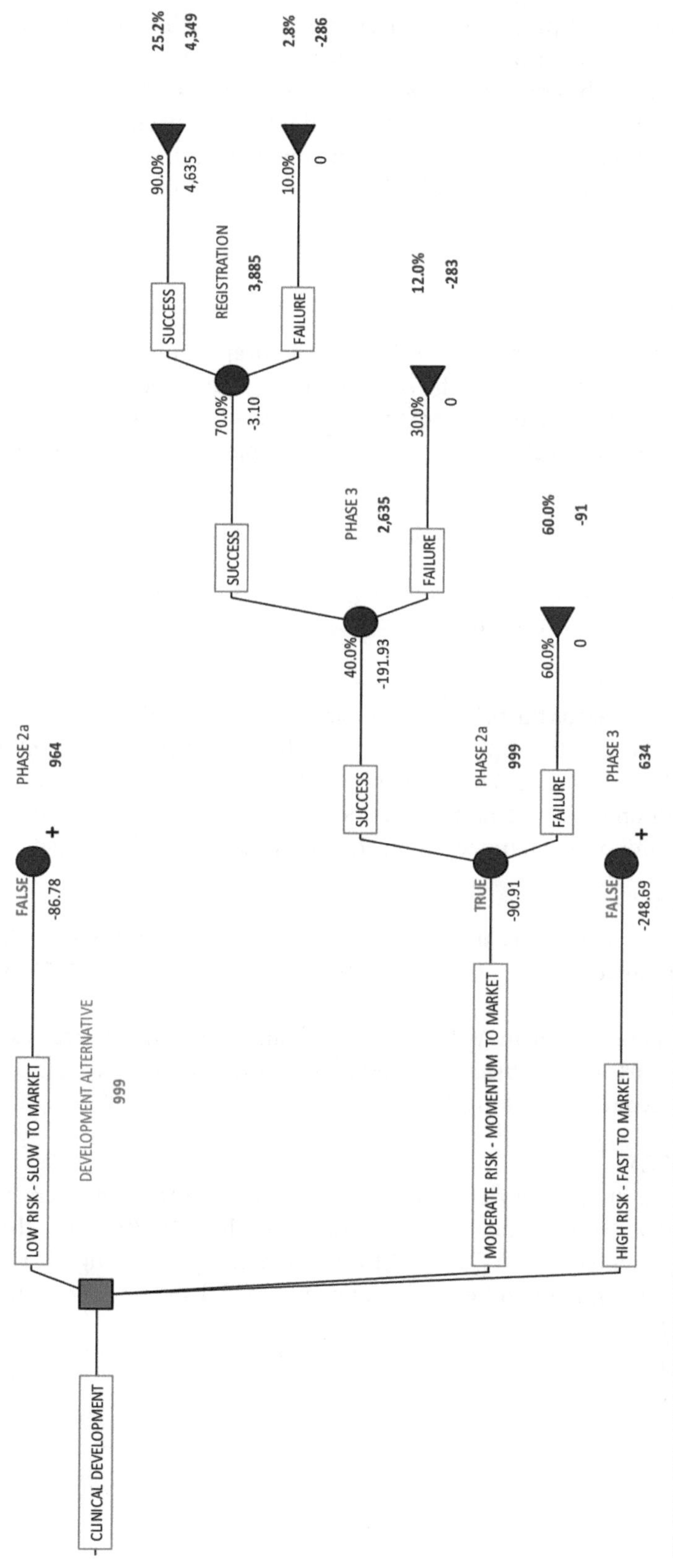

Figure 7.11 Decision tree of dominant strategic alternative—moderate risk.

Resource requirements and projected financial outcomes for the 10 projects—each with three strategic alternatives—are summarized in WAV Table 7.10. *Diverse Therapies R Us* evaluates its portfolio against three financial objectives: (a) Present value (PV) of risk-adjusted sales (eSales), (b) Risk-adjusted eNPV, and (c) PV of gross profit (GP). Additionally, the portfolio spans three therapeutic areas: central nervous system, metabolic diseases, and oncology (ONC).

Annual budget constraints for the portfolio are defined as follows:

- Year 1: $750M
- Year 2: $850M
- Year 3: $950M

Before optimization can be performed, a critical structural requirement must be enforced: no more than one strategic alternative may be selected for any given project. While each project offers three possible alternatives, these alternatives are mutually exclusive, and management is not obligated to advance every project. This requirement is implemented through a constraint applied at the project level:

$$\text{Low-risk}_i + \text{Moderate-risk}_i + \text{High-risk}_i \leq 1$$

This formulation ensures that, at most, one strategic pathway is selected per project while allowing projects to be excluded entirely from the optimal portfolio.

Maximization of PV eSales

When PV eSales is designated as the primary financial objective, and the optimization is constrained by the year 1 budget, the resulting optimal portfolio achieves a value of $17,399M. The associated tracking metrics are: (a) eNPV: $9,627M and (b) PV GP: $54,515M. These results are shown in Table 7.11. Notably, the solution selects no more than one strategic alternative per project, with risk levels varying across the portfolio to maximize aggregate sales value under the budget constraint.

Maximization of eNPV

When the optimization objective is shifted to eNPV, the optimal portfolio achieves a maximum value of $9,885M. The corresponding tracking metrics are: (a) PV eSales: $17,372M and (b) PV GP: $51,640M.

These outcomes, summarized in WAV Table 7.12, align with theoretical expectations: when eNPV is weighted at 100% in the objective function, its portfolio value exceeds that obtained when eNPV is treated as a secondary tracking metric.

Maximization of PV GP

When PV GP becomes the primary objective, the optimized portfolio delivers a maximum gross profit value of $57,701M, with tracking metrics of: (a) PV eSales: $16,146M and (b) eNPV: $8,489M.

These results are presented in WAV Table 7.13. As expected, prioritizing PV GP yields the highest possible gross profit, while the other financial measures fall short of the values achieved when optimized individually.

Table 7.11 Portfolio optimization subject to year 1 budget constraint—maximization of PV eSales.

PROJECT	DECISION VARIABLES	YEAR 1 INVESTMENT REQUIRED ($M)	YEAR 2 INVESTMENT REQUIRED ($M)	YEAR 3 INVESTMENT REQUIRED ($M)	PRESENT VALUE RISK-ADJUSTED SALES (PV eSALES) ($M)	RISK-ADJUSTED NPV (eNPV) ($M)	PRESENT VALUE GROSS PROFIT ($M)
CNS A - LOW RISK	0	50	50	95	1,252	964	4,026
CNS A - MODERATE RISK	1	100	100	100	1,298	999	5,098
CNS A - HIGH RISK	0	125	115	115	981	579	5,716
CNS B - LOW RISK	0	40	60	80	313	125	1,007
CNS B - MODERATE RISK	0	80	120	100	973	719	3,824
CNS B - HIGH RISK	1	100	100	90	981	641	5,716
CNS C - LOW RISK	0	60	60	60	2,128	896	3,802
CNS C - MODERATE RISK	1	70	70	90	2,223	983	4,852
CNS C - HIGH RISK	0	90	90	95	1,687	615	5,457
MD D - LOW RISK	0	65	75	85	532	73	951
MD D - MODERATE RISK	0	75	85	90	1,667	696	3,639
MD D - HIGH RISK	1	110	120	135	1,687	542	5,457
MD E - LOW RISK	0	45	50	100	1,755	1,421	5,645
MD E - MODERATE RISK	1	80	90	100	1,884	1,549	7,402
MD E - HIGH RISK	0	75	125	115	1,435	1,033	8,358
MD F - LOW RISK	0	50	60	70	1,053	786	3,387
MD F - MODERATE RISK	0	75	75	125	1,319	1,042	5,181
MD F - HIGH RISK	1	65	85	110	1,435	1,079	8,358
ONC G - LOW RISK	0	40	50	80	2,520	1,112	4,503
ONC G - MODERATE RISK	1	65	100	125	2,612	1,162	5,700
ONC G - HIGH RISK	0	70	100	120	1,975	750	6,390
ONC H - LOW RISK	0	60	50	125	1,260	450	2,252
ONC H - MODERATE RISK	1	55	75	95	2,089	927	4,560
ONC H - HIGH RISK	0	85	85	95	1,975	768	6,390
ONC I - LOW RISK	1	55	75	95	1,170	872	3,762
ONC I - MODERATE RISK	0	85	125	75	1,170	898	4,598
ONC I - HIGH RISK	0	100	90	100	872	544	5,079
ONC J - LOW RISK	1	40	50	60	2,020	873	3,609
ONC J - MODERATE RISK	0	70	125	100	2,012	857	4,392
ONC J - HIGH RISK	0	75	125	125	1,497	482	4,843
ALL		740	865	1,000	17,399	9,627	54,515

Across all three optimizations, a consistent principle emerges: when a portfolio is optimized for a single objective, the resulting value of that objective is maximized because it carries an implicit weight of 100%. While tracking metrics may approach the values achieved under alternative objective functions, they cannot exceed them.

Because clinical development unfolds over multiple years, portfolio selection based on a single-year budget can produce misleading results. A portfolio that appears optimal under short-term constraints may become infeasible or suboptimal when downstream funding requirements are considered.

To illustrate this effect, the *Diverse Therapies R Us* portfolio is re-optimized under three years of sequential budget constraints, with separate optimizations conducted for:

1. PV eSales (see WAV Table 7.14)
2. eNPV (see WAV Table 7.15)
3. PV GP (see WAV Table 7.16)

The results demonstrate systematic differences between short- and multi-horizon budget optimization:

- PV eSales optimization
    - Year 1 constraint only: $17,399M; years 1–3 constraints: $17,288M
- eNPV optimization
    - Year 1 constraint only: $9,885M; years 1–3 constraints: $9,835M
- PV GP optimization
    - Year 1 constraint only: $57,701M; years 1–3 constraints: $55,367M

These findings, summarized in Table 7.17, confirm that multi-period constraints modestly but consistently reduce attainable portfolio value, reflecting the cumulative impact of downstream capital commitments.

This case study underscores three critical insights for biopharma portfolio governance:

1. Portfolio optimization can explicitly accommodate strategic flexibility by treating development alternatives as decision variables rather than assuming binary project choices
2. The choice of financial objective materially influences portfolio composition and value realization
3. Multi-year optimization provides a more realistic and defensible basis for strategic decision making than single-year budget analyses

Table 7.17 Comparison of portfolio optimizations subject to single- and multi-year budget constraints.

PROJECT	MAX: PV eSALES YEAR 1 CONSTRAINT	MAX: eNPV YEAR 1 CONSTRAINT	MAX: PV GP YEAR 1 CONSTRAINT	MAX: PV eSALES YEAR 1 - 3 CONSTRAINTS	MAX: eNPV YEAR 1 - 3 CONSTRAINTS	MAX: PV GP YEAR 1 - 3 CONSTRAINTS
CNS A - LOW RISK	0	0	1	0	0	1
CNS A - MODERATE RISK	1	1	0	1	1	0
CNS A - HIGH RISK	0	0	0	0	0	0
CNS B - LOW RISK	0	0	0	0	0	0
CNS B - MODERATE RISK	0	1	0	0	1	0
CNS B - HIGH RISK	1	0	1	1	0	1
CNS C - LOW RISK	0	0	0	0	0	1
CNS C - MODERATE RISK	1	1	1	1	1	0
CNS C - HIGH RISK	0	0	0	0	0	0
MD D - LOW RISK	0	0	0	0	0	0
MD D - MODERATE RISK	0	1	0	1	1	1
MD D - HIGH RISK	1	0	1	0	0	0
MD E - LOW RISK	0	0	0	0	0	0
MD E - MODERATE RISK	1	1	0	1	1	0
MD E - HIGH RISK	0	0	1	0	0	1
MD F - LOW RISK	0	0	0	0	0	0
MD F - MODERATE RISK	0	0	0	0	0	0
MD F - HIGH RISK	1	1	1	1	1	1
ONC G - LOW RISK	0	0	0	1	1	0
ONC G - MODERATE RISK	1	1	0	0	0	0
ONC G - HIGH RISK	0	0	1	0	0	1
ONC H - LOW RISK	0	0	0	0	0	0
ONC H - MODERATE RISK	1	1	0	1	1	0
ONC H - HIGH RISK	0	0	1	0	0	1
ONC I - LOW RISK	1	0	1	1	0	0
ONC I - MODERATE RISK	0	1	0	0	1	0
ONC I - HIGH RISK	0	0	0	0	0	1
ONC J - LOW RISK	1	1	0	1	1	1
ONC J - MODERATE RISK	0	0	1	0	0	0
ONC J - HIGH RISK	0	0	0	0	0	0
PV eSALES ($M)	17,399	17,372	16,146	17,288	17,281	15,741
eNPV ($M)	9,627	9,885	8,489	9,731	9,835	8,244
PV GP ($M)	54,515	51,640	57,701	51,500	50,443	55,367

7.4 APPLICATION OF DETERMINISTIC MULTIPLE OBJECTIVE OPTIMIZATION FOR R&D PORTFOLIO SELECTION

Up to this point, all portfolio optimization exercises—whether conducted across a single year or multiple time periods, and whether constrained by one or several resource limitations—have been framed around the maximization of a single financial objective. From an MP perspective, these formulations fall within the class of SOLP problems, even though they are implemented as integer linear programs due to the binary nature of project selection.

In practice, however, executive decision makers rarely evaluate R&D portfolios through the lens of a single metric. Instead, they are typically concerned with the simultaneous achievement of multiple financial objectives, such as maximizing risk-adjusted sales (PV eSales), maximizing economic value creation (eNPV), and maximizing profitability (PV GP). Addressing this reality requires a fundamental shift in modeling approach—from SOLP to MOLP. As discussed previously in Chapter 3, MOLP frameworks require two essential elements:

1. Explicit weighting of each objective to reflect executive preferences and strategic priorities
2. Normalization of weights, such that the sum of all objective weights equals 100%

By introducing weighted objectives, the optimization no longer seeks to maximize a single outcome but instead searches for a portfolio that best balances competing goals under stated constraints.

To operationalize MOLP in a portfolio context, the MINIMAX Q technique (described in detail in Chapter 3) is applied. This method reformulates the problem such that the objective is to minimize the maximum weighted deviation of each financial metric from its theoretical maximum, where that maximum is defined as the value obtained when the objective is optimized individually at a 100% weight.

A MINIMAX Q model for *Diverse Therapies R Us*, subject to the year 1 budget constraint, is shown in Table 7.18.

In this formulation, PV eSales is weighted at 34.0%, eNPV is weighted at 33.0%, and PV GP is weighted at 33.0%. Under these weights, the maximum deviation from any objective's individual optimum is 4.52%, resulting in an optimal portfolio with the following aggregate values (see Table 7.19): (a) PV eSales: $17,240M, (b) eNPV: $9,459M, and (c) PV GP: $56,108M.

A fundamental property of MOLP is evident here: when an objective carries less than a 100% weight, its portfolio value cannot exceed the value achieved when it is optimized independently.

Table 7.18 Application of MINIMAX Q technique to portfolio optimization subject to year 1 budget constraint.

	PV eSales YEAR 1	eNPV YEAR 1	PV GP YEAR 1
Optimal Solutions @100% Weights	**17,399**	**9,885**	**57,701**
Current Difference	0.92%	4.31%	2.76%
MiniMax Weights	34.00%	33.00%	33.00%
Weighted Difference	0.31%	1.42%	0.91%
Max Weighted Difference (Q)	1.42%		

Table 7.19 MOLP portfolio optimization subject to year 1 budget constraint—objectives weighted at 34% (PV eSales); 33% (eNPV); 33% (PV GP).

PROJECT	PV eSales:eNPV:PV GP 34%:33%:33% YEAR 1 CONSTRAINT	YEAR 1 INVESTMENT REQUIRED ($M)	YEAR 2 INVESTMENT REQUIRED ($M)	YEAR 3 INVESTMENT REQUIRED ($M)	PRESENT VALUE RISK-ADJUSTED SALES (PV eSALES) ($M)	RISK-ADJUSTED NPV (eNPV) ($M)	PRESENT VALUE GROSS PROFIT ($M)
CNS A - LOW RISK	1	50	50	95	1,252	964	4,026
CNS A - MODERATE RISK	0	100	100	100	1,298	999	5,098
CNS A - HIGH RISK	0	125	115	115	981	579	5,716
CNS B - LOW RISK	0	40	60	80	313	125	1,007
CNS B - MODERATE RISK	0	80	120	100	973	719	3,824
CNS B - HIGH RISK	1	100	100	90	981	641	5,716
CNS C - LOW RISK	0	60	60	60	2,128	896	3,802
CNS C - MODERATE RISK	1	70	70	90	2,223	983	4,852
CNS C - HIGH RISK	0	90	90	95	1,687	615	5,457
MD D - LOW RISK	0	65	75	85	532	73	951
MD D - MODERATE RISK	0	75	85	90	1,667	696	3,639
MD D - HIGH RISK	1	110	120	135	1,687	542	5,457
MD E - LOW RISK	0	45	50	100	1,755	1,421	5,645
MD E - MODERATE RISK	1	80	90	100	1,884	1,549	7,402
MD E - HIGH RISK	0	75	125	115	1,435	1,033	8,358
MD F - LOW RISK	0	50	60	70	1,053	786	3,387
MD F - MODERATE RISK	0	75	75	125	1,319	1,042	5,181
MD F - HIGH RISK	1	65	85	110	1,435	1,079	8,358
ONC G - LOW RISK	0	40	50	80	2,520	1,112	4,503
ONC G - MODERATE RISK	1	65	100	125	2,612	1,162	5,700
ONC G - HIGH RISK	0	70	100	120	1,975	750	6,390
ONC H - LOW RISK	0	60	50	125	1,260	450	2,252
ONC H - MODERATE RISK	0	55	75	95	2,089	927	4,560
ONC H - HIGH RISK	1	85	85	95	1,975	768	6,390
ONC I - LOW RISK	0	55	75	95	1,170	872	3,762
ONC I - MODERATE RISK	1	85	125	75	1,170	898	4,598
ONC I - HIGH RISK	0	100	90	100	872	544	5,079
ONC J - LOW RISK	1	40	50	60	2,020	873	3,609
ONC J - MODERATE RISK	0	70	125	100	2,012	857	4,392
ONC J - HIGH RISK	0	75	125	125	1,497	482	4,843
ALL		750	875	975	17,240	9,459	56,108
CONSTRAINT		750	850	950			

This principle is confirmed by comparing the MOLP results with the single-objective solutions summarized earlier in Table 7.18.

The MOLP solution is obtained efficiently using Lumivero's Evolver, requiring fewer than 20,000 trials. An efficient frontier analysis of the MINIMAX Q formulation (see Figure 7.14) demonstrates that the maximum deviation across objectives is only 1.42%, indicating a well-balanced compromise solution across competing financial metrics.

A broader set of MOLP results, reflecting alternative executive preference structures, is summarized in WAV Table 7.20. These findings illustrate that changes in objective weights can materially alter both portfolio composition and financial outcomes, reinforcing the importance of transparency and alignment when defining executive preferences.

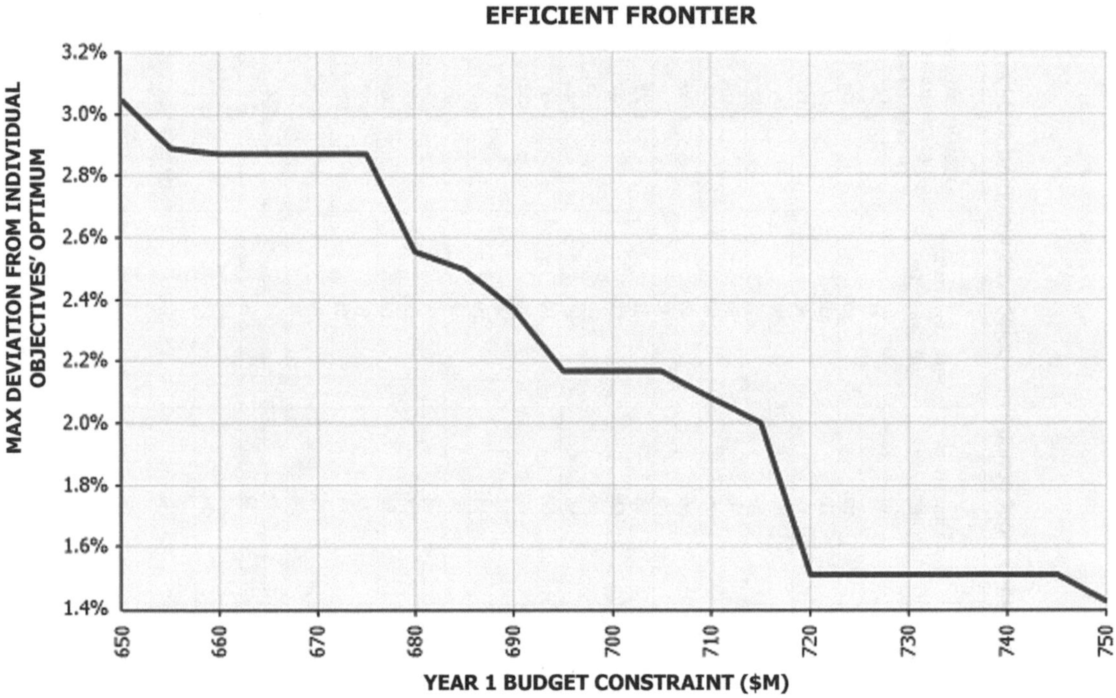

Figure 7.14 Efficient frontier analysis of year 1 budget constraint with objectives weighted at 34% (PV eSales); 33% (eNPV); 33% (PV GP).

7.5 APPLICATION OF STOCHASTIC OPTIMIZATION FOR R&D PORTFOLIO SELECTION

All linear programming–based portfolio optimizations described thus far share a common assumption: all input data are known (or believed to be known) with certainty. Annual resource requirements, risk-adjusted financial values, and budget constraints are treated as fixed point estimates. While analytically convenient, this deterministic representation is rarely realistic.

As discussed in Chapter 6, most portfolio inputs—capital requirements, FTE demands, and projected cash flows—are inherently uncertain. A more faithful representation of reality requires these parameters to be expressed as ranges or probability distributions, rather than single values.

To address uncertainty explicitly, the analysis now shifts from deterministic to stochastic optimization. Returning to the *Innovative Therapies R Us* case study, each base (deterministic) estimate for annual resource requirements is bounded by a low value (typically ≤10th percentile) and a high value (typically ≥90th percentile). These bounds are then converted into probability distributions, as shown in WAV Table 7.21, enabling optimization under uncertainty. Once uncertainty is introduced, the optimization problem becomes a stochastic problem, where simulation and optimization are performed simultaneously.

Single-Objective Stochastic Optimization (SOSO)

Using Lumivero's RISKOptimizer, an *@RISK* software component, portfolio optimization is conducted through integrated Monte Carlo or Latin Hypercube simulation and search. Even if sufficient resources existed to fund every project, uncertainty in downstream variables—such as future sales driving NPVs—means that portfolio eNPV is no longer a fixed value, but a distribution of possible outcomes.

Subject to the year 1 budget constraint, a stochastic optimization with 5,000 simulation trials yields an eNPV distribution ranging from $4.691B to $5.099B, with an 85% probability of exceeding the deterministic eNPV of $4.820B (see Figure 7.15). The mean eNPV of the optimized portfolio is $4.887B, notably higher than the deterministic estimate.

Minor variations between runs are expected due to the stochastic nature of the simulation. The authors employ 5,000 iterations as a practical balance between computational efficiency and solution stability, achieving convergence within approximately 2% of the mean and standard deviation of key outputs.

The stochastic optimization identifies the same optimal project set as in the deterministic solution (see Table 7.3), excluding projects C, G, and H. Among the 2^{10} (1,024) feasible project combinations evaluated, this portfolio yields the highest expected eNPV distribution. All other combinations, while feasible, are economically suboptimal. Sensitivity analysis (see Figure 7.16) indicates that the mean portfolio eNPV is most influenced by uncertainty in projects F, D, and J, and least influenced by projects C, G, and H, consistent with their exclusion from the optimal solution.

In addition to financial outcomes, stochastic optimization provides distributions for required resources. For the year 1-only constraint scenario (see WAV Figure 7.17 A–C): (a) budget range: $1.408B–$1.582B, (b) function A FTE range: 7.0–9.2, and (c) function B FTE range: 11.7–14.0. These distributions provide decision makers with a probabilistic understanding of resource exposure rather than single-point estimates.

When the portfolio is optimized subject to years 1–3 budget constraints, the optimal solution excludes projects A, C, and H. The resulting eNPV distribution spans $4.481B to $4.897B, with a mean of $4.683B and an 85% likelihood of exceeding the deterministic eNPV of $4.620B (see WAV Figure 7.18). Corresponding distributions of year 1 resource requirements for budget and FTEs can be generated (data not shown).

Finally, when all year 1 resource constraints are enforced simultaneously, the optimal portfolio omits projects B, F, H, and I. The resulting eNPV distribution ranges from $4.087B to $4.473B, with a mean of $4.288B (see WAV Figure 7.19).

Collectively, these analyses demonstrate that stochastic portfolio optimization:

- Preserves the rigor of linear programming
- Explicitly incorporates uncertainty in both value and resource requirements
- Provides probabilistic insight into downside and upside exposure
- Enables executives to make portfolio decisions that are robust, transparent, and aligned with risk tolerance

In doing so, stochastic optimization represents a natural and necessary evolution of deterministic portfolio management methodologies in biopharma R&D. Where material uncertainty exists in

PROJECT	DECISION VARIABLES
A	1
B	1
C	0
D	1
E	1
F	1
G	0
H	0
I	1
J	1

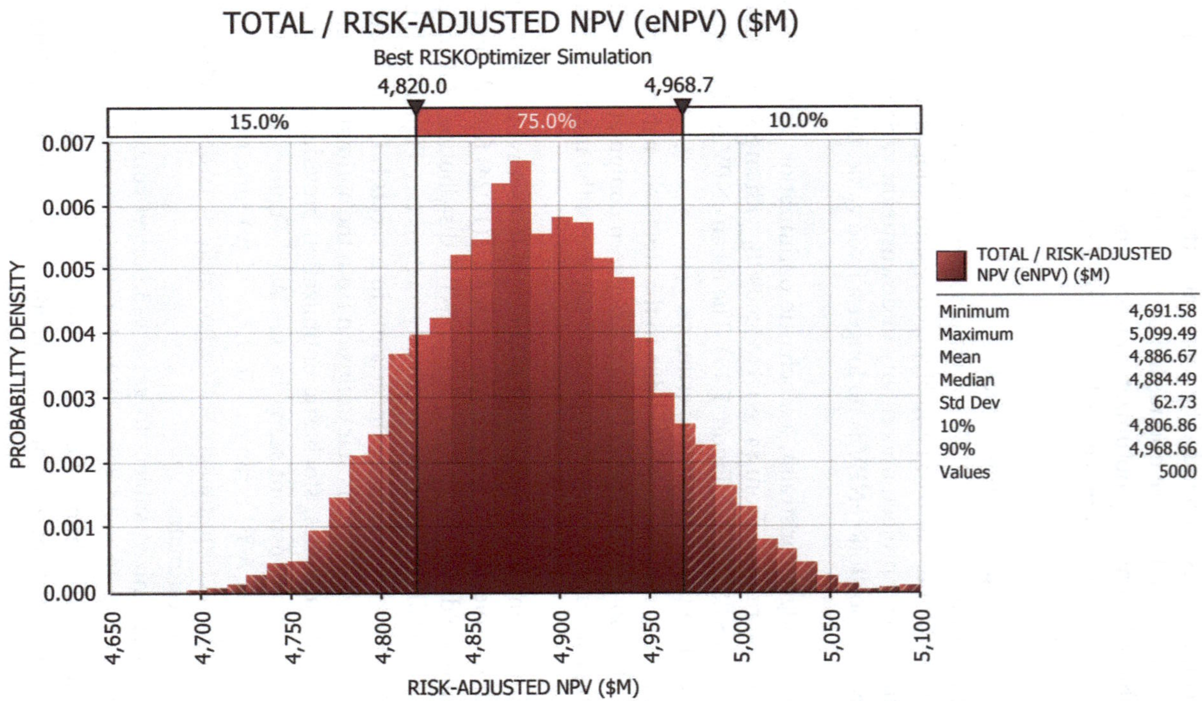

Figure 7.15 Optimal portfolio and distribution of eNPVs from stochastic portfolio optimization subject to year 1 budget constraint.

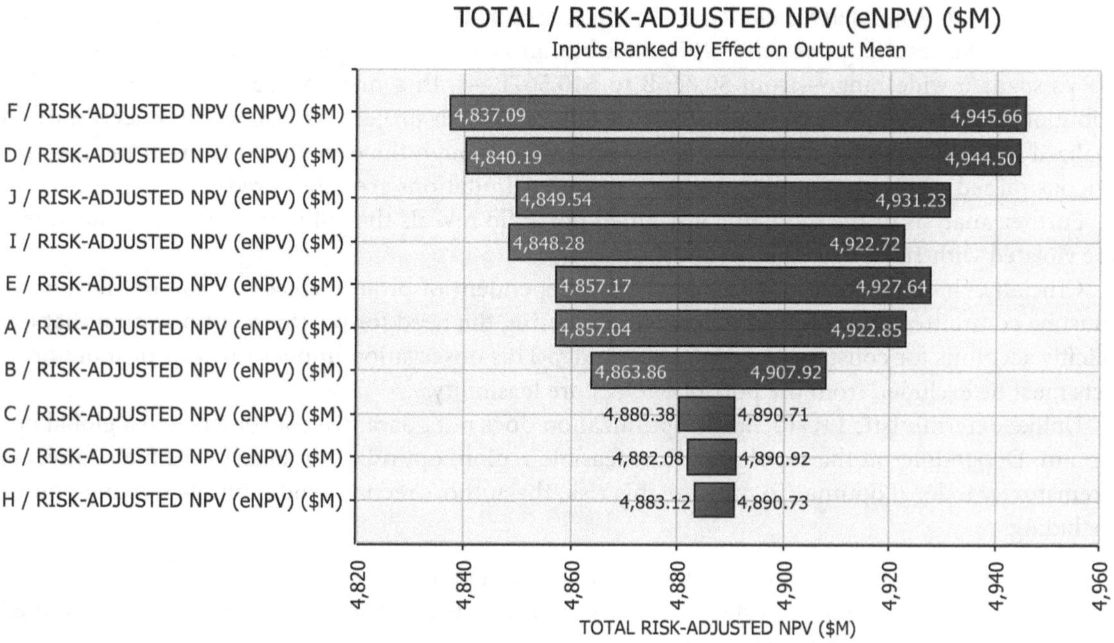

Figure 7.16 Sensitivity analysis to mean portfolio eNPV from stochastic optimization subject to year 1 budget constraint.

project data—particularly when large capital investments must be committed over extended timelines—the authors strongly recommend the use of stochastic portfolio optimization rather than deterministic approaches.

However, all stochastic optimizations conducted thus far rest on a critical simplifying assumption: once a project is selected into the portfolio, it is implicitly assumed to succeed and deliver either its full deterministic value or a probability-weighted realization of its uncertain commercial value. In practice, this assumption is often violated.

In high-risk industries—such as oil and gas exploration, mining, and drug discovery and development—portfolio inclusion does not imply project success. Many projects fail outright and contribute no economic value despite consuming substantial resources. Consequently, project risk must be explicitly embedded in the analytic model, not merely inferred through discounted or risk-adjusted value estimates.

This concept is illustrated in the expanded dataset shown in WAV Table 7.22, where project-specific risk is introduced explicitly via a Bernoulli distribution, representing the probability of technical and regulatory success. Commercial uncertainty is modeled independently using triangular NPV distributions defined by minimum, modal, and maximum values.

Under this formulation:

- Each project has two discrete outcome states: (a) success, in which 100% of the realized NPV is captured and (b) failure, in which 0% of the NPV is realized
- Commercial uncertainty affects the magnitude of NPV conditional on success, while project risk determines whether any value is realized at all

This combined representation captures both binary technical risk and continuous commercial uncertainty, resulting in a more realistic depiction of portfolio value outcomes.

When all 10 projects are included in an unconstrained portfolio and simulated under these assumptions, the resulting portfolio NPV distribution is shown in Figure 7.20. The distribution of NPVs spans a wide range—from \$0.866B to \$10.567B—with a mean value of \$6.317B. This distribution reflects the probabilistic success or failure of each project, as well as the uncertainty in realized NPV for projects that succeed. Importantly, although the expected value is attractive, this unconstrained portfolio is not feasible once resource limitations are considered.

Further analysis of the same unconstrained portfolio reveals that all year 1 resource constraints are violated with 100% probability (not shown).

Crucially, these resource requirements are independent of project success or failure. Resources must be committed regardless of outcome, reinforcing the need for portfolio optimization that explicitly accounts for constraints under uncertainty. This observation implies that one or more projects must be excluded from the portfolio to restore feasibility.

Unlike deterministic LP, stochastic optimization does not guarantee convergence to a global optimum. Depending on the topology of the feasible region, optimization algorithms may converge prematurely to local optima. To mitigate this risk, the authors recommend multiple starting points, including:

1. Initializing all decision variables first to 0 and then to 1
2. Applying constraints to statistical measures (e.g., mean resource usage), when constrained variables themselves are probabilistic
    - In the examples presented, budget and FTE constraints are applied to the mean of total resource usage
3. Ensuring a sufficient number of simulation iterations for output stability
    - Tools such as *@RISK* provide diagnostics to determine convergence within specified confidence levels
4. Allowing adequate optimization time
    - Although some solutions emerged within ~200 trials, all optimizations were run for at least 250 trials to ensure robustness

When stochastic optimization is conducted subject to year 1 budgetary and human resource constraints, the resulting optimal portfolio yields the highest mean NPV among all feasible portfolios. As shown in Figure 7.21, the distribution of portfolio NPVs for this solution ranges from \$0, where all selected projects fail, to \$7.288B where all selected projects succeed, with a mean NPV of \$4.287B. Of course, the associated resource distributions for this portfolio (not shown) demonstrate 0% probability of exceeding any of the three constraints. The resulting optimal portfolio consists of six projects, excluding projects B, F, H, and I.

When the stochastic portfolio optimization is then repeated subject to all budgetary and human resource constraints across years 1–3 (not shown), the resulting optimal portfolio is identical to the portfolio identified under year 1 constraints alone.

At the conclusion of the optimization, the mean NPV and standard deviation of NPV for each evaluated portfolio can be plotted, producing an efficient frontier of feasible solutions (see Figure 7.22). In this analysis, over 110 portfolios were evaluated. The frontier illustrates the trade-off between (a) reward, measured by mean NPV, and (b) risk, measured by the standard deviation of NPV. The portfolio with the highest mean NPV (~\$4.29B) exhibits a standard deviation of approximately \$1.42B (triangle marker). Beyond this level of risk, no additional portfolios satisfy all constraints.

An alternative portfolio (square marker) has a mean NPV of ~\$4.28B and a lower standard deviation of ~\$1.38B. This alternative represents a 2.5% reduction in risk for a 0.2% reduction in value, making it a viable option depending on executive risk tolerance. The composition of these two portfolios is compared in Table 7.23.

PROJECT	DECISION VARIABLES
A	1
B	1
C	1
D	1
E	1
F	1
G	1
H	1
I	1
J	1

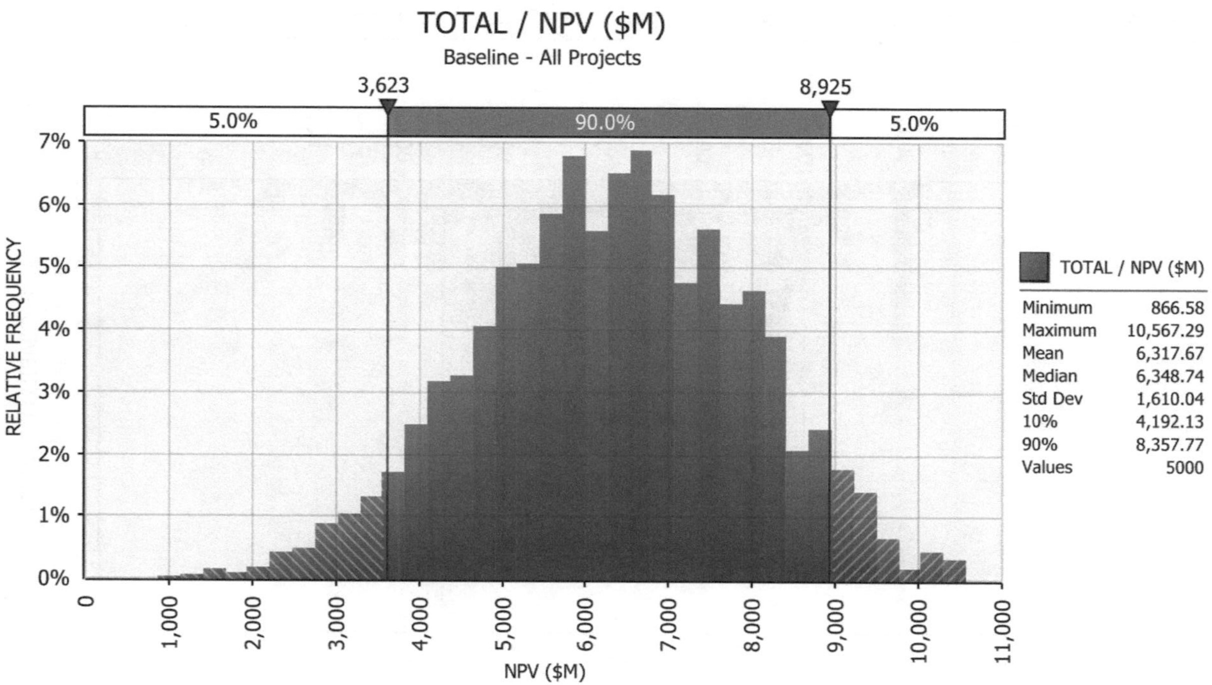

Figure 7.20 Distribution of NPVs for stochastic portfolio without resource constraints.

PROJECT	DECISION VARIABLES
A	1
B	0
C	1
D	1
E	1
F	0
G	1
H	0
I	0
J	1

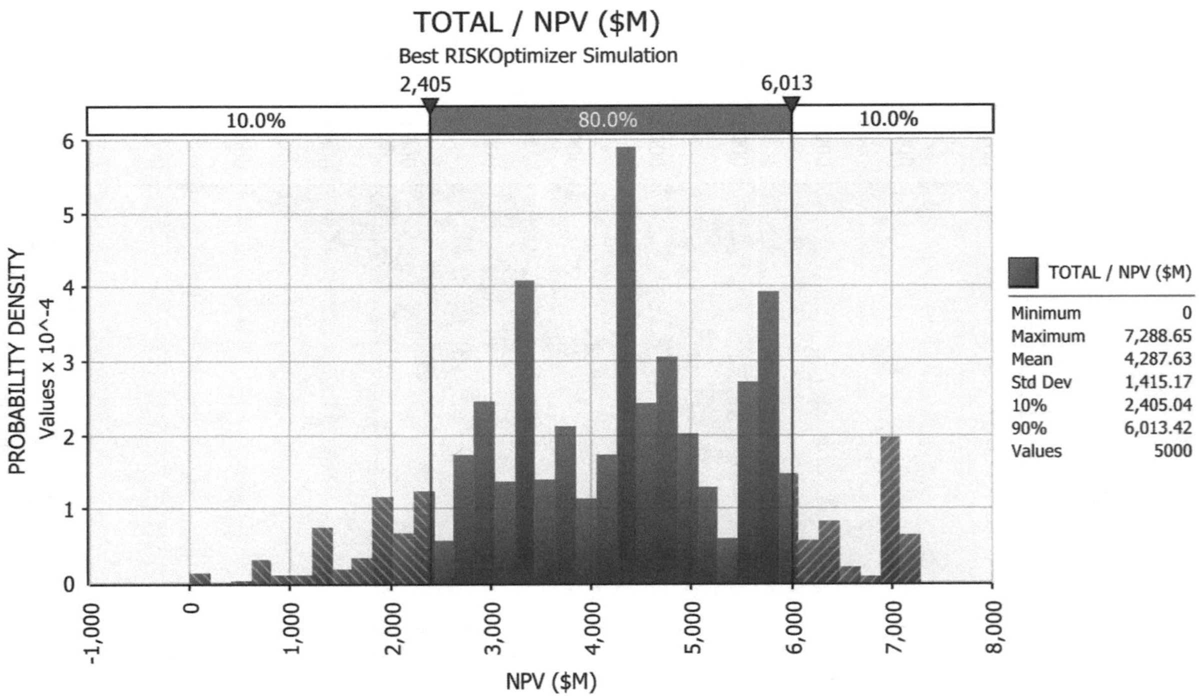

Figure 7.21 Optimal portfolio and distribution of NPVs from stochastic portfolio optimization with project risk and commercial uncertainty subject to year 1 budgetary and human resource constraints.

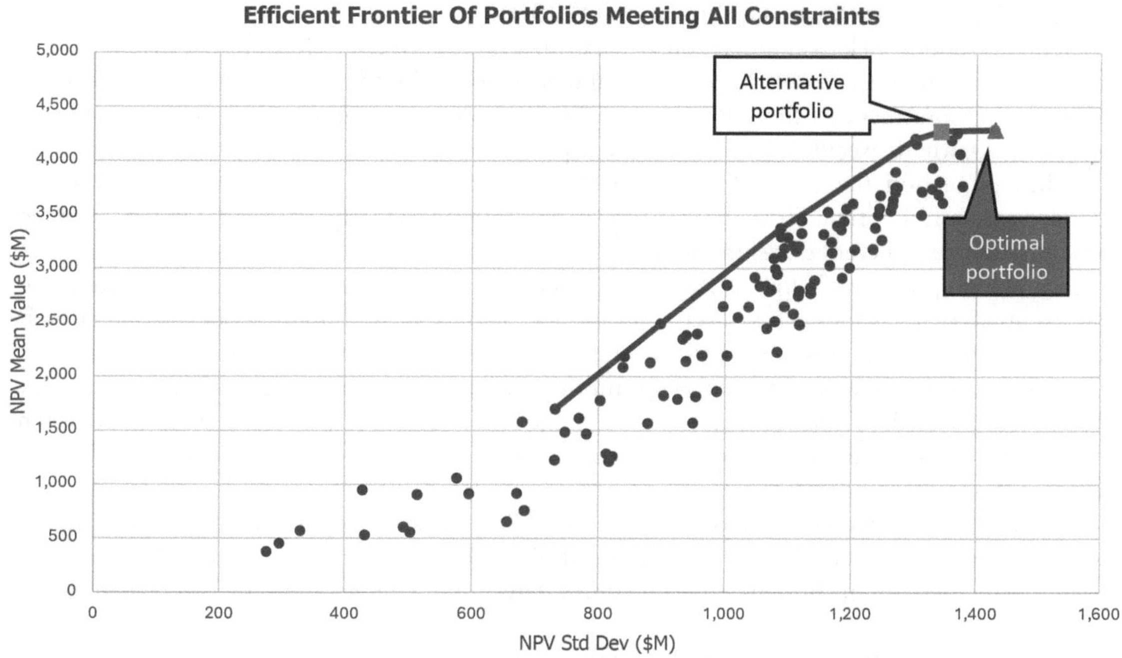

Figure 7.22 Efficient frontier of portfolios meeting all constraints.

Table 7.23 Summary of project selection for stochastic portfolio optimization.

	Optimal Portfolio	Alternative Portfolio
Mean NPV	$4.288B	$4.278B
Std Dev NPV	$1.428B	$1.379B
PROJECT	Included?	Included?
A	1	0
B	0	0
C	1	1
D	1	1
E	1	1
F	0	1
G	1	0
H	0	0
I	0	1
J	1	1

Both portfolios consistently exclude projects B and H, while differing primarily in the inclusion of projects A and G versus F and I. Given the minimal loss in expected value, the final portfolio decision may hinge on strategic preferences, scientific alignment, or organizational risk appetite rather than purely on financial metrics.

Multiple-Objective Stochastic Optimization (MOSO)

When organizations seek to simultaneously maximize value across multiple objectives and manage risk, no single solution will dominate across all objectives, and trade-offs become unavoidable. As before, the authors advocate the application of MOSO under uncertainty, using techniques such as MINIMAX Q, where executive preferences are expressed explicitly through objective weights.

In the next section, the *Diverse Therapies R Us* case study is revisited, and MOLP optimization is extended into a fully probabilistic framework, integrating uncertainty, risk, and competing objectives into a single, coherent decision model.

As discussed previously, *Diverse Therapies R Us* evaluates three mutually exclusive strategic development alternatives for each project: low-risk, moderate-risk, and high-risk development pathways. Each alternative represents a distinct sequence of clinical phases, timelines, risk exposure, and financial outcomes. To extend portfolio optimization into a stochastic framework, the authors explicitly incorporate probabilities of technical and regulatory success (PTS/PRS) at each development phase. These probabilities are modeled sequentially, such that progression through later phases is conditional on success in earlier phases.

For each project alternative, phase-specific PTS/PRS values are applied to construct a chain of Bernoulli distributions, one for each development phase. In each iteration of a Latin Hypercube or Monte Carlo simulation, the Bernoulli outcomes determine whether the project advances to the next phase.

A key structural feature of this model is phase dependency:

- The Bernoulli outcome for Phase IIb, Phase III, or regulatory approval can only equal 1 (success) if the Bernoulli outcome of the immediately preceding phase is also equal to 1
- If any phase fails (Bernoulli outcome = 0), all subsequent phases are forced to fail in that iteration

This conditional logic ensures that the simulated development pathway mirrors real-world attrition dynamics in clinical development. The PTS/PRS values used for each phase and the structure of the Bernoulli distributions are summarized in WAV Table 7.24.

The investment requirements for each project over years 1–3 are presented in WAV Table 7.25. Given that these investment magnitudes are relatively small when compared to the scale of NPV and PV Sales, the authors elected not to model uncertainty in investment requirements. Instead, the modeling assumption is that:

- The full forecasted investment for each selected project alternative is committed and expended over the three-year horizon, regardless of development success or failure
- PTS/PRS values are not applied to investment, reflecting the reality that most costs are incurred before clinical failure is observed

This approach focuses the stochastic modeling effort on value uncertainty, where it is most consequential for portfolio outcomes.

Diverse Therapies R Us recognizes material uncertainty in its estimates of (a) PV Sales, (b) NPV, and (c) PV GP. Consequently, the uncertainty assumptions are as follows:

- PV Sales and NPV may vary between 90% and 130% of their base-case estimates for all strategic alternatives
- PV GP for moderate- and high-risk alternatives may also vary between 90% and 130% of the base case
- PV GP for low-risk alternatives has limited upside and is constrained to 90%–100% of the base case

These bounds are assumed to represent the absolute minimum and maximum outcomes. Each financial metric is modeled using a three-point triangular distribution, implemented in *@RISK*. The full specification of these distributions is shown in WAV Table 7.26.

The optimization model applies annual investment constraints across all three years as follows: year 1: $750M; year 2: $850M; year 3: $950M. As before, the model enforces mutual exclusivity across strategic alternatives by imposing the constraint that no more than one alternative may be selected per project.

To ensure that simulation outputs are stable and reproducible, a formal convergence test was performed using *@RISK*. The test evaluated convergence for both the mean and standard deviation of key outputs, using 2% tolerance and 95% confidence level. The convergence test determined that a minimum of 5,300 iterations is required. To provide an additional margin of stability, all Latin Hypercube or Monte Carlo simulations were run using 10,000 iterations. SOSOs were conducted independently for (a) PV Sales, (b) NPV, and (c) PV GP, where each SOLP assigns a 100% weight to the selected objective. A preference weighting of 34% (PV Sales), 33% (NPV), and 33% (PV GP) is applied to MOSOs. The resulting project selections and mean outcomes are summarized in Table 7.27.

Table 7.27 Project selection and mean outcomes for SOLP and MOLP optimizations.

PROJECT	MAX PV SALES YEAR 1-3 CONSTRAINT	MAX NPV YEAR 1-3 CONSTRAINT	MAX PV GP YEAR 1-3 CONSTRAINT	34%:33%:33% YEAR 1-3 CONSTRAINT
CNS A - LOW RISK	1	1	0	1
CNS A - MODERATE RISK	0	0	1	0
CNS A - HIGH RISK	0	0	0	0
CNS B - LOW RISK	0	0	0	0
CNS B - MODERATE RISK	0	0	0	0
CNS B - HIGH RISK	1	1	1	1
CNS C - LOW RISK	0	0	0	0
CNS C - MODERATE RISK	1	1	1	1
CNS C - HIGH RISK	0	0	0	0
MD D - LOW RISK	0	0	0	0
MD D - MODERATE RISK	1	1	1	1
MD D - HIGH RISK	0	0	0	0
MD E - LOW RISK	0	0	0	0
MD E - MODERATE RISK	1	1	1	1
MD E - HIGH RISK	0	0	0	0
MD F - LOW RISK	0	0	0	0
MD F - MODERATE RISK	0	0	0	0
MD F - HIGH RISK	1	1	1	1
ONC G - LOW RISK	0	0	1	1
ONC G - MODERATE RISK	1	1	0	0
ONC G - HIGH RISK	0	0	0	0
ONC H - LOW RISK	0	0	0	0
ONC H - MODERATE RISK	1	1	1	1
ONC H - HIGH RISK	0	0	0	0
ONC I - LOW RISK	0	0	0	1
ONC I - MODERATE RISK	1	1	1	0
ONC I - HIGH RISK	0	0	0	0
ONC J - LOW RISK	1	1	1	0
ONC J - MODERATE RISK	0	0	0	1
ONC J - HIGH RISK	0	0	0	0
PV SALES MEAN ($M)	18,550.6	18,550.6	18,480.5	18,468.0
NPV MEAN ($M)	11,473.6	11,473.6	11,459.6	11,398.6
PV GP MEAN ($M)	12,977.5	12,977.5	13,319.0	13,153.9

Because the optimization is stochastic, the MINIMAX Q metric itself is a distribution. Accordingly, the MOLP objective is to minimize the mean value of this distribution.

As summarized in Table 7.27, several important observations emerge:

- The same project combination maximizes PV sales and NPV under SOLP
- The MOLP solution is not a simple intersection of the SOLP portfolios
- The MOLP portfolio includes projects unique to each SOLP solution, as well as projects not selected by any SOLP, such as (a) ONC I—low risk and (b) ONC J—moderate risk

This illustrates a defining feature of MOLP: the optimal compromise portfolio may include assets that are individually suboptimal under any single objective but collectively improve balance across objectives.

As expected, when objectives are weighted rather than maximized individually:

- The mean value of each financial metric in the MOLP solution is lower than its corresponding SOLP maximum
- This reduction reflects the explicit trade-offs required to satisfy multiple objectives simultaneously

Stochastic MOLP enables decision makers to move beyond *winner-take-all* objective optimization and instead identify balanced, preference-consistent portfolios that remain robust under uncertainty. This approach is particularly well-suited to biopharma R&D, where risk, uncertainty, and competing strategic objectives are unavoidable and must be addressed explicitly rather than implicitly.

8

PROJECT PRIORITIZATION

Some of the figures and tables in this chapter are too large or complex to properly format in printed form. Figures and tables marked with a "WAV" label can be downloaded from the Web Added Value Download Resource Center at www.jrosspub.com/wav for proper viewing.

At its core, prioritization denotes an explicit mechanism for establishing precedence—that is, determining which initiatives are entitled to proceed ahead of others. In a portfolio management context, prioritization extends beyond a simple ordering exercise and instead represents a structured assessment of relative urgency and importance across a collection of projects (and programs). Each project's position within a rank-ordered list reflects its comparative contribution to organizational objectives when evaluated against a defined set of criteria.

In biopharma portfolio management, prioritization is most commonly operationalized through formal methodologies that integrate quantitative metrics, qualitative judgments, or a combination of both. The outcome is a ranked list of assets or initiatives that informs funding, resourcing, and sequencing decisions under conditions of constrained capital and organizational capacity.

Single-Attribute Prioritization Based on Financial Value

The most elementary form of project prioritization relies on a single quantitative attribute, typically the net present value (NPV) of projected future cash flows. Under this approach, projects are evaluated independently and ranked according to their estimated economic contribution, implicitly assuming that financial value is the dominant—or sole—decision criterion.

As introduced in Chapter 7, *Innovative Therapies R Us* manages a portfolio of 10 development projects but lacks sufficient financial and human resources to advance all of them simultaneously. For simplicity, resource constraints are expressed in terms of total budget, function A full-time equivalents (FTEs), and function B FTEs (see Table 7.2). In the prioritization example presented here, the objective function is defined as the maximization of NPV, replacing the earlier use of expected NPV (eNPV).

If maximization of NPV is assumed to be the sole organizational objective, a straightforward prioritization procedure would involve sorting projects in descending order of NPV. Projects are then selected sequentially until one or more resource constraints are violated. The resulting cutoff defines the boundary between funded projects and those that are deferred or terminated.

Applying this method to the data in Table 8.1 yields the ranking shown in Table 8.2. Under this prioritization, the top five projects—D, J, A, E, and B—are selected, while the remaining five projects are excluded. The resulting portfolio is feasible (because it violates none of the three constraints), requiring a total budget of \$1.170B, 4.8 function A FTEs, and 9.3 function B FTEs, and is projected to generate \$6.326B in NPV.

Table 8.1 *Innovative Therapies R Us*—resource requirements and NPV for a portfolio of 10 projects.

PROJECT	DECISION VARIABLES	FUNCTION A FTEs REQUIRED YEAR 1	FUNCTION B FTEs REQUIRED YEAR 1	FUNCTION B FTEs REQUIRED YEAR 3	BUDGET REQUIRED ($M) YEAR 1	NPV ($M)
A	1	0.70	2.40	1.20	200	1,300
B	1	1.60	2.50	1.30	175	1,000
C	1	0.90	1.60	0.90	250	1,000
D	1	0.80	1.80	0.90	275	1,377
E	1	0.90	0.70	0.70	300	1,286
F	1	0.60	1.60	1.50	150	733
G	1	1.50	0.50	0.50	125	643
H	1	1.60	2.00	1.50	145	577
I	1	2.40	1.80	1.20	160	875
J	1	0.80	1.90	1.10	220	1,364
TOTAL		11.80	16.80	10.80	2,000	10,154
CONSTRAINT		7.00	10.00	12.00	1,500	

Table 8.2 *Innovative Therapies R Us*—project prioritization on the basis of NPV.

PROJECT	DECISION VARIABLES	FUNCTION A FTEs REQUIRED YEAR 1	FUNCTION B FTEs REQUIRED YEAR 1	BUDGET REQUIRED ($M) YEAR 1	NPV ($M)
D	1	0.80	1.80	275	1,377
J	1	0.80	1.90	220	1,364
A	1	0.70	2.40	200	1,300
E	1	0.90	0.70	300	1,286
B	1	1.60	2.50	175	1,000
C	0	0.90	1.60	250	1,000
I	0	2.40	1.80	160	875
F	0	0.60	1.60	150	733
G	0	1.50	0.50	125	643
H	0	1.60	2.00	145	577
TOTAL		4.80	9.30	1,170	6,326
CONSTRAINT		7.00	10.00	1,500	

While this solution satisfies all stated constraints and delivers substantial economic value, it raises a critical question: Does this rank-based selection represent the *best* portfolio the organization can pursue? More formally, is this prioritization-derived solution *optimal* with respect to the stated objective of NPV maximization?

8.1 COMPARISON OF PORTFOLIO SELECTION BY PRIORITIZATION AND OPTIMIZATION

The answer becomes evident when the prioritization outcome is compared with the portfolio generated through deterministic linear programming, as presented in Chapter 7. The optimal solution, shown in Table 8.3, achieves a total NPV of $7.060B—an incremental increase of $734M relative to the prioritization-based portfolio.

Notably, the composition of the optimized portfolio differs materially from that produced by simple prioritization. The optimization algorithm excludes projects B, I, G, and H, while selecting the remaining six projects. This divergence highlights a fundamental limitation of rank-order prioritization: selecting projects sequentially based on a single attribute does not guarantee that the resulting portfolio maximizes value under multiple simultaneous constraints.

Using the same underlying project data (see Table 8.1), reliance on a suboptimal selection methodology results in the destruction of $734M in potential NPV—more than 10% of total achievable portfolio value. This comparison underscores an essential distinction: project prioritization can produce a *feasible* portfolio, but feasibility alone does not imply optimality. Without an explicit search for the best possible combination of projects—such as that enabled by mathematical optimization—organizations remain unaware of the opportunity cost embedded in their decision processes.

The prioritization approach previously described is explicitly single-objective, focused solely on maximizing NPV. This class of methods can be referred to as *single-objective decision analysis* (SODA). In practice, however, biopharma organizations rarely make portfolio decisions based solely on financial value. Strategic alignment, risk exposure, development cost, timing, and organizational capabilities are frequently considered alongside economic return.

To address decision contexts characterized by multiple, often competing objectives, organizations employ *multiple-objective decision analysis* (MODA). MODA leverages utility theory to integrate heterogeneous quantitative and qualitative criteria into a unified analytic framework. Each criterion—referred to as an attribute—is transformed into a common unit of measurement known as *utils*, a construct first introduced in Chapter 6. Utils represent the relative attractiveness or usefulness of a project with respect to a specific objective. By converting diverse attributes into commensurable utility scores and aggregating them using explicitly defined weights, MODA enables simultaneous consideration of multiple objectives.

Table 8.3 *Innovative Therapies R Us*—portfolio optimization on the basis of maximizing NPV.

PROJECT	DECISION VARIABLES	FUNCTION A FTEs REQUIRED YEAR 1	FUNCTION B FTEs REQUIRED YEAR 1	BUDGET REQUIRED ($M) YEAR 1	NPV ($M)
D	1	0.80	1.80	275	1,377
J	1	0.80	1.90	220	1,364
A	1	0.70	2.40	200	1,300
E	1	0.90	0.70	300	1,286
B	0	1.60	2.50	175	1,000
C	1	0.90	1.60	250	1,000
I	0	2.40	1.80	160	875
F	1	0.60	1.60	150	733
G	0	1.50	0.50	125	643
H	0	1.60	2.00	145	577
TOTAL		4.70	10.00	1,395	7,060
CONSTRAINT		7.00	10.00	1,500	

MODA is particularly valuable in situations where no single criterion dominates decision making. When applied rigorously, it offers several advantages:

- Explicit incorporation of decision-makers' objectives, preferences, and risk tolerances
- Systematic integration of subject matter expert (SME) judgment for both numerical and categorical inputs
- Accommodation of uncertainty through discretization techniques such as the extended Swanson–Megill method (10th, 50th, and 90th percentiles) or the extended Pearson–Tukey method (5th, 50th, and 95th percentiles)
- Transparent decomposition of how each project contributes to overarching portfolio objectives
- Explicit representation of trade-offs among weighted objectives
- Support for sensitivity analyses to assess the impact of attribute weights on prioritization outcomes
- Facilitation of alignment and resolution among stakeholders with differing perspectives
- Relative ease of understanding, adaptation, and iterative refinement

Effective application of MODA requires not only sound analytics but also a disciplined and transparent process. One such process—Multiple-Objective Project Prioritization (MOPP^{SM})—is depicted in Figure 8.1 and provides a structured pathway for implementation.

MOPP^{SM} relies on close collaboration between the portfolio management function and a steering or governance committee charged with guiding prioritization decisions. The process begins with a clear articulation of *why* prioritization is being undertaken, whether at an enterprise-wide level or within specific therapeutic areas, sectors, or business units.

When executed rigorously, the MOPP^{SM} framework unfolds as an eight-step sequence:

1. **Objective definition:** Governance determines the primary objective(s) of the prioritization exercise (e.g., maximizing financial growth subject to a defined risk tolerance).
2. **MODA hierarchy construction:** An analytic hierarchy is developed (see Figure 8.2) that links primary objectives to directly measurable secondary attributes.
3. **Attribute definition:** Each attribute (e.g., NPV) is defined with explicit numerical or categorical thresholds corresponding to varying levels of attractiveness.
4. **Utility function development:** Utility curves (see Figures 8.3 and 8.5 and WAV Figures 8.4 and 8.6) are constructed, typically on a 1–10 raw utility scale, to reflect collective risk preferences and returns to scale.
5. **Weight assignment:** Weights are assigned to objectives and attributes, and an initial MODA model is assembled (see Figure 8.7).
6. **Asset scoring:** SMEs score each project against the defined attributes, followed by validation of the resulting utility scores.
7. **Model testing and refinement:** The model is stress-tested to ensure logical consistency in project rankings and to eliminate anomalous results. Iterative adjustments—most often to attribute weights or utility functions—may be required (as indicated by steps 1–4 in Figure 8.1).
8. **Model closeout:** The finalized MODA model is approved for portfolio selection and subsequent prioritization cycles.

Experience suggests that while prioritization initiatives often have well-defined starting points, they frequently lack clear termination criteria. Repeated or perpetual reprioritization is often symptomatic of weak decision discipline and can create confusion and instability for project and program teams.

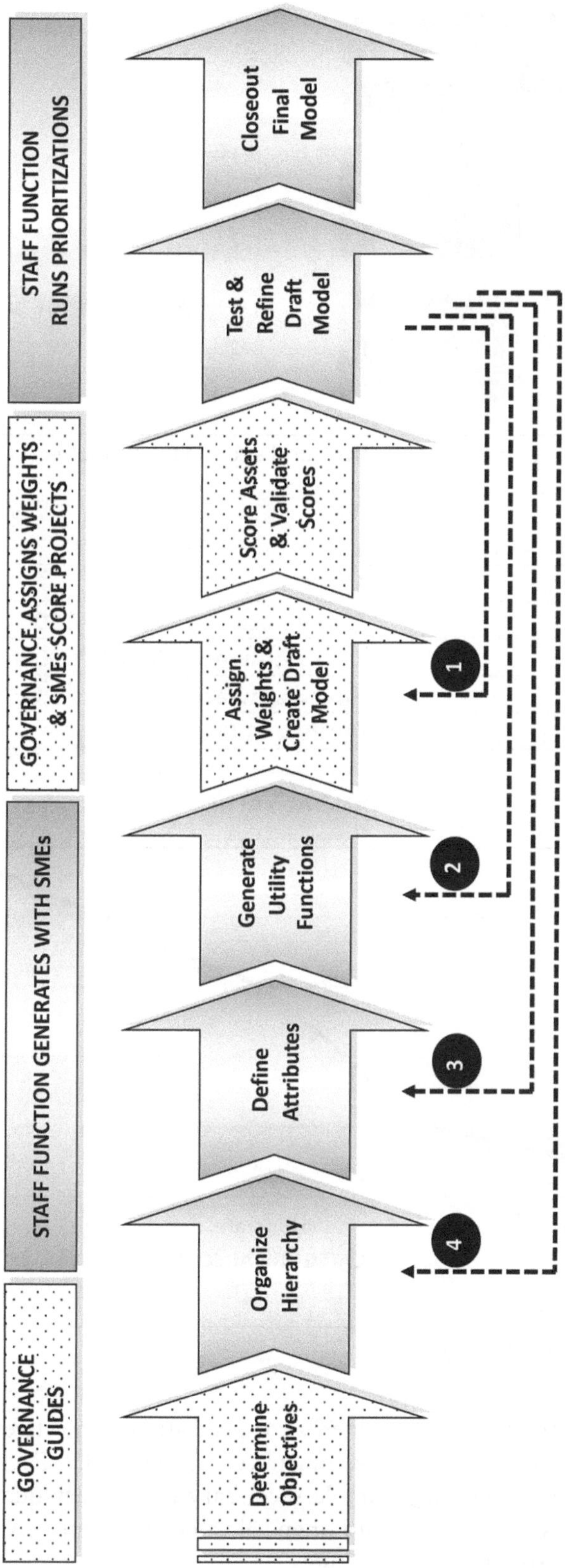

Figure 8.1 MOPP℠—a process that supports the MODA prioritization methodology.

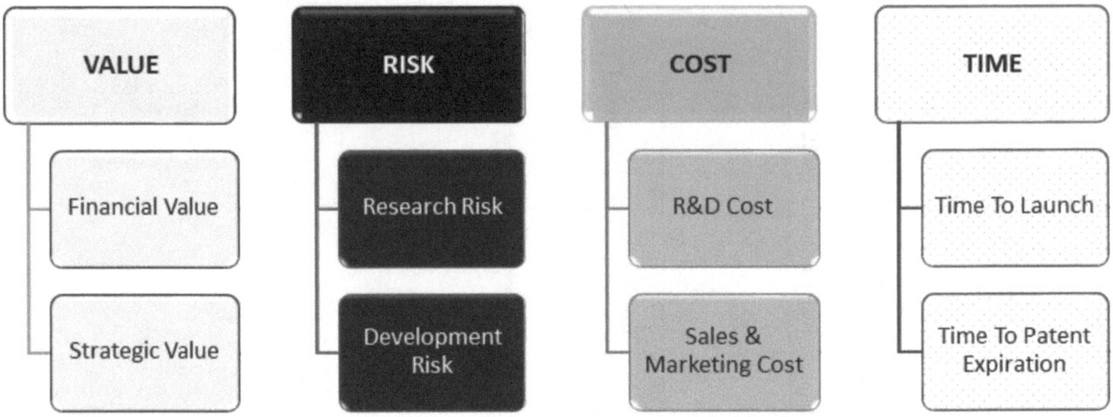

Figure 8.2 MODA hierarchy of primary objectives and secondary attributes.

FINANCIAL VALUE	SCORE	DEFINITION
Very Low	1	NPV ≤ $600M
Low	2	NPV between $601M - $800M
Moderate	6	NPV between $801M - $900M
High	8	NPV between $901M - $1,000M
Very High	10	NPV >$1,000M

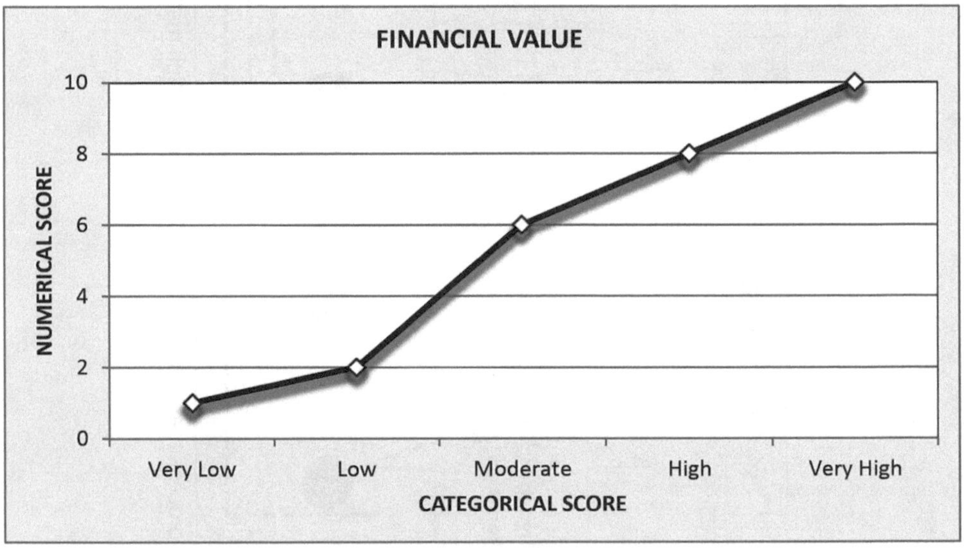

Figure 8.3 Utility function for financial value.

In the absence of a dominant decision criterion, organizational objectives typically cluster into four categories: *value, risk, cost,* and *time,* as illustrated in Figure 8.2. Each primary objective is quantified through a set of directly measurable secondary (or tertiary/quaternary) attributes. In the example shown, value is decomposed into financial value and strategic value, while each of the remaining objectives is measured by two attributes.

R&D COST	SCORE	DEFINITION
Very Low	10	≤ $500M
Low	7	between $501M - $600M
Moderate	5	between $601 - $700M
High	3	between $701M - $800M
Very High	1	> $800M

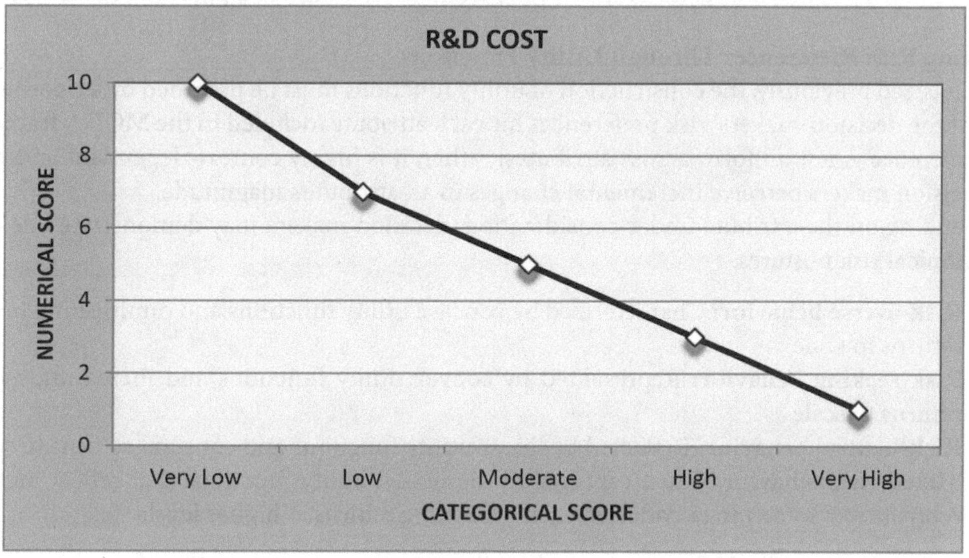

Figure 8.5 Utility function for research and development (R&D) cost.

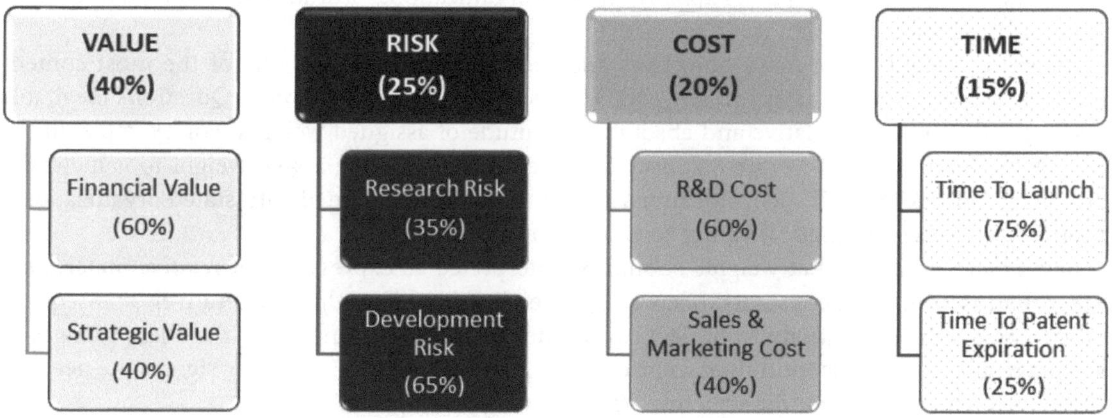

Figure 8.7 MODA hierarchy of weighted primary objectives and secondary attributes.

Although no fixed rule governs the number of attributes in a MODA hierarchy, several guiding principles apply:

- Attributes should be collectively exhaustive and mutually exclusive, covering all critical decision concerns without redundancy

- Attributes should exhibit mutual preferential independence (i.e., preference for one attribute level should not depend on the level of another attribute)
- Attributes must be operable and readily understood by stakeholders
- Parsimony is essential: smaller, well-constructed hierarchies are easier to communicate, measure, and weight meaningfully

When these principles are respected, MODA-based prioritization provides a robust, transparent, and strategically aligned foundation for portfolio decision making in biopharma organizations.

Capturing Risk Preferences Through Utility Functions
As emphasized previously, the construction of utility functions must be preceded by a rigorous elicitation of the decision-makers' risk preferences for each attribute included in the MODA framework. Risk preference is not uniform across attributes; rather, it is highly context-dependent and reflects how decision makers perceive incremental changes in an attribute's magnitude.

Depending on the attribute under consideration, decision makers may demonstrate one of several canonical risk postures:

- **Risk-averse behavior:** Characterized by concave utility functions and diminishing marginal returns to scale
- **Risk-seeking behavior:** Represented by convex utility functions and increasing marginal returns to scale
- **Risk-neutral behavior:** Reflected by linear utility functions and constant returns to scale
- **Mixed risk behavior:** Often captured by sigmoidal utility functions that reflect increasing returns over some range, followed by diminishing returns at higher levels

Figures 8.3 and 8.5 and WAV Figures 8.4 and 8.6 illustrate representative utility functions reflecting common risk preferences across the four primary objective domains generally used in project prioritization exercises. These utility functions translate raw numerical or categorical attribute values into normalized utility scores that explicitly encode organizational attitudes toward risk, trade-offs, and non-linear value perception.

Assigning weights to primary objectives and secondary attributes is one of the most contentious—and consequential—steps in any MODA-based prioritization exercise. Questions inevitably arise regarding both the relative and absolute magnitude of assigned weights. For example, in the MODA model depicted in Figure 8.7, how can one justify assigning a 40% weight to value and a 25% weight to risk? Even if such allocations appear directionally aligned with stated organizational priorities, the precise numerical values remain open to scrutiny.

Best practice dictates that weights should be interpreted as expressions of *relative importance*, rather than as immutable constants. Accordingly, weights should be adjusted iteratively and sequentially to assess their sensitivity to project rank order before being finalized. If modest changes in weights result in large discontinuities in prioritization outcomes, this instability signals the need for further refinement of the model structure or underlying assumptions.

An alternative—and often complementary—approach to weight determination is the use of the *Analytic Hierarchy Process* (AHP), discussed in greater detail later in this chapter. AHP relies on systematic pairwise comparisons to elicit the relative importance of objectives and attributes. Using a predefined preference scale (see WAV Table 8.5), decision makers compare each objective against every other objective (e.g., value vs. risk, value vs. cost, value vs. time), resulting in a comparison matrix such as that shown later in Table 8.8. From this matrix, a consistent set of weights is mathematically derived.

Once weights are assigned at the objective level, the same pairwise comparison process is repeated to derive weights for the associated secondary attributes. In principle, AHP can also be applied at the project level—comparing each project to every other project—but this quickly becomes impractical for portfolios of nontrivial size. The usefulness of AHP in the assignment of weights to primary objectives and secondary attributes deserves its own detailed discussion under the subsequent heading: *A Matter of Weights—The Utility of AHP.*

For this reason, the authors recommend a hybrid approach in which AHP is used, where necessary, to derive defensible weights for objectives (and attributes), while MODA is used to score projects against each attribute using utility functions. This division of labor balances methodological rigor with operational feasibility.

Regardless of the method used to assign weights, the scoring of projects follows a consistent logic. Each project's numerical or categorical attribute value must first be translated into a raw utility score and subsequently into a weighted utility (WU) score. As illustrated in Table 8.4, Project A has a financial value of $1,300M, corresponding to a maximum raw utility score of 10. Its strategic value is assessed as *high*, corresponding to a raw utility score of 8. Because no weights have yet been applied, these values are referred to as *raw* utility scores.

The next step is to apply weights. In the MODA hierarchy shown in Figure 8.7, financial value and strategic value carry relative weights of 60% and 40%, respectively, within the primary objective of value, which itself has an overall weight of 40%. This results in absolute weights of 24% (0.60 × 0.40) for financial value and 16% (0.40 × 0.40) for strategic value.

Multiplying project A's raw utility scores by these absolute weights yields weighted utility scores of 2.4 for financial value and 1.3 for strategic value, as shown in WAV Table 8.6.

This procedure is repeated for all remaining attributes in the MODA hierarchy. The resulting weighted utility scores are summed to generate a weighted aggregate utility (WAU) score for each project (see WAV Table 8.7). These WAU scores form the basis for project rank ordering, as illustrated in Figure 8.8.

When all year 1 constraints are imposed—most notably the binding constraint on function B FTEs—the feasible portfolio consists of projects A, B, E, C, and J. This portfolio is valued at $5,949M in NPV. While this outcome reflects a balanced consideration of multiple objectives, it is economically inferior to portfolios selected using (a) simple NPV rank ordering ($6,326M) and (b) NPV optimization ($7,060M).

The principal strength of MODA-based prioritization lies not solely in the final rank order, but in the richness of insight derived from relative comparisons across attributes and projects. For example, Figure 8.8 suggests that projects A and E exhibit comparable financial value attractiveness, whereas project A's strategic value contribution appears approximately fourfold greater than that of project E. Such comparisons often provide a compelling rationale for prioritizing outcomes and facilitating acceptance among decision makers.

Further insight can be gained by examining portfolio diversification across key objective dimensions. Figure 8.9 illustrates the distribution of selected projects across value, risk, and cost, while WAV Figure 8.10 does so across value, risk, and time.

Table 8.4 Conversion of numerical and categorical values to raw utility scores.

PROJECT	FINANCIAL VALUE ($M)	FINANCIAL VALUE (U)	STRATEGIC VALUE (CATEGORICAL)	STRATEGIC VALUE (U)	RESEARCH RISK (PTS)	RESEARCH RISK (U)	DEVELOPMENT RISK (PTS)	DEVELOPMENT RISK (U)
A	1,300	10	HIGH	8	0.65	1	0.75	10
B	1,000	8	VERY HIGH	10	0.70	1	0.70	9
C	1,000	8	LOW	3	0.80	6	0.65	6
D	1,377	10	MODERATE	5	0.75	3	0.60	3
E	1,286	10	VERY LOW	2	0.70	1	0.70	9
F	733	2	MODERATE	5	0.80	6	0.75	10
G	643	2	VERY HIGH	10	0.90	10	0.50	1
H	577	1	VERY HIGH	10	0.75	3	0.60	3
I	875	6	VERY LOW	2	0.85	9	0.65	6
J	1,364	10	MODERATE	5	0.60	1	0.55	1

PROJECT	R&D COST ($M)	R&D COST (U)	SALES & MARKETING COST ($M)	SALES & MARKETING COST (U)	TIME TO LAUNCH (YEARS)	TIME TO LAUNCH (U)	TIME TO PATENT EXPIRATION (YEARS)	TIME TO PATENT EXPIRATION (U)
A	625	5	175	5	5	10	12	6
B	550	7	325	2	6	9	14	8
C	675	5	100	10	7	7	9	2
D	875	1	250	3	8	7	13	6
E	850	1	125	7	6	9	15	10
F	450	10	125	7	9	3	8	2
G	725	3	225	3	11	1	10	2
H	435	10	275	2	10	3	11	4
I	760	3	200	5	11	1	14	8
J	570	7	150	7	5	9	9	2

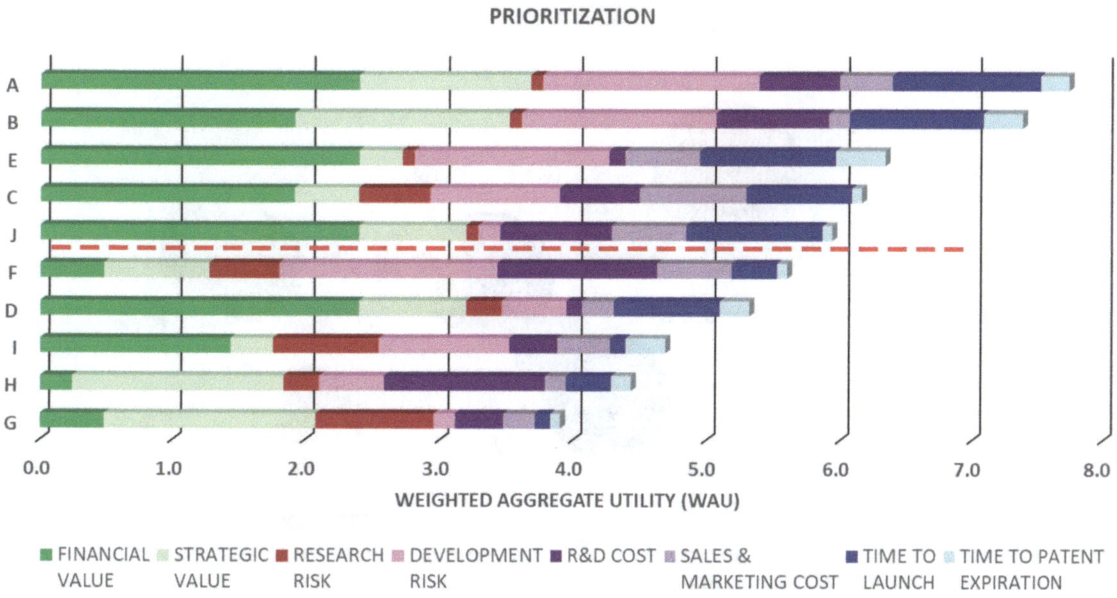

Figure 8.8 Project prioritization using MODA and based on weighted aggregate utility scores.

Despite its many advantages, MODA—like all prioritization techniques—has inherent limitations:

- **Project dependencies are not captured:** For example, if project E (above the constraint line in Figure 8.8) depends on project I (below the constraint line), the selected portfolio becomes infeasible. Such dependencies are more appropriately handled using binary integer logic, as discussed in Chapter 7.
- **Project selection logic is not supported:** MODA cannot enforce rules such as selecting only one project from a mutually exclusive subset (e.g., projects B, E, and C competing within the same market).
- **Portfolio optimality is not guaranteed:** Prioritization does not reveal whether projects below the constraint line could displace one or more projects above the line to create a superior portfolio.
- **The efficient frontier is not identified:** As a result, prioritization provides no direct estimate of the value left unrealized by selecting a suboptimal combination of projects. The gap between any such portfolio and the true optimal portfolio represents value destruction attributable to an inferior decision methodology.

Finally, it bears repeating that many prioritization exercises have well-defined starting points but ill-defined endpoints. This lack of closure often results in repeated reprioritization cycles, organizational instability, and erosion of confidence among project teams. While project data inevitably evolve over time, such changes rarely justify a wholesale reversal of prior decisions. Highly disciplined organizations address this challenge by making a sufficient set of prioritization decisions up front and then reprioritizing *only at the margin* as resources are released through project completion or termination.

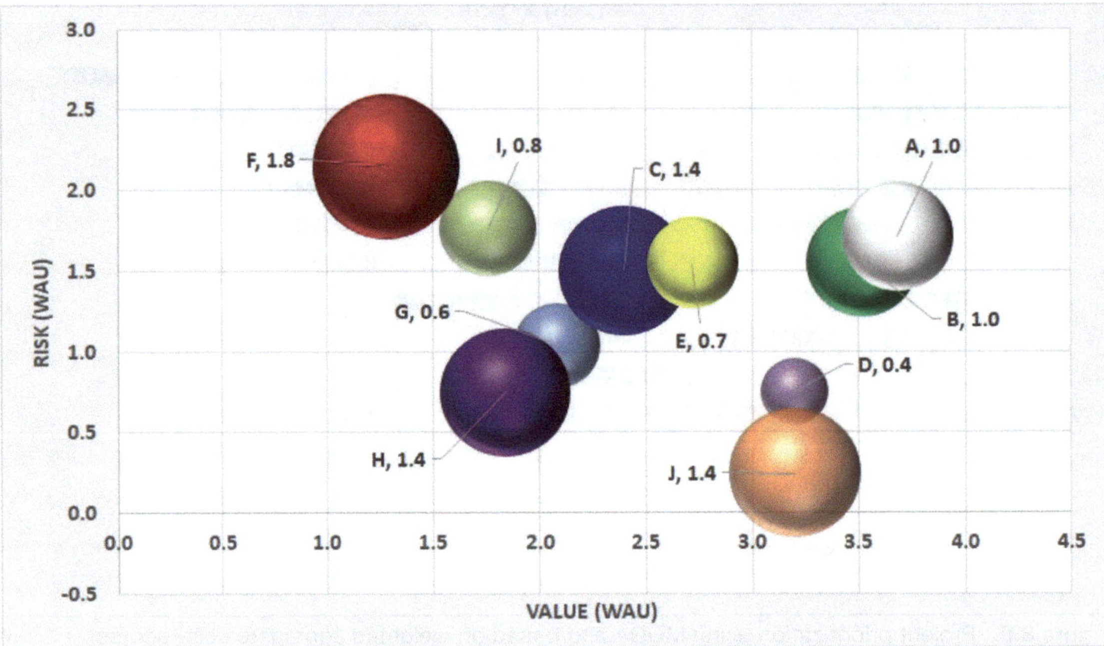

Figure 8.9 Representation of potential portfolio by value (X-axis), risk (Y-axis), and cost (Z-axis).

8.2 PRIORITIZATION IN RESEARCH: CHALLENGES, TENSIONS, AND EMERGING PRACTICE

While asset prioritization in clinical development is now a standard and largely institutionalized practice across the biopharma industry, the application of similarly structured prioritization frameworks to discovery and early research portfolios remains far more controversial. This reluctance is driven primarily by the profound uncertainty inherent in research-stage assets, where biological mechanisms, translational relevance, and eventual clinical utility are only partially understood.

Proponents of research-stage prioritization argue that finite scientific, financial, and organizational resources must be deployed with rigor, transparency, and defensibility—even in the face of uncertainty. From this perspective, systematic prioritization is an essential lever for reducing downstream attrition, improving the quality of candidates entering development, and increasing the probability that early investments mature into clinically viable assets. In contrast, opponents contend that scientific discovery and innovation are fundamentally non-linear, serendipitous processes that defy structured management. They argue that imposing formal portfolio selection methodologies risks suppressing creativity and prematurely terminating programs that might otherwise yield transformative therapies. This position is often supported by retrospective examples of successful drugs that survived only because individual scientists or leaders intervened against a formal decision process.

However, such arguments frequently overlook the substantial costs associated with unstructured decision making—particularly the impact of cognitive and emotional biases. Chief among these is the tendency to continue funding projects simply because investment has already begun, conflating sunk costs with rational forward-looking decision criteria. In many organizations, sunk costs are tacitly accepted as *the cost of doing business*, rather than being recognized appropriately as the cost of acquiring information. This bias materially increases the likelihood of continuing low-quality assets into later, more expensive phases of R&D.

As discussed in Chapter 1, the exceptionally high cost of successful drug development is widely attributed to elevated attrition rates, particularly those arising from Type I (false positive) errors that advance weak assets into late-stage development. These errors are often tolerated—implicitly or explicitly—because organizations seek to avoid the perceived reputational and strategic consequences of Type II (false negative) errors, in which potentially valuable assets are terminated prematurely.

This asymmetry in error tolerance creates a structural bias that makes it difficult, time-consuming, and expensive to translate early scientific success into sustainable development success. While translational medicine has gained broad acceptance in early clinical development, the industry continues to lack a reproducible, analytically grounded process for *translational research*—one that systematically improves the quality and durability of assets progressing from early discovery through late research and, ultimately, into first-in-human studies.

As highlighted in Chapter 1, several organizations—including AstraZeneca, Eli Lilly, and Pfizer—have demonstrated that sustained, phase-specific attention to translational risk factors can materially improve the quality of assets entering mid-clinical development. These efforts are explicitly designed to mitigate failure in what is often referred to as the *valley of clinical death*, where attrition rates are historically highest.

In this context, Wehling (2009) introduced a structured, MODA-type scoring methodology to assess the *translatability* of research projects into viable development assets. In this framework, five feasibility parameters—functionally equivalent to primary objectives—are decomposed into 13 directly measurable secondary attributes. Each attribute is explicitly weighted, as shown in WAV Figure 8.11. Notably, attributes related to biomarkers for efficacy and safety prediction receive the highest aggregate weight (37%), followed closely by attributes capturing human evidence (31%). This weighting scheme reflects a deliberate emphasis on translational confidence rather than scientific novelty alone.

Wehling (2009) further identifies three additional feasibility domains that, while not directly predictive of translational success, remain critical to overall asset viability:

- **Chemical tractability**—supported by lead identification and lead optimization metrics
- **Intellectual property**—supported by patent strength, freedom to operate, and patent expiry considerations
- **Area of clinical interest**—supported by measures of unmet medical need, competitive intensity, and reimbursement or payer cost pressures

Taken together, these domains enable a holistic, yet structured, evaluation of research-stage assets and provide a defensible basis for prioritization under uncertainty.

In a subsequent report, Wendler and Wehling (2017) clarified that the original 2009 translatability model was developed specifically for cardiovascular drug research. Importantly, they demonstrated that the same underlying secondary attributes—when reweighted appropriately—could be applied across a broad range of therapeutic areas, including oncology; psychiatry; antiviral, antibacterial, and antifungal indications; and monogenic disorders. This adaptability underscores a key strength of MODA-based approaches. While the attributes themselves may be largely universal, their relative importance can—and should—be tailored to reflect therapeutic-area-specific risks, regulatory expectations, and clinical development paradigms.

Building on the empirical experience of AstraZeneca, Pfizer, and Eli Lilly, PPVC has applied the MOPP^{SM} process to develop MODA-based prioritization models for research portfolios that share conceptual alignment with the frameworks proposed by Wendler and Wehling (2017). These models are explicitly designed to support disciplined, transparent decision making in environments characterized by high uncertainty and limited data.

In these implementations, for example, the risk dimension is decomposed into five secondary attributes:

(1) Biological evidence
(2) Pharmacokinetics and pharmacodynamics (PK/PD)
(3) Safety liabilities
(4) Patient identification and stratification feasibility
(5) Manufacturability

Each of these secondary attributes is, in turn, supported by two directly measurable tertiary attributes, enabling granular assessment while maintaining traceability to higher-level objectives.

Through this structured decomposition of translational risk, MODA-based prioritization provides a pragmatic compromise between unstructured scientific judgment and overly rigid algorithmic selection—offering a reproducible, defensible mechanism to improve the quality of assets progressing from research into development without extinguishing innovation at its source.

8.3 A MATTER OF WEIGHTS—THE UTILITY OF AHP

Determination and Validation of Objective and Attribute Weights

The assignment of hierarchical weights to primary objectives and their associated attributes is rarely a neutral or uncontested exercise. Indeed, weight elicitation is often one of the most debated elements of a MODA-based prioritization effort. As stated earlier, in the MODA hierarchy shown in Figure 8.7, how can one objectively justify assigning a weight of 40% to value and 25% to risk? Even if such allocations appear directionally aligned with stated organizational priorities, the question remains whether the *exact* values—40% and 25%—are appropriate. Best practice recognizes that weights should be interpreted as expressions of *relative importance* rather than as precise or immutable quantities. Accordingly, weights should be assigned initially to reflect broad preference relationships and then adjusted iteratively to assess the sensitivity of project rank ordering to incremental changes. If modest perturbations in weights lead to disproportionate changes in prioritization outcomes, this instability signals the need for refinement before finalization.

An alternative and more structured approach to determining objective and attribute weights is the AHP, originally developed by T. L. Saaty (1980). AHP employs systematic pairwise comparisons to elicit the relative importance of decision criteria and includes an explicit mechanism to assess internal consistency. An overview of the AHP methodology, as described by Ragsdale (2017), is shown in Figure 8.12. Its application to the MODA hierarchy depicted in Figure 8.7 proceeds as follows.

Step 1: Determine Primary Objectives

The primary objectives for the R&D portfolio under consideration are defined as value, risk, cost, and time.

Step 2: Conduct Pairwise Comparisons of Objectives

Using the AHP pairwise preference-value scale (see WAV Table 8.5), each objective is compared against every other objective. The resulting matrix of comparisons is shown in Table 8.8.

In the context of biopharma R&D prioritization, experience suggests that value is typically the most important objective and is considered less than twice as important as risk. Risk, in turn, is viewed as slightly more important than cost, which itself is marginally more important than time. These heuristics are encoded directly into the pairwise comparisons in Table 8.8.

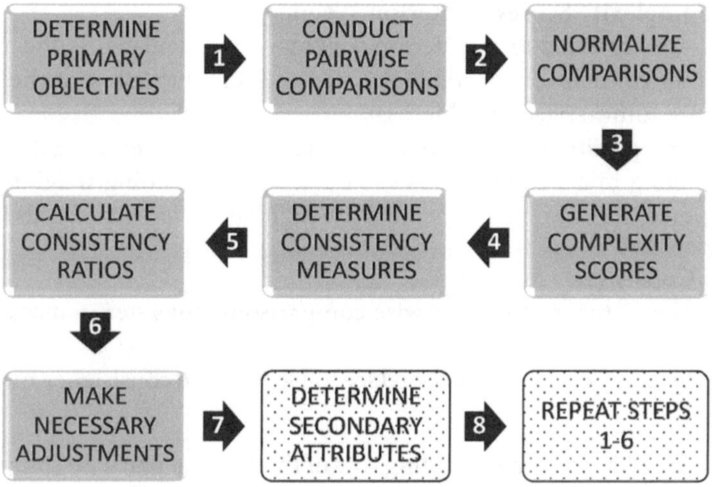

Figure 8.12 Outline of the Analytic Hierarchy Process (AHP).

For example, because value is judged to be *moderately preferred* relative to risk, the corresponding entry in the first row and second column is assigned a value of 3.000. The reciprocal comparison—risk relative to value—appears as 0.333 in the second row and first column. The final row of Table 8.8 reports the aggregate of the pairwise comparison scores for each objective.

Step 3: Normalize the Pairwise Comparisons

The pairwise comparison matrix is normalized by dividing each cell value in Table 8.8 by the column total for the corresponding objective. The normalized results are shown in Table 8.9.

As an illustration, the normalized score in the first row and third column is calculated as 4.000 divided by 7.500, yielding a value of 0.533.

Table 8.8 Matrix of MODA model objectives used to calculate weights by pairwise comparison in the AHP.

	VALUE	RISK	COST	TIME
VALUE	1.000	3.000	4.000	5.000
RISK	0.333	1.000	2.000	3.000
COST	0.250	0.500	1.000	2.000
TIME	0.200	0.333	0.500	1.000
TOTAL	**1.783**	**4.833**	**7.500**	**11.000**

Table 8.9 Normalized comparison of MODA model objectives.

	VALUE	RISK	COST	TIME
VALUE	0.561	0.621	0.533	0.455
RISK	0.187	0.207	0.267	0.273
COST	0.140	0.103	0.133	0.182
TIME	0.112	0.069	0.067	0.091

Step 4: Generate Complexity Scores (Criterion Weights)

Complexity scores—also referred to as criterion weights—are calculated as the arithmetic mean of the normalized values in each row of Table 8.9. These averages represent the relative weight of each objective and are summarized in Table 8.10.

For example, the complexity score for the value objective is the average of the normalized values in the first row, resulting in a weight of 0.542 (54.2%). Corresponding weights for the remaining objectives are computed in the same manner and are highlighted as shaded cells in Table 8.10.

Step 5: Determine Consistency Measures

To assess the internal coherence of the pairwise comparisons, consistency measures are calculated for each objective, as shown in Table 8.11.

For each row in the normalized comparison (see Table 8.9), the consistency measure is computed as follows:

1. Multiply each cell in the corresponding row of the original pairwise comparison matrix (see Table 8.8) by the respective complexity score from Table 8.10
2. Sum these products to obtain an aggregate score
3. Divide the aggregate score by the complexity score for that objective

For example, the aggregate score for the value objective is calculated as:

$$(1 \times 54.2\%) + (3 \times 23.3\%) + (4 \times 14.0\%) + (5 \times 8.5\%) = 2.224$$

Dividing this aggregate by the complexity score for value (54.2%) yields a consistency measure of 4.102. In a perfectly consistent decision framework, each consistency measure would equal the number of objectives under consideration—in this case, four. The small deviations observed in Table 8.11 are typical and reflect the inherent imprecision in human judgment.

Step 6: Calculate Consistency Index and Consistency Ratio

Overall consistency is assessed using the consistency index (CI) and the consistency ratio (CR), which compare observed inconsistency against a benchmark derived from random judgments.

Table 8.10 Complexity scores of MODA model objectives.

	VALUE	RISK	COST	TIME	Complexity Scores
VALUE	0.561	0.621	0.533	0.455	0.542
RISK	0.187	0.207	0.267	0.273	0.233
COST	0.140	0.103	0.133	0.182	0.140
TIME	0.112	0.069	0.067	0.091	0.085

Table 8.11 Consistency measures of MODA model objectives.

	VALUE	RISK	COST	TIME	Complexity Scores	Consistency Measures
VALUE	0.561	0.621	0.533	0.455	0.542	4.102
RISK	0.187	0.207	0.267	0.273	0.233	4.061
COST	0.140	0.103	0.133	0.182	0.140	4.018
TIME	0.112	0.069	0.067	0.091	0.085	4.024

The calculations are defined as:

$$CI = \frac{\lambda - n}{n - 1}$$

$$CR = \frac{CI}{RI}$$

where:

- n is the number of objectives
- λ is the average consistency measure across all objectives
- RI is the random index (see WAV Table 8.12) corresponding to n (0.90 for four objectives)

For this case:

$$CI = \frac{4.051 - 4}{3} = 0.017$$

$$CR = \frac{0.017}{0.90} = 0.019$$

A commonly accepted rule of thumb is that a CR less than or equal to 0.10 indicates acceptable consistency. The resulting CR of 0.019 demonstrates that the AHP component of the MODA model is highly consistent.

Step 7: Adjust Weights as Necessary

The AHP-derived weights for the objectives are: (1) value: 54.2%, (2) risk: 23.3%, (3) cost: 14.0%, and (4) time: 8.5%.

These values differ from the heuristic weights initially assumed by decision makers in Figure 8.7, necessitating adjustment of the pairwise preferences. It is noteworthy that classical AHP restricts pairwise comparisons to integer values on the 1–9 scale defined previously in WAV Table 8.5.

The authors note, however, that if non-integer pairwise preferences were permitted, the matrix shown in Table 8.13A would produce complexity scores (see Table 8.13B) closely aligned with the heuristic judgments reflected in Figure 8.7. Under this modified scheme, the consistency ratio would be calculated as 0.0038.

Step 8: Extend the AHP to Secondary Attributes

Steps 1 through 6 are then repeated for the secondary attributes associated with each primary objective in the MODA hierarchy (see Figure 8.7) in order to determine their relative weights.

AHP offers several advantages, including conceptual simplicity, structured elicitation of preferences, and an explicit check for internal consistency. However, it also exhibits notable limitations when applied to project prioritization (Goodwin and Wright 2010):

- **Scalability constraints:** The number of required comparisons grows rapidly with the number of projects and attributes. For example, prioritizing seven projects across seven attributes requires 168 pairwise comparisons.
- **Rank reversal:** The introduction of new projects can alter the ranking of existing projects due to normalization effects inherent in AHP.

Table 8.13 A and B Matrix of objectives used to calculate weights in the AHP—application of non-integer pairwise preferences.

A

	VALUE	RISK	COST	TIME
VALUE	1.000	1.750	2.000	2.250
RISK	0.571	1.000	1.250	1.500
COST	0.500	0.800	1.000	1.500
TIME	0.444	0.667	0.667	1.000
TOTAL	**2.516**	**4.217**	**4.917**	**6.250**

B

Normalized Comparisons

	VALUE	RISK	COST	TIME	Complexity Scores	Consistency Measures
VALUE	0.397	0.415	0.407	0.360	0.395	4.014
RISK	0.227	0.237	0.254	0.240	0.240	4.013
COST	0.199	0.190	0.203	0.240	0.208	4.008
TIME	0.177	0.158	0.136	0.160	0.158	4.007

- **Arbitrary numerical mapping:** Several authors have pointed to the fact that verbal preference descriptors (e.g., *moderately preferred*) are mapped to fixed numerical values (e.g., 3) without empirical validation.
- **Restricted preference scale:** The 1–9 scale cannot accommodate extreme preference ratios. For instance, if A is moderately to strongly preferred to B (4), and B is similarly preferred to C (4), strict transitivity would imply that A should be 16 times more important than C—a value not representable within the scale.

Given these considerations, it is recommended that AHP be used selectively—primarily to derive defensible weights for primary objectives (and, if necessary, secondary attributes)—while MODA is used to score projects against each attribute using utility functions.

Irrespective of the method used to determine weights, each project's attribute-level utility score must be multiplied by its corresponding absolute weight to generate a weighted utility score. These weighted scores are then aggregated across all attributes to produce a weighted aggregate score for each project, leading to a rank order or prioritization.

9

BEST PRACTICES IN PORTFOLIO MANAGEMENT

Some of the figures and tables in this chapter are too large or complex to properly format in printed form. Figures and tables marked with a "WAV" label can be downloaded from the Web Added Value Download Resource Center at www.jrosspub.com/wav for proper viewing.

9.1 KEY CONCEPTS

Capability Maturity Model (CMM)
The CMM is a formal framework that describes how organizational capabilities evolve through progressive stages of discipline and sophistication within a defined domain, such as research and development (R&D), software engineering, or enterprise portfolio management. Originating at the Software Engineering Institute (SEI) and later integrated as CMMI under ISACA stewardship, the model provides a structured approach to institutionalizing repeatable, continuously improving processes. By embedding these processes across functions, the CMM enhances predictability, governance, and alignment between strategy and execution.

Portfolio Analysis
Portfolio analysis is the coordinated evaluation of projects and programs managed collectively to achieve time-phased strategic and financial objectives under resource constraints. In biopharma, it enables the comparison of heterogeneous investments across therapeutic areas, development stages, and risk profiles, supporting informed trade-offs among value, risk, and capital efficiency.

Risk, Resource, and Stakeholder Management

- **Risk management** is a continuous discipline encompassing identification, quantification, monitoring, and treatment of uncertainty across the portfolio. This includes efforts to mitigate risks where feasible, eliminate risks where possible, and actively manage both controllable and uncontrollable sources of uncertainty.
- **Resource management** governs the dynamic allocation of constrained financial, human, and infrastructure assets through mechanisms such as stage-gate reviews. In a biopharma enterprise, disciplined resource management is essential to ensuring that capital and capabilities are continuously directed toward the most value-accretive opportunities as technical and regulatory outcomes unfold.
- **Stakeholder management** ensures structured engagement of internal and external parties whose alignment and support are critical to success.

Together, these disciplines underpin credible and sustainable portfolio management. Portfolio management is inherently a continuous, multi-period decision process operating under uncertainty. Its objective is to support high-quality strategic decisions that balance near-term execution with long-term value creation. Accordingly, this chapter presents a portfolio management CMM and associated best practices to guide organizations seeking higher maturity. The chapter is divided into two major sections:

1. **Portfolio management CMM**—a diagnostic and benchmarking framework applicable to R&D and enterprise portfolios
2. **Portfolio management best practices**—practical guidance across portfolio analysis, risk, resource, and stakeholder management

9.2 CAPABILITY MATURITY MODEL

CMMs are widely used to guide systematic process improvement from the project level to the enterprise. Beyond software development, CMM principles are now standard in enterprise project and portfolio management (EPPM). This chapter proposes a portfolio-specific CMM to assess and benchmark maturity across core portfolio competencies.

A key barrier to portfolio excellence is not analytical technique, but sustained investment in people, processes, governance, and decision frameworks. These investments are characterized using a five-level CMM:

- **Level 1:** *Project and program office*: Activity is limited to data collection and reporting. Decision authority resides with senior management, portfolio reviews are infrequent, and decision making is largely ad hoc and reactive.
- **Level 2:** *Basic portfolio management*: A centralized portfolio function emerges but remains reporting-focused. Data aggregation improves visibility, yet analytical depth, role clarity, and continuous governance mechanisms are limited.
- **Level 3:** *Standard portfolio management*: Portfolio management becomes strategy-anchored and decision-support oriented. Enterprise data ownership is established, skilled practitioners are deployed, and semi-annual portfolio reviews and structured stage gates enable continuous oversight.
- **Level 4:** *Advanced portfolio management*: The focus shifts to active value maximization through integrated strategic planning and formal portfolio optimization using holistic decision frameworks (e.g., CREOPM™). Multiple optimized portfolios define the efficient frontier, supporting transparent trade-off discussions. Reviews occur quarterly with monthly stage gates.
- **Level 5:** *Center of Excellence* (COE): Portfolio management is fully institutionalized as a strategic enterprise capability. A COE sets standards, integrates strategy and portfolio optimization, embeds advanced methodologies enterprise-wide, and sustains continuous governance and talent development.

Regardless of level, effective EPPM rests on four enabling pillars: (1) robust project and program management, (2) disciplined financial management, (3) credible resource management, and (4) structured, analytically sound decision making.

9.2.1 Portfolio Management Competencies

A rigorous CMM requires assessment across multiple competencies rather than a single maturity label. Organizational readiness is evaluated across four primary domains:

1. **Strategy**—Assesses alignment between portfolio decisions and enterprise strategy, including strategic alignment, business case standards, planning integration, value realization, and cross-functional collaboration
2. **Process**—Evaluates the structure and transparency of portfolio activities, including project selection, decision criteria, integration with planning and stage-gate processes, portfolio reviews, and continuous improvement
3. **People (Organization)**—Focuses on organizational structure, portfolio expertise, executive sponsorship, resource management capability, and effectiveness of communication and decision framing
4. **Technology**—Assesses data quality, analytical tools, and the reporting infrastructure supporting portfolio decisions

Each domain comprises sub-competencies independently assessed using structured diagnostic questions. Domain maturity is calculated as the average of sub-competency scores. Detailed maturity definitions for each domain are provided in WAV Tables 9.1 through 9.4.

9.2.2 Application of CMM Diagnosis for EPPM

The CMM framework can be applied as a structured diagnostic tool to assess current and aspirational EPPM maturity. PPVC's proprietary diagnostic defines five EPPM levels—informal, defined, governed, managed, and optimized—as illustrated in Figure 9.1 and described in WAV Table 9.5.

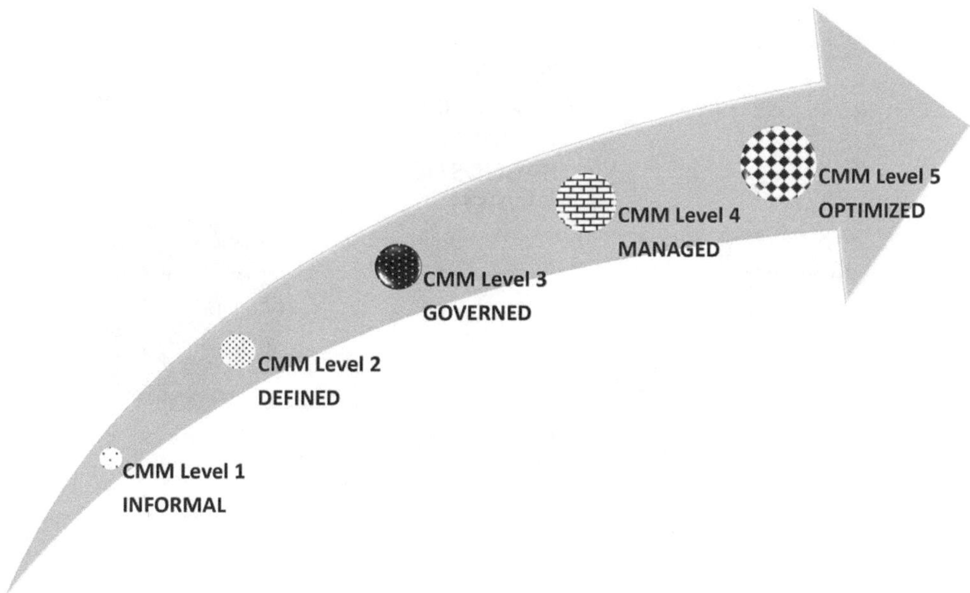

Figure 9.1 EPPM CMM levels.

PPVC's Nonagonal EPPM Diagnostic Model

PPVC operationalizes its diagnostic through a nonagonal wheel comprising nine primary competency dimensions supported by 38 measurable sub-dimensions as illustrated in Figure 9.2:

- Executive sponsorship and governance
- Strategic alignment
- Project management
- Resource management
- Financial management
- Methodologic and analytic proficiency
- Technology management
- Portfolio review and selection process
- Stakeholder management

A structured five-step process (see Figure 9.3) integrates stakeholder input and quantitative scoring to establish current- and future-state maturity profiles.

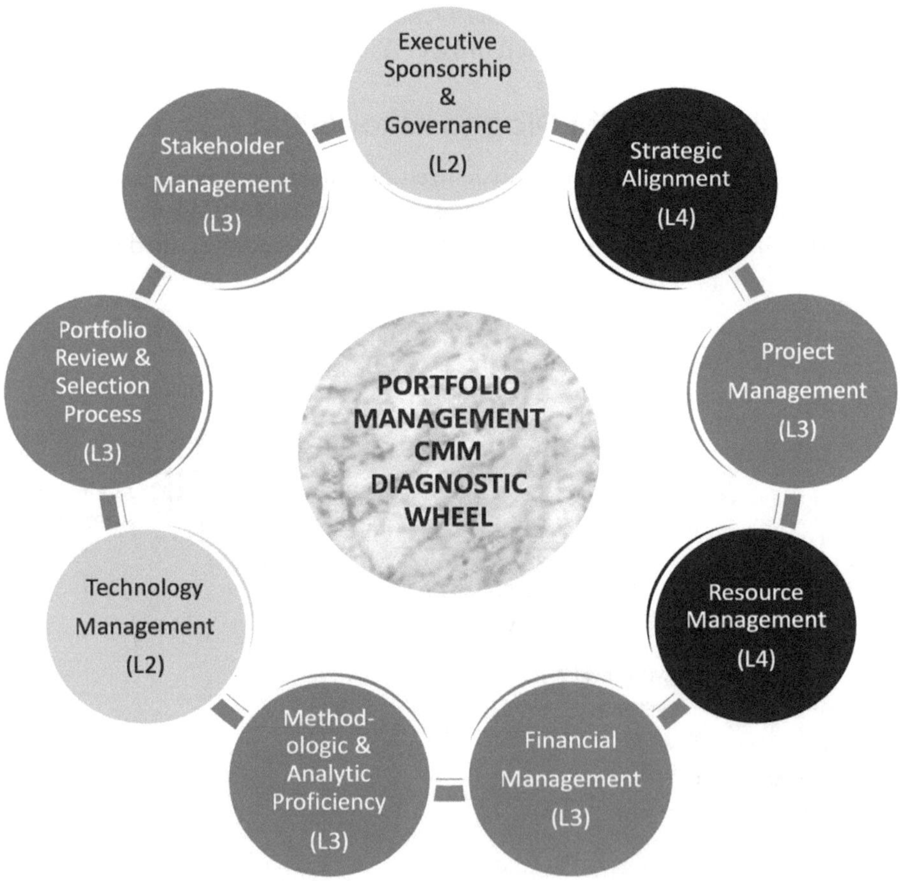

Figure 9.2 PPVC's nonagonal EPPM diagnostic wheel.

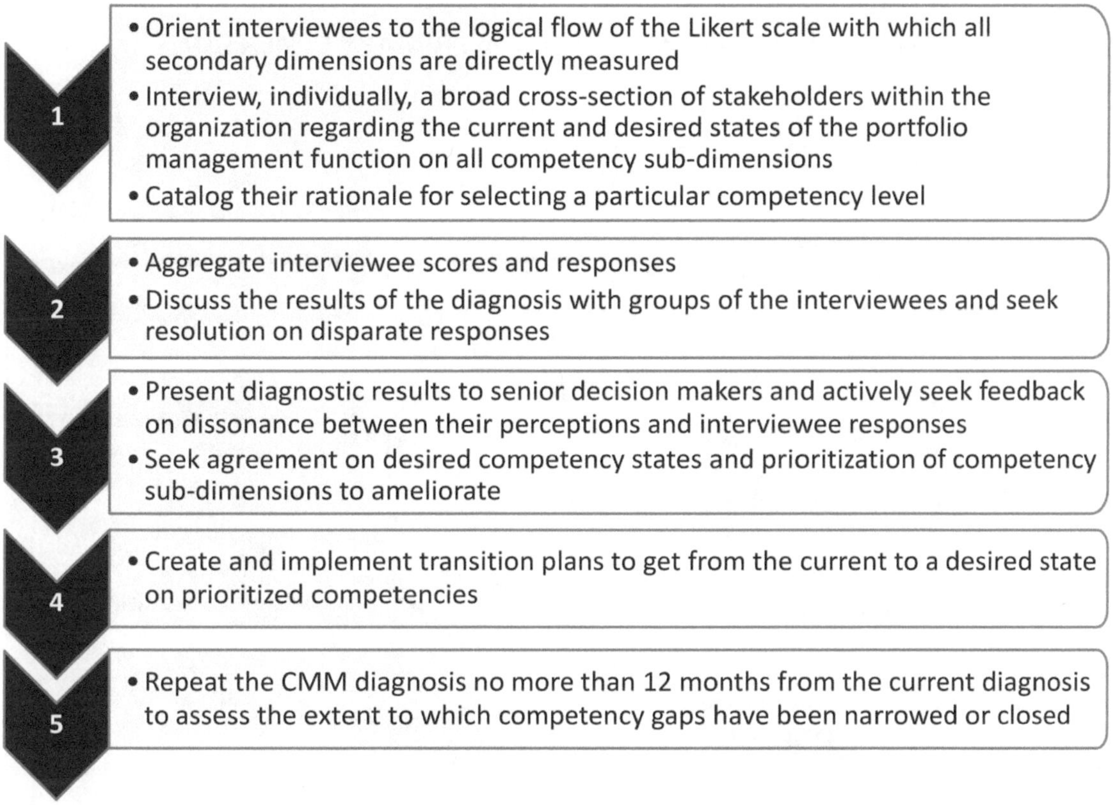

Figure 9.3 PPVC's five-step process for diagnosing and improving EPPM maturity in organizations.

Illustrative diagnostics (see Figure 9.4) demonstrate how organizations identify capability gaps at the primary-dimension level, enabling targeted, prioritized improvement initiatives in transitioning from a current to a future state.

Maturity assessment is iterative rather than transformational. Periodic reassessment enables progress tracking and recalibration. Because EPPM competencies are interdependent, extreme variance across domains is rare, and selective over-optimization yields limited benefit. Organizations should therefore pursue pragmatic, incremental improvements aligned with strategic needs and cultural readiness.

Depending on its commitment to sustaining and enhancing portfolio management maturity, an organization may occupy any one of the competitive positions illustrated in Figure 9.5. These positions reflect relative capability in the ongoing *maturity race* and underscore the strategic implications of underinvestment—or sustained leadership—in EPPM.

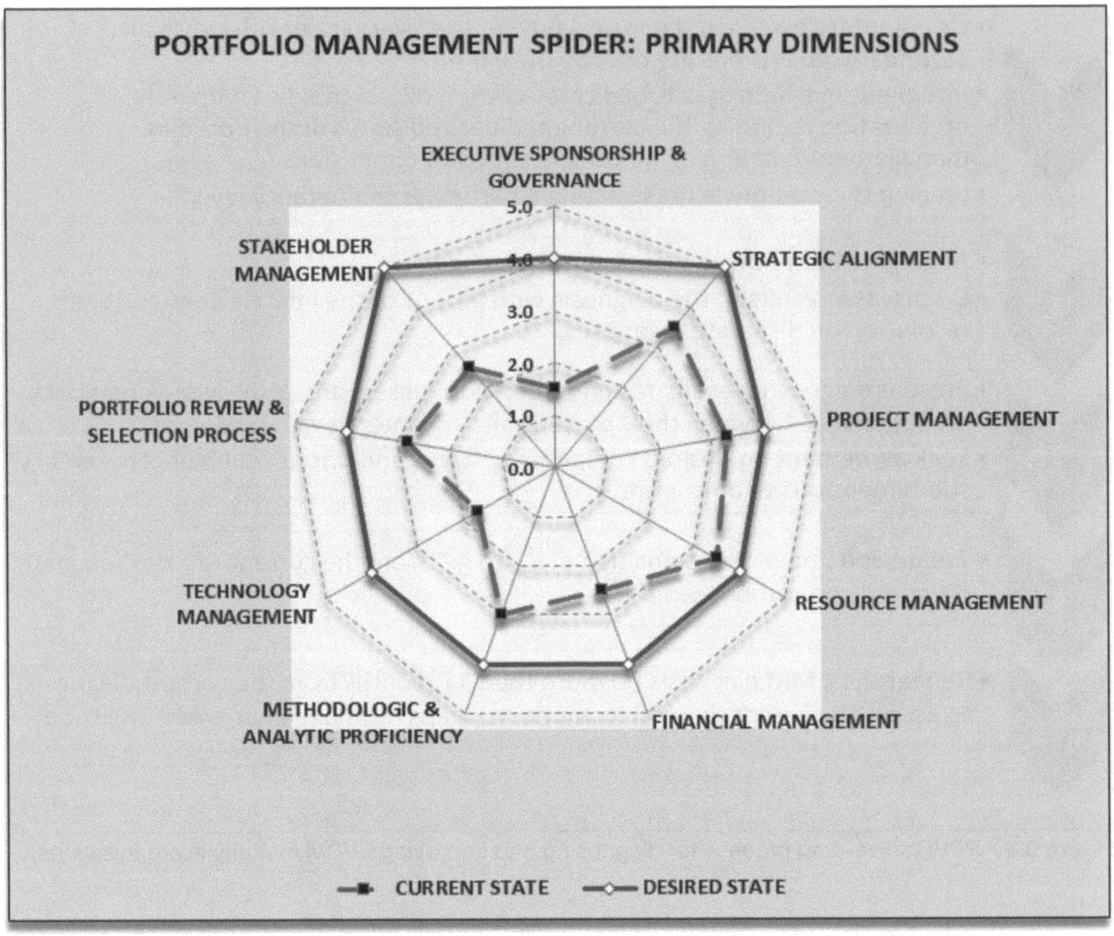

Figure 9.4 Diagnosis of current and desired EPPM states.

9.2.3 Building DA&PM Functions in Biopharma

Over a ten-year period spanning 1997 to 2007, one of the authors was directly responsible for designing, implementing, sustaining, and advancing decision analysis and portfolio management (DA&PM) capabilities within three major biopharma organizations—Bayer, Bristol-Myers Squibb (BMS), and Johnson & Johnson. Each organization presented a distinct cultural context, governance philosophy, and structural configuration, which materially influenced how portfolio management capabilities were established and matured.

As illustrated in Figure 9.6, these differences resulted in varying levels of CMM attainment. Notably, in the case of BMS—where a *greenfield* opportunity existed to build the function from the ground up—the highest level of maturity was achieved and maintained. The following case studies examine how DA&PM maturity was either sustained or elevated within each organization, highlighting both critical success factors and structural impediments encountered along the way.

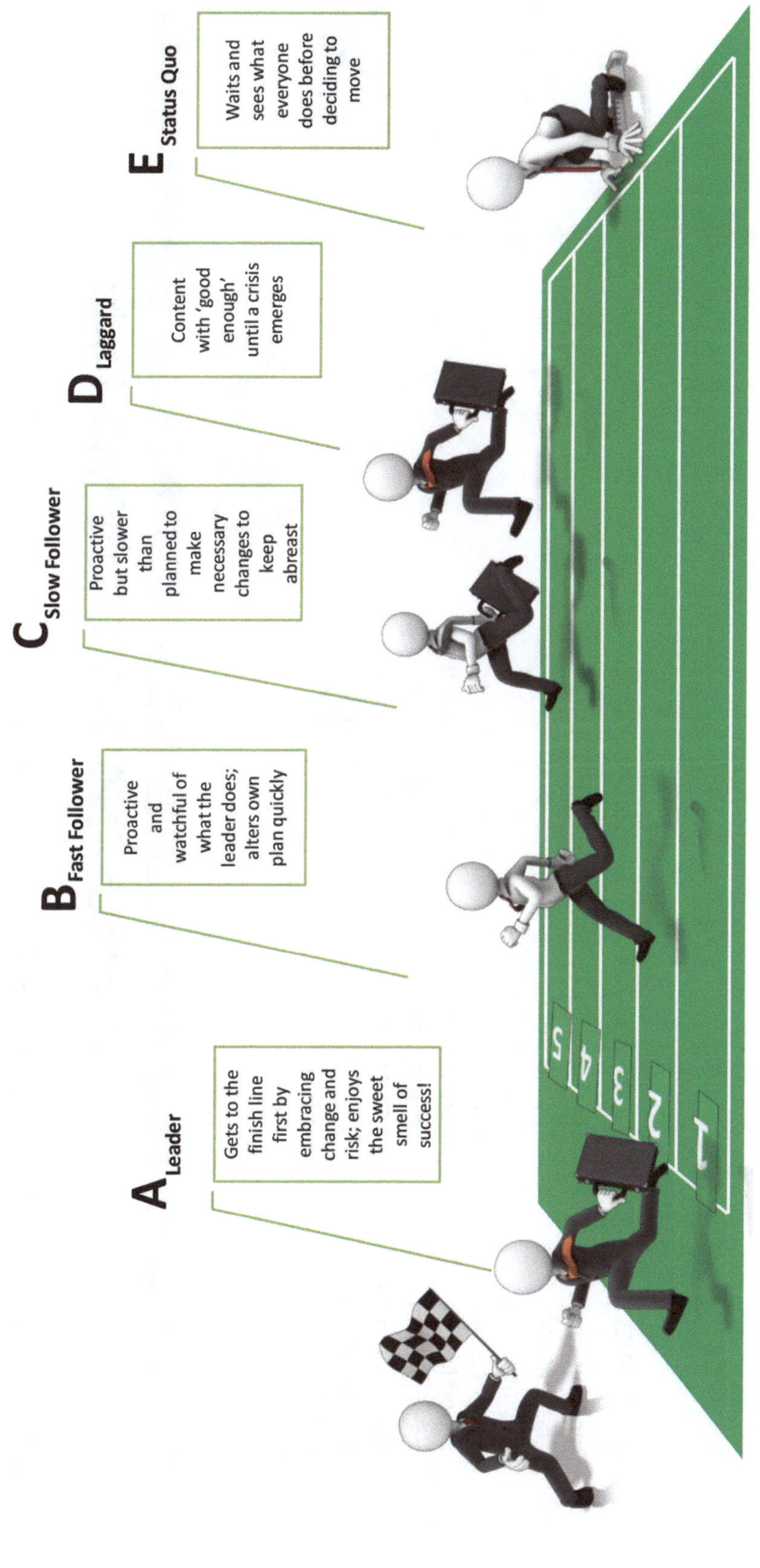

Figure 9.5 Competitive positioning in the EPPM competency maturity race.

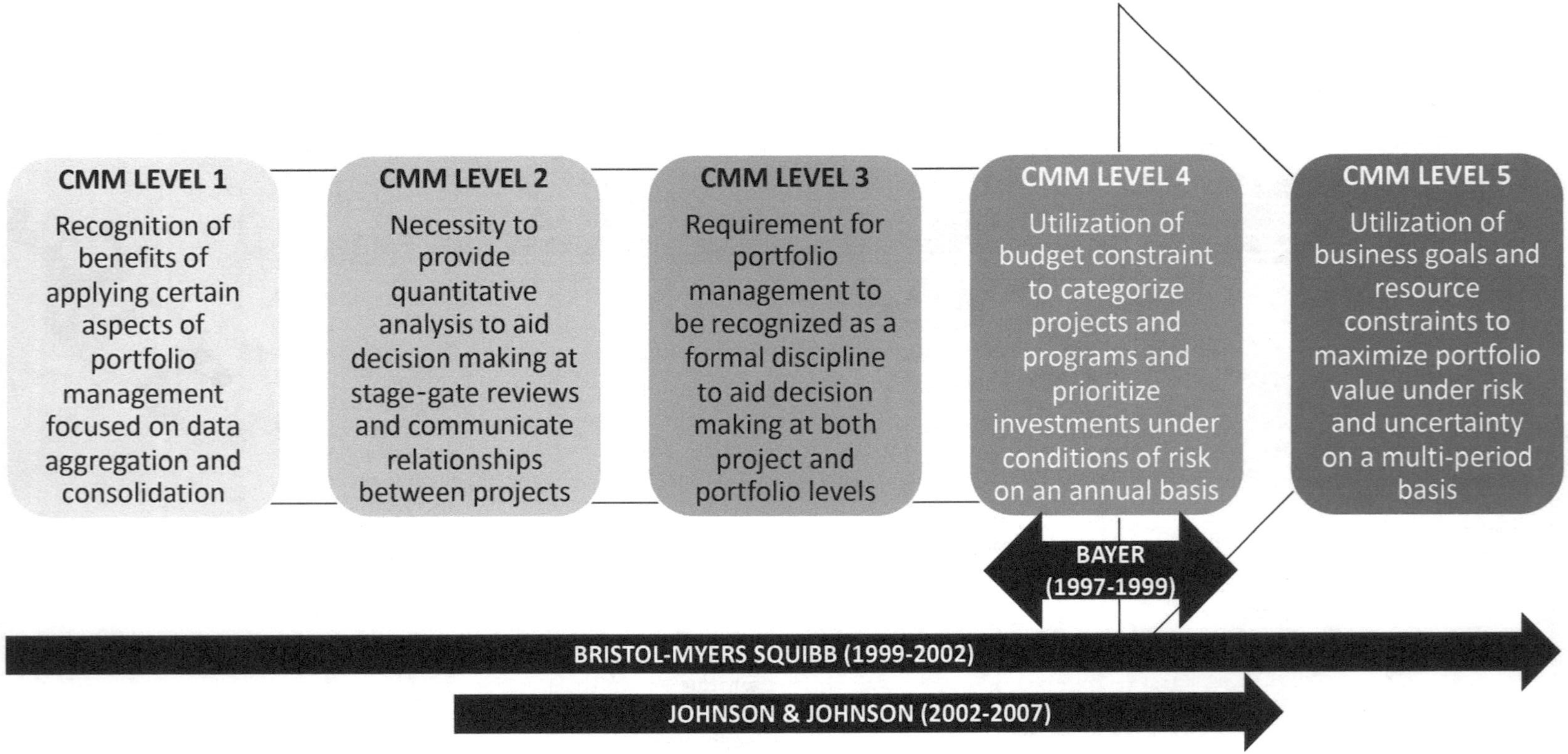

Figure 9.6 DA&PM capability maturation—Bayer, BMS, and Johnson & Johnson.

9.2.3.1 *Bayer (1997–1999): Regionally Decentralized DA&PM*

During the late 1990s, Bayer operated a regionally decentralized DA&PM model, with dedicated functions located in Germany, the United States (with separate functions for pharmaceuticals and biotechnology), and Japan. While each region maintained control over its own budgets and local decision support, overall functional leadership and final decision authority for cross-regional investments resided at corporate headquarters in Germany.

Bayer distinguished itself early as an industry leader with the adoption of formal decision analysis, alongside peers such as Eli Lilly and Pfizer. Beginning in the mid-1980s, the organization made sustained investments in DA capability development, supported by extensive training from the Strategic Decisions Group. As a result, each regional DA&PM function possessed the technical competence required to support stage-gate investment decisions using rigorous analytical methods.

Beyond analytical execution, each regional function assumed responsibility for educating a broad range of stakeholders—project teams, functional leaders, and governance bodies—through internal workshops and training sessions. This emphasis on education proved foundational, enabling contributors from sales forecasting, finance, resource management, and other functions to understand not only the mechanics of DA but also the importance of their inputs to transparent, defensible portfolio evaluations.

By 1997, the biotechnology business unit required dedicated leadership to sustain and further develop its DA capability. At that time, the corporate function was staffed with four senior leaders and four direct reports covering eight therapeutic areas (see Figure 9.7). Given the comparatively smaller size of the biotechnology portfolio in the United States, that function operated with one senior leader and two analysts. Each regional team worked closely with local counterparts in finance, project management, resource management, sales forecasting and market research, and licensing and business development. Importantly, finance leaders across regions maintained strong coordination with headquarters to support both regional and global budget allocation decisions.

While regional functions were empowered to recommend the continuation or termination of locally funded projects, final decisions on assets involving shared regional investment were made centrally. This governance structure, while ensuring consistency, often introduced delays and generated frustration at the regional level. Nevertheless, through sustained internal education, close collaboration between regional teams—particularly between biotechnology and pharma in the United States—and strict adherence to common processes and methodologies, Bayer successfully maintained a Level 4 CMM maturity during the 1997–1999 period (see Figure 9.8). Despite geographic dispersion and language barriers, DA&PM at Bayer functioned as an integrated, highly disciplined system characterized by standardized tools, templates, and models applied consistently across the global organization.

9.2.3.2 *BMS (1999–2002): Centralized DA&PM*

Long before the late 1990s, BMS was already a highly successful biopharma company. However, leadership recognized that critical R&D investment decisions were often protracted, contentious, and insufficiently documented, leaving little institutional memory to inform future choices. In response to a clear mandate from the head of R&D, the organization embarked on a deliberate effort to build a centralized DA&PM function.

The primary objective was to improve decision quality and efficiency at the most capital-intensive and risk-laden point in development—clinical proof-of-concept, previously referred to as the *valley of clinical death*. Success at this juncture would justify extending the function's remit to late-stage clinical development and, ultimately, to preclinical portfolios.

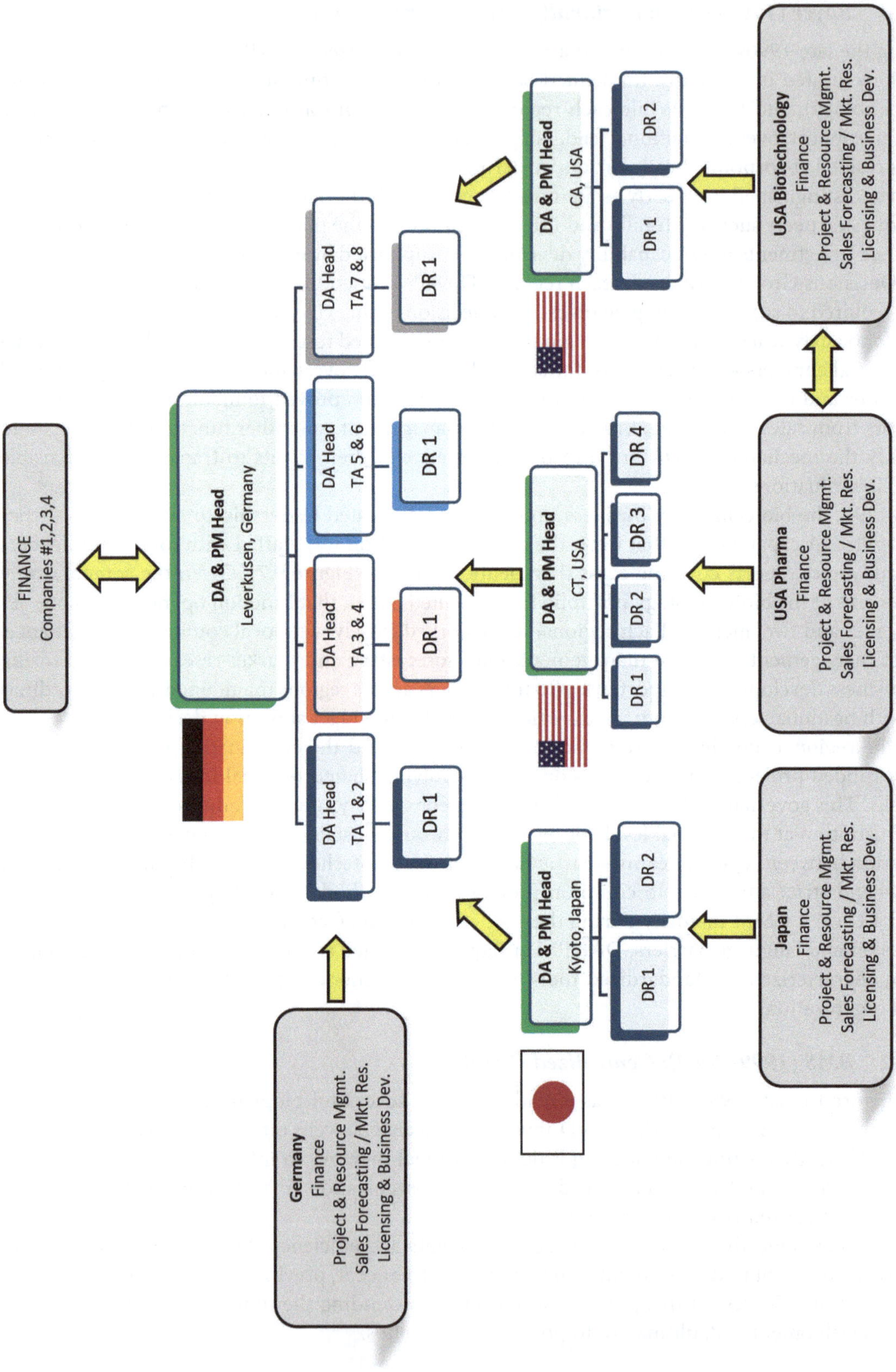

Figure 9.7 Bayer—decentralized, regional DA&PM functions.

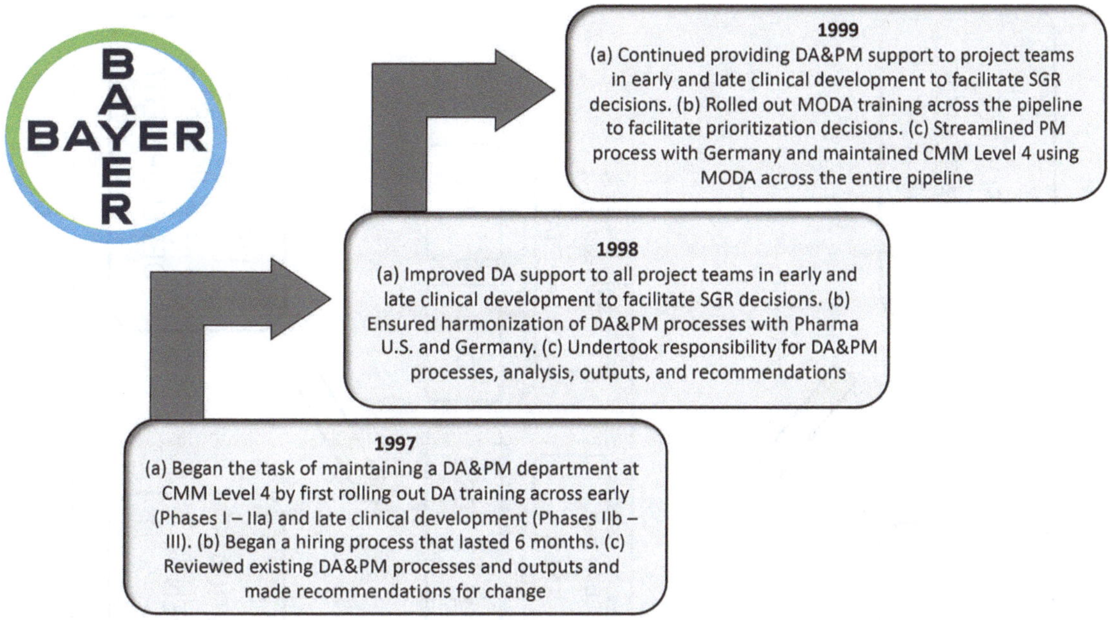

Figure 9.8 Maintenance of DA&PM capabilities at Bayer (1997–1999). SGR = stage-gate review.

Over the first 18 months, an extensive recruitment effort resulted in the formation of a sizable team comprising four internal hires and eight external recruits (see Figure 9.9). Given the scarcity of formally trained decision analysts, the hiring strategy emphasized analytical aptitude over functional pedigree. Consequently, the team included individuals with backgrounds in clinical research, clinical development, finance, sales forecasting and market research, and licensing and business development. This diversity of perspectives proved invaluable, as complex development decisions were examined through multiple analytical lenses, fostering deeper insight and more robust debate.

The function was structured around four senior DA leaders responsible for five therapeutic areas, with one also serving as head of portfolio management. The centralized organizational model facilitated seamless communication with other core functions, enabling the establishment of a centralized data repository—*a single source of truth*—from which all portfolio analyses were derived.

During its first year, the DA&PM function focused on rationalizing existing methodologies, harmonizing evaluation practices, and standardizing decision outputs. Concurrently, an intensive education program was rolled out targeting three critical stakeholder groups: asset teams, enabling functions (e.g., finance, resource management, sales forecasting), and senior governance bodies. This broad-based engagement was instrumental in embedding DA as an accepted and valued component of R&D governance.

By the end of year 1, the function had delivered several tangible benefits (see Figure 9.10). These included standardized evaluation processes; adoption of a limited and consistent set of financial decision metrics: expected net present value (eNPV), expected productivity index (ePI), expected internal rate of return (eIRR); improved strategic optionality; and a dramatic reduction in portfolio review duration—from two days prior to 1999 to four hours by 2001.

Building on this success, the function expanded its remit in year 2 to include late-stage clinical development and, subsequently, preclinical portfolios. The introduction of portfolio optimization as a superior alternative to project prioritization was readily accepted by executive leadership,

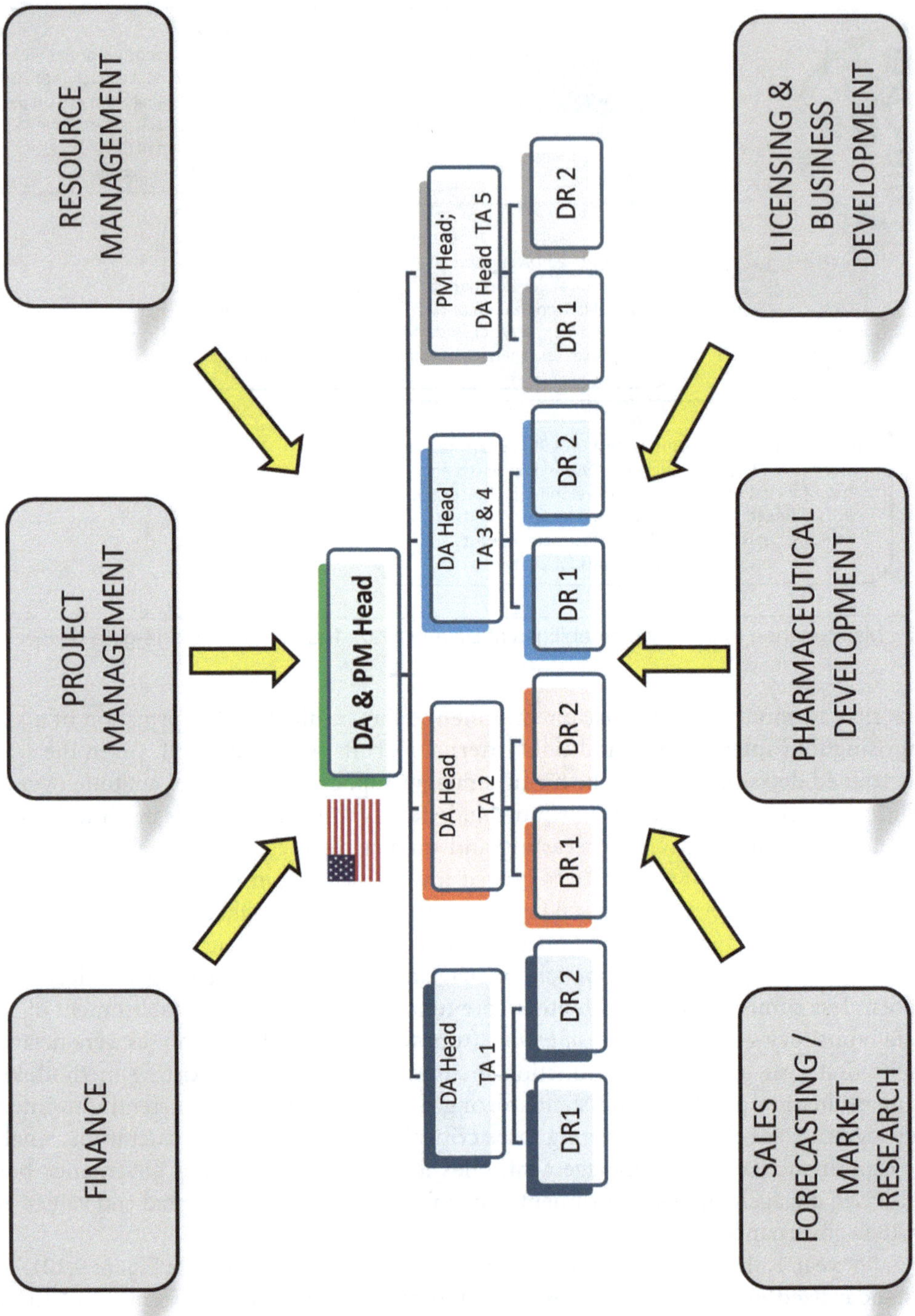

Figure 9.9 BMS—centralized DA&PM function.

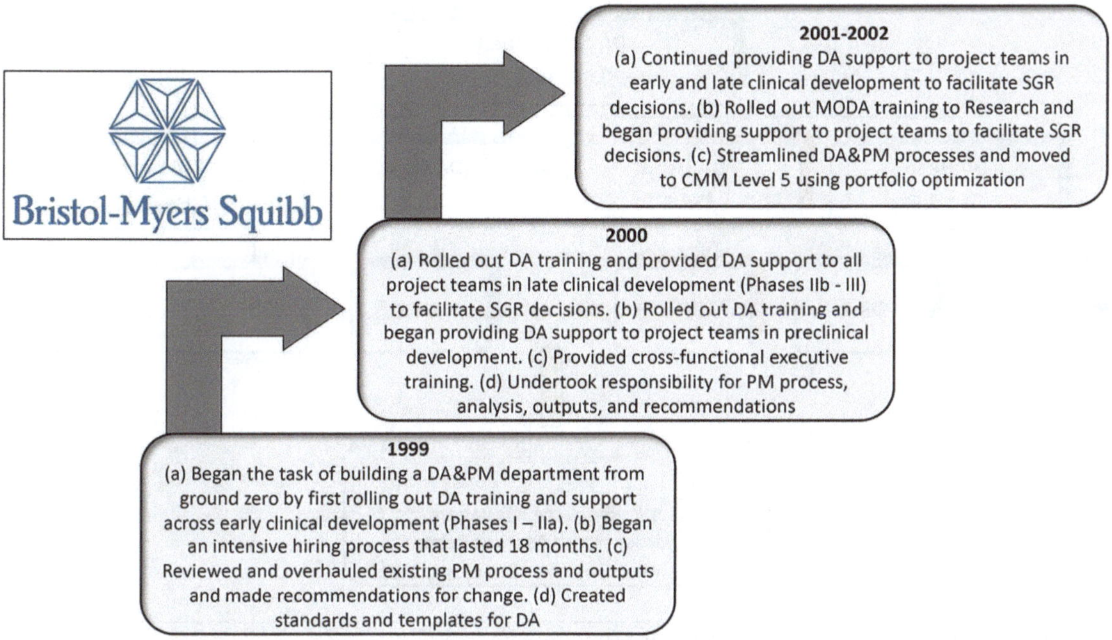

Figure 9.10 Building DA&PM capabilities at BMS (1999–2002).

culminating in a clear mandate at the end of year 2: "Show us the optimal portfolio and help us decide."

By year 3, and consistent with CMM criteria—including deterministic portfolio optimization—BMS's DA&PM function was assessed to have achieved Level 5 maturity. While extensive training and engagement were essential contributors, one factor was decisive: visible, sustained executive sponsorship at the highest level of R&D leadership. Without this commitment, the function would likely have been relegated to an advisory role with limited strategic impact.

9.2.3.3 Johnson & Johnson (2002–2007): Decentralized DA&PM by Operating Company

During the early 2000s, Johnson & Johnson's pharmaceutical business operated through five largely autonomous operating companies: Alza, Centocor, PRD, Scios, and Tibotec. Each entity maintained independent budgets and decision authority. DA&PM capabilities varied widely, ranging from modest dedicated functions to reliance on finance-led evaluation processes.

In 2002, PRD—the largest operating company—initiated the development of a formal DA&PM function. Within 12 months, a robust organization was established, comprising four senior DA leaders (one serving concurrently as head of portfolio management) and 10 analysts responsible for five therapeutic areas (see Figure 9.11). Strong linkages were established with finance, project management, and resource management, supported by a centralized data repository, although occasional data synchronization issues persisted.

Drawing on prior experience, PRD rapidly advanced to a higher maturity. A key early objective was to assess and selectively adopt best practices from across the five operating companies, followed by a multi-year effort to harmonize processes, methodologies, and standards (see Figure 9.12). Resistance from more established functions was not unexpected, but sustained education and engagement eventually enabled alignment.

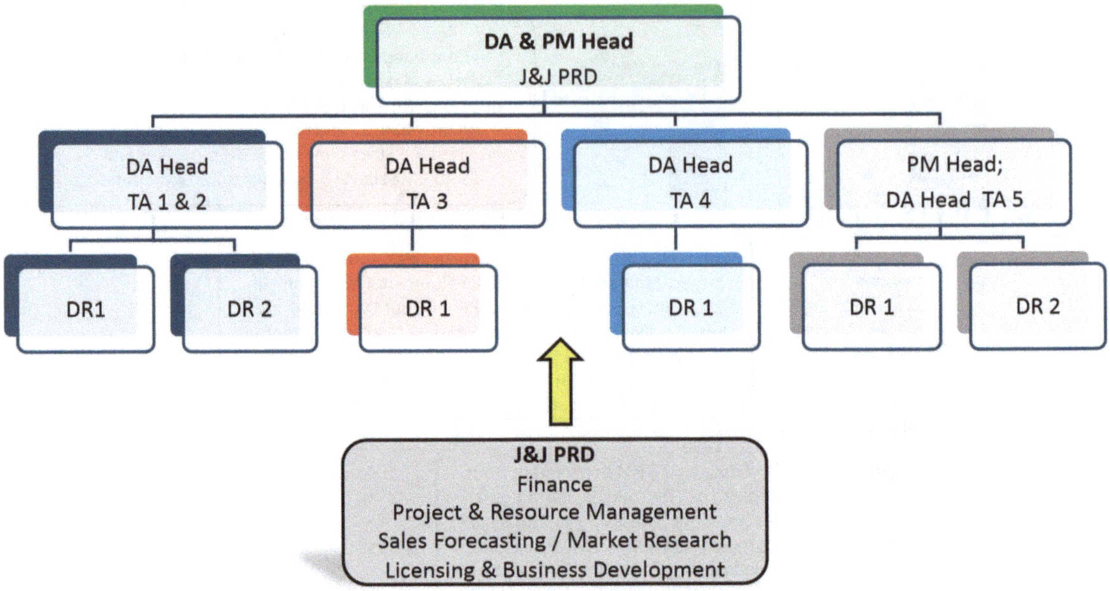

Figure 9.11 Johnson & Johnson PRD—DA&PM function.

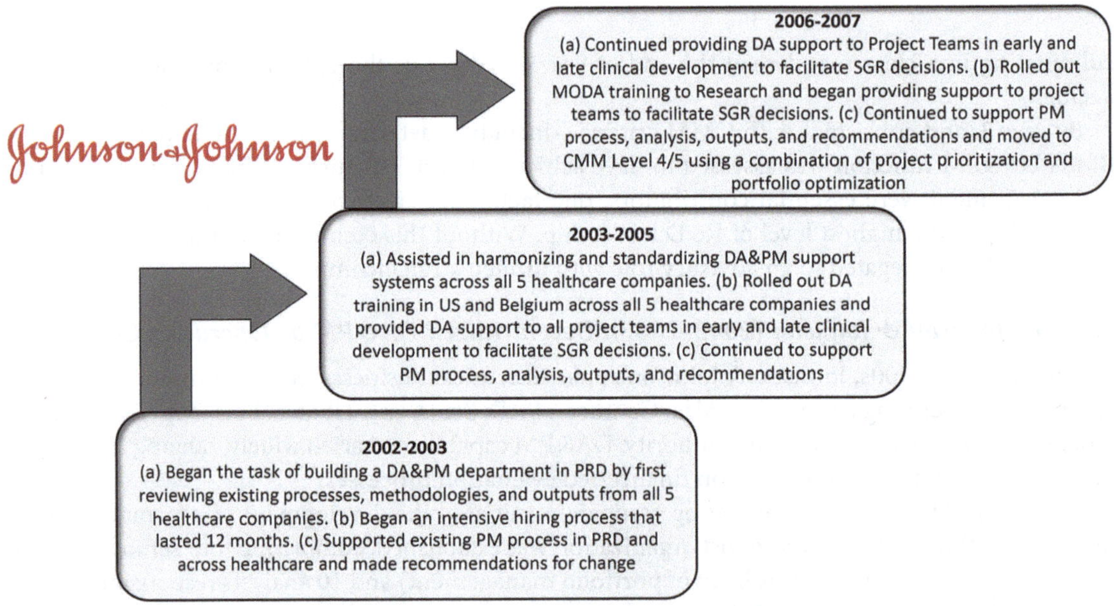

Figure 9.12 Building and harmonizing DA&PM capabilities at Johnson & Johnson (2002–2007).

By 2006–2007, PRD had implemented Multiple Objective Decision Analysis-based prioritization for research portfolios and adopted deterministic portfolio optimization for development-stage portfolio selection. However, other operating companies did not embrace optimization approaches. As a result, PRD gradually reverted to prioritization-based selection in 2007, leading to a slight regression in assessed maturity—from Level 5 to a position between Levels 4 and 5.

Collectively, these case studies illustrate that while technical capability and analytical rigor are essential, sustained maturity in DA&PM ultimately depends on organizational structure, executive sponsorship, cultural readiness, and disciplined integration across functions.

9.3 ROLES AND RESPONSIBILITIES IN DA&PM

An effective DA&PM function—whether centralized or decentralized—depends on close operational coordination with project management, resource management, sales forecasting/market research, and finance. When these functions are aligned, information flows consistently and transparently, enabling defensible asset- and portfolio-level decisions.

Achieving this coordination requires explicit resolution of several foundational questions:

1. Who produces required inputs?
2. Who is accountable for their integrity and timeliness?
3. Who delivers information to whom?
4. When and where does accountability ultimately reside for decisions?

When these issues are unresolved, organizations experience data ownership conflicts, multiple versions of assumptions, and inconsistent asset valuations, eroding governance confidence in analytic outputs.

Lacking clear, shared answers to these questions, organizations frequently experience so-called *data and information ownership* issues—often a symptom of entrenched functional silos or fiefdoms. In such environments, data are selectively shared, competing versions of assumptions proliferate, and multiple *sets of books* emerge. The downstream consequences are predictable: inconsistent asset valuations, conflicting portfolio narratives, and diminished confidence among governance bodies in the analytic outputs presented to them.

RACI as a Mechanism to Eliminate Ambiguity

A proven approach to reducing this dysfunction is the systematic application of a RACI framework. By clearly defining who is ***responsible*** and ***accountable*** for each data element and analysis—before specifying who must be ***consulted*** or ***informed***—organizations establish clarity around data provenance, validation, and reporting expectations (see WAV Table 9.6). The objective of RACI is behavioral alignment, not bureaucracy: it enables DA&PM to function as an integrator and steward of decision quality rather than a competing analytical authority. Several principles are evident in this illustrative RACI:

- The DA&PM function is typically *accountable* for integration, synthesis, and analytic coherence, but not for generating most primary data
- Functional experts (e.g., project management, sales forecasting, finance) retain *accountability* for the accuracy of their respective inputs
- Governance bodies remain *accountable* for final decisions, reinforcing the separation between analysis and decision rights

When reinforced through training and leadership sponsorship, a RACI framework reduces inter-functional friction, accelerates decision cycles, and improves the credibility of portfolio recommendations.

9.4 PORTFOLIO ANALYSIS AND PRESENTATION

In mature organizations, portfolio analysis is an integrated management discipline supporting strategy execution, resource allocation, and risk management. It typically serves three interrelated imperatives:

1. Formal annual or semi-annual portfolio review to assess viability, balance, and strategic alignment
2. Marginal portfolio analysis to guide resource reallocation as projects progress, terminate, or initiate
3. Portfolio impact analysis linked to stage-gate decisions, assessing enterprise-level consequences of individual project outcomes

Once an appropriate decision frame is established, core portfolio management responsibilities include validated data collection, integrated potential and risk-adjusted evaluation, forward-looking assessment of strategic and financial objectives, quantification of value gaps, and translation of results into actionable recommendations. Without an appropriate decision frame, the chasm between ineffective and effective portfolio analysis and decision making can become pronounced (see WAV Figure 9.13).

No portfolio analysis should begin or end without explicit articulation of strategic intent. A clear definition of objectives provides a principled basis for prioritization and resource allocation and mitigates resistance to difficult trade-offs. Organizations that avoid defining intent in the name of *opportunism* often struggle to execute asymmetric investment decisions.

A practical entry point for portfolio review is a SWOT assessment (strengths, weaknesses, opportunities, and threats) (see WAV Figure 9.14), which integrates internal capabilities and external forces to contextualize portfolio recommendations within broader strategic realities.

According to Tufte (2001), portfolio information is commonly communicated through quantitative displays, relational diagrams, reference tables, and conceptual frameworks. Cooper, Edgett, and Kleinschmidt (2001) document commonly used portfolio graphics, dominated by risk–reward plots where reward is expressed via net present value (NPV), internal rate of return (IRR), return on investment (ROI), or peak sales, and risk via probabilities of technical or commercial success (see Table 9.7).

Pipeline throughput modeling quantifies the number of projects required at each development stage to support one or more launches, based on assumed success probabilities (see Chapter 5). Extending this analysis to multiple launches allows organizations to compare required pipeline depth with annual launch targets and identify gaps between expected and risk-adjusted outcomes.

Launch profiles, as described in Chapter 5, provide a time-based view of potential and risk-adjusted product launches across the planning horizon. Coupling launch timing with estimates of the probability of technical and regulatory success highlights risk concentration and informs partnering, licensing, and acquisition strategies.

Simulation techniques further enhance analysis by estimating probabilities of achieving launch and growth targets over time. Risk-adjusted growth trajectories and probability-based assessments of exceeding or missing targets provide a more realistic view of portfolio viability than deterministic projections alone.

Strategic alternatives can be compared probabilistically across value creation, downside risk, and cost exposure (see Figure 9.15), recognizing that no single portfolio typically dominates across all dimensions. Rarely does a single portfolio dominate across all dimensions, necessitating informed trade-offs. In this graphic, portfolio A has the highest probability of exceeding $20B in value and the

Table 9.7 Graphical displays for portfolio analysis (Cooper et al. 2001).

RANK	TYPE	AXIS 1	AXIS 2	USE
1	RISK & REWARD	NPV, IRR, ROI, PEAK SALES	PROBABILITY OF TECHNICAL OR COMMERCIAL SUCCESS	+++++
2	INNOVATION	TECHNICAL INNOVATION	MARKET INNOVATION	++
3	EASE & ATTRACTIVENESS	TECHNICAL FEASIBILITY	MARKET ATTRACTIVENESS	++
4	STRENGTH & ATTRACTIVENESS	COMPETITIVE POSITION	MARKET ATTRACTIVENESS	++
5	COST & TIMING	COST TO IMPLEMENT OR LAUNCH	TIME TO IMPACT	+
6	STRATEGIC & BENEFIT	STRATEGIC FOCUS OR FIT	NPV, IRR, ROI, BUSINESS INTENT	+
7	COST & BENEFIT	CUMULATIVE INVESTMENT COSTS	CUMULATIVE REWARD	+

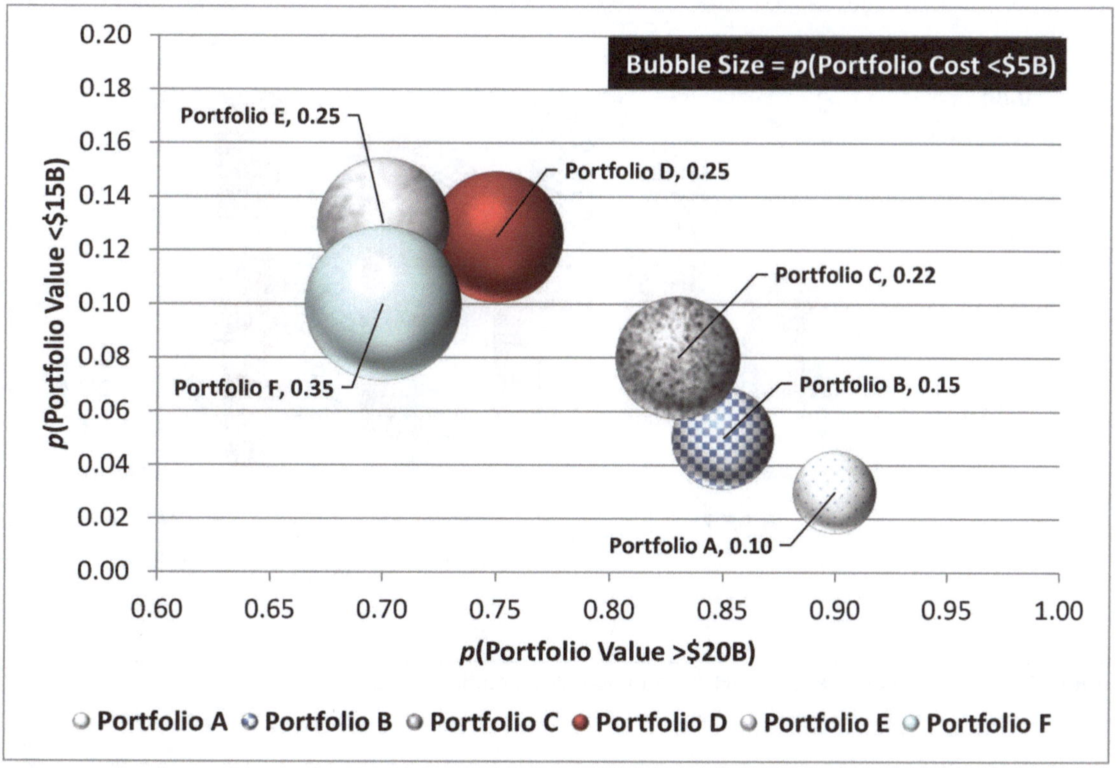

Figure 9.15 Analysis of portfolio strategies along three risk dimensions.

lowest probability of falling below $15B. Portfolio A also has the lowest chance of exceeding $5B in cumulative cost. On the other hand, portfolio D has a 75% likelihood of exceeding $20B in value, the highest likelihood (13%) of yielding less than $15B in value, and the second highest probability (25%) of exceeding $5B in cumulative cost.

Full outcome distributions derived from simulation are visualized using cumulative distribution functions (see WAV Figure 9.16), enabling evaluation of upside potential relative to downside thresholds and supporting informed trade-offs when enhanced upside is accompanied by increased risk. Assuming an upper corporate target of exceeding $8.5B and a lower threshold of not delivering less than $7.0B, which of the three portfolio strategies is best? Portfolio A has a range of outcomes between $6.5B and $9.25B, with a 5.3% probability of exceeding the $8.5B target. The corresponding range of values for portfolio B is between $5.0B and $10.0B, with a 26.6% likelihood of exceeding the target. For portfolio C, the range is between $2.5B and $12.5B, with 45.2% likelihood of exceeding the $8.5B target. However, while portfolio C has the highest likelihood of exceeding the target, it also has the highest likelihood (27.9%) of falling below the lower threshold of $7.0B. By comparison, portfolios A and B are 11.7% and 20.4% likely to fall below the $7.0B threshold.

Trend analyses of portfolio value by business unit (BU) or therapeutic area reveal the sustainability of value creation and highlight areas requiring incremental investment (see Figure 9.17).

Finally, portfolio waterfall analyses provide a longitudinal view of value added and lost over time, attributing changes in risk-adjusted NPV to launches, terminations, transitions, partnerships, and strategic interventions (see Figure 9.18). This type of analysis can serve as a direct indicator of the value contributed by a DA&PM discipline itself.

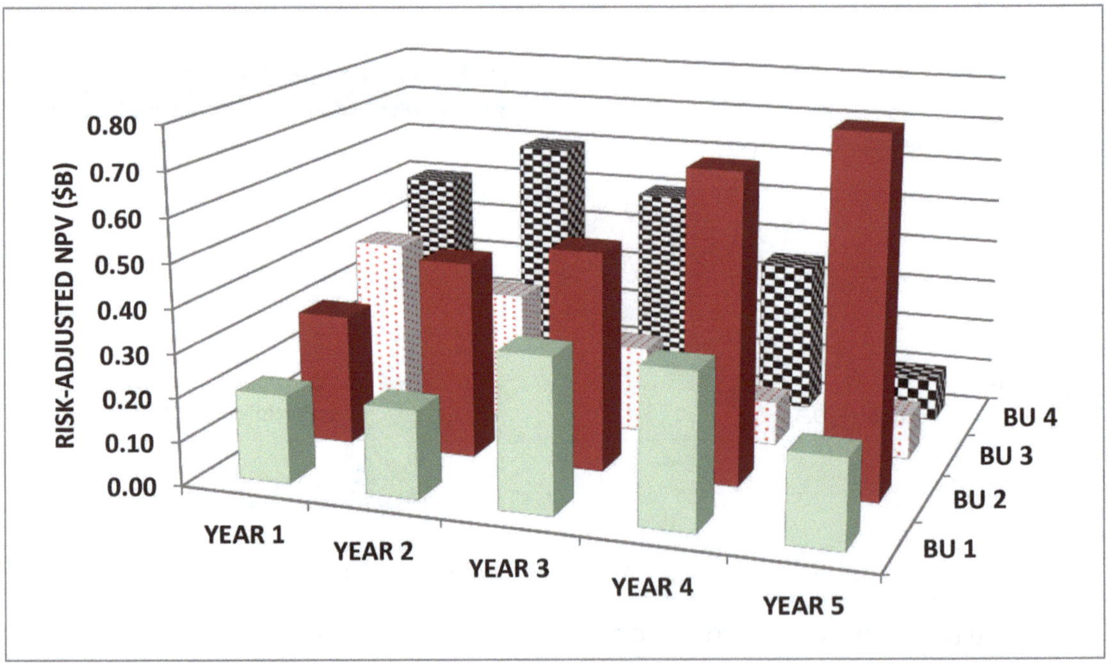

Figure 9.17 Annual business unit (BU) portfolio value creation trend analysis.

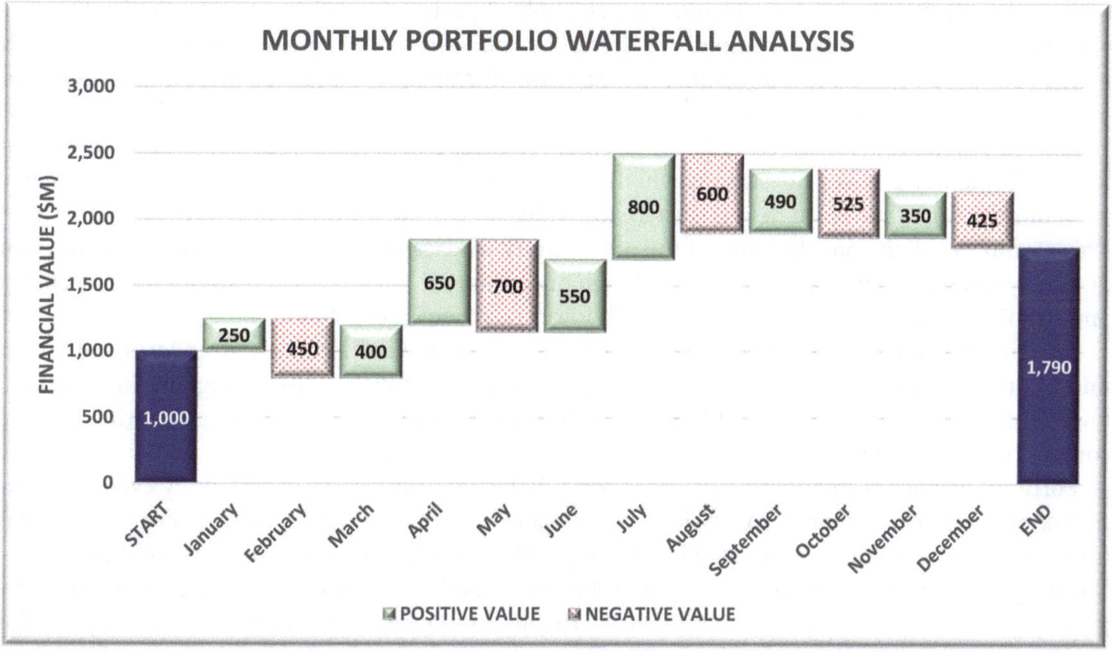

Figure 9.18 Portfolio waterfall analysis.

9.5 RISK MANAGEMENT

In biopharma enterprises, risk management is often misconstrued as the ability to control or eliminate uncertainty. A more precise and analytically useful framing distinguishes controllable operational risks from uncontrollable states of nature and separates the probability of an event from the magnitude of its consequences.

Operational risks—such as schedule adherence, cost control, and execution efficiency—are largely amenable to managerial intervention through resourcing, sourcing strategies, or process optimization. By contrast, risks intrinsic to biomedical innovation—particularly drug efficacy and safety—are fundamentally uncertain and cannot be directly controlled. These risks resolve only as development progresses, independent of organizational intent or investment levels. For such uncertainties, management focuses on anticipating and mitigating downstream impact rather than preventing occurrence.

As established in Chapter 5, risk is defined by the probability of an event, while impact (or consequence) reflects the range and severity of possible outcomes. Conflating probability and impact degrades decision quality and distorts resource allocation. Accordingly, the terminology of *positive* and *negative* risk is avoided; risks are value-neutral, while their realized consequences may be favorable or unfavorable and often asymmetric across value dimensions.

The development history of sildenafil illustrates this distinction. Although the compound failed to demonstrate sufficient efficacy for its original cardiovascular indications—an anticipated development risk—it later achieved commercial success through recognition of an unanticipated pharmacological effect in the treatment of erectile dysfunction. This outcome was not the result of a *positive risk*, but of effective exploitation of impact following resolution of uncertainty.

For these reasons, the commonly used notion of *risk impact* is insufficient at the portfolio level. Enterprise portfolio risk management must evaluate impact in strategic, financial, and temporal terms rather than at the level of isolated projects. Robust project-level risk management is necessary but not sufficient; portfolio decisions are driven by the objective of maximizing enterprise value, which may require reallocating resources away from technically successful initiatives if doing so improves portfolio-level outcomes.

From a portfolio perspective, risk management is inseparable from strategic prioritization and temporal balance. The key question is whether the current configuration of assets and investments positions the organization to deliver sustainable value across short-, mid-, and long-term horizons. This requires explicit consideration of the opportunity landscape, ranging from early-stage internal discovery programs with long horizons and high uncertainty to late-stage acquisitions or in-licensing opportunities offering nearer-term value with lower risk (see Figure 9.19).

Absent deliberate intervention, portfolios tend to drift toward low-risk, near-term initiatives, undermining future growth. Maintaining temporal balance often requires foregoing incremental, lower-risk opportunities in favor of higher-risk innovation programs capable of generating durable competitive advantage.

Portfolio balance is frequently misunderstood. Balance is not achieved simply by even distribution of projects or resources across risk–value categories, nor by assigning uniform growth expectations to all segments. Such symmetry is arbitrary without explicit strategic and temporal objectives. Meaningful balance is context-dependent and must be evaluated relative to enterprise goals, competitive positioning, and capital constraints. A critical discipline in this process is marginal portfolio decision making in which incremental or freed resources are redeployed to maximize portfolio-level outcomes.

Portfolio risk can be further examined by assessing the relative contributions of BUs to value creation and resource utilization, particularly when mid-term viability is a concern (see Figure 9.20).

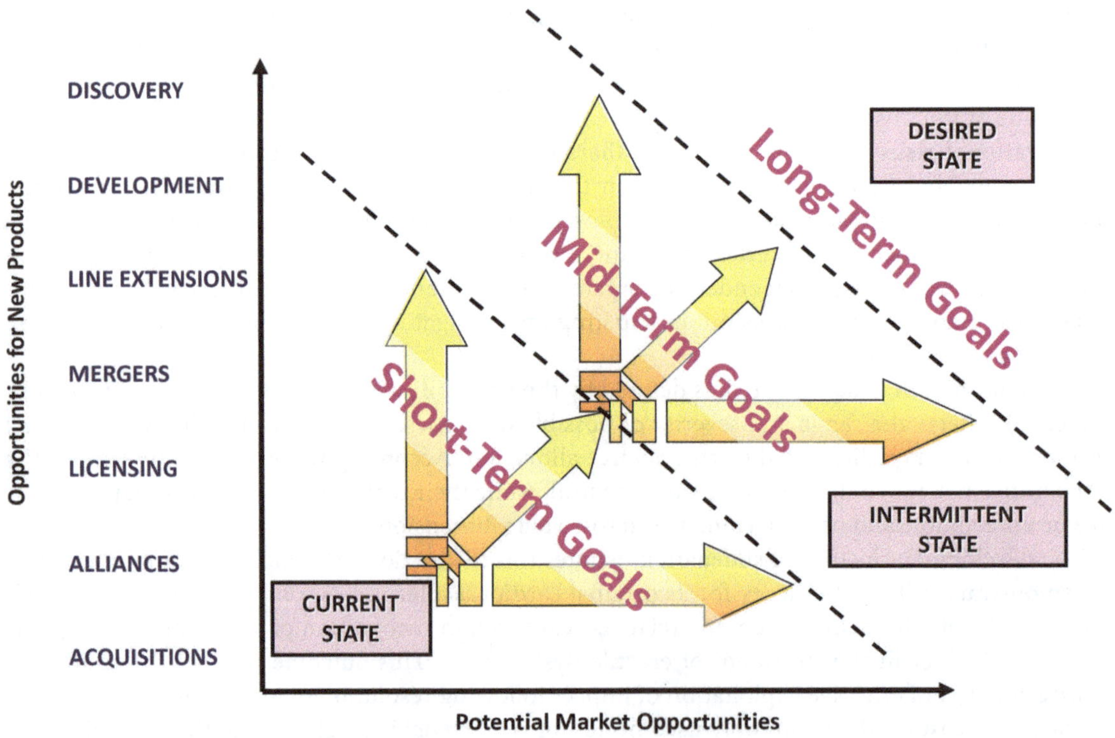

Figure 9.19 Temporal pursuit of portfolio goals.

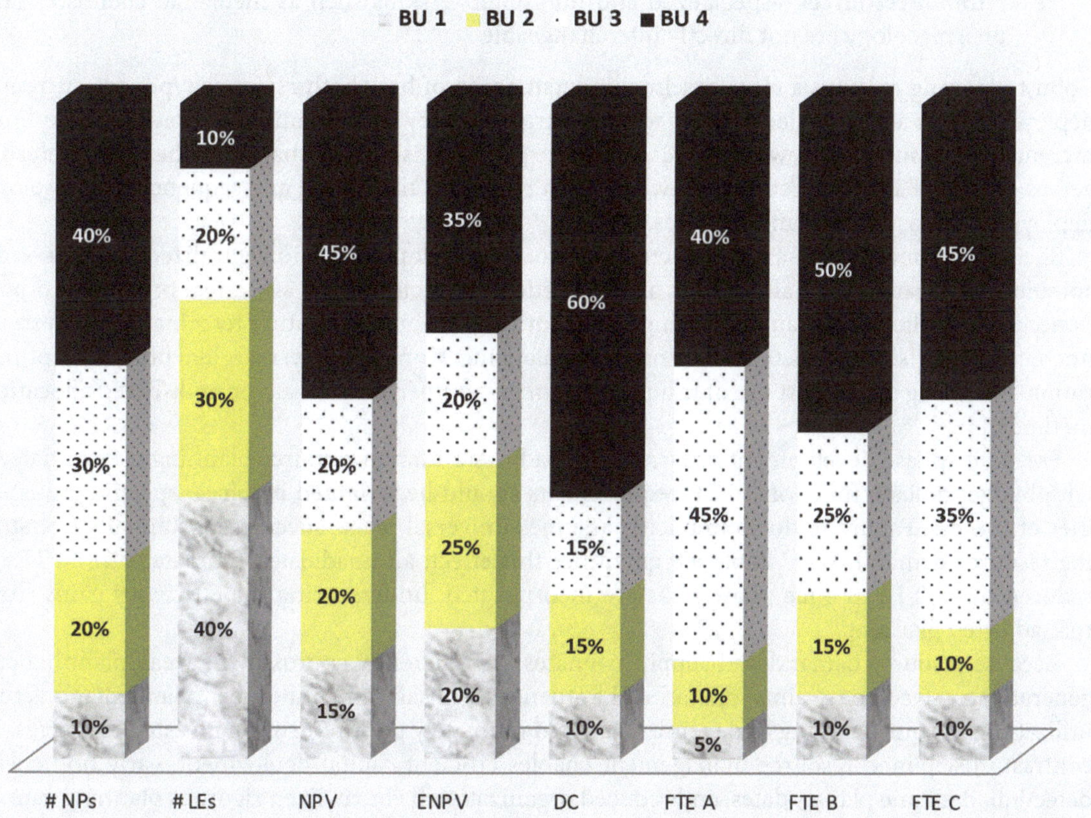

Figure 9.20 Contribution of BUs to value creation and resource utilization metrics.

Metrics typically include new products, line extensions, NPV, eNPV, direct costs, and functional full-time equivalents (FTEs).

Such analyses reveal implicit risk-return profiles. A BU generating substantial NPV but consuming disproportionate resources while contributing relatively little to eNPV may carry an unfavorable risk load. Conversely, a BU that delivers balanced contributions to NPV and eNPV with moderate resource consumption reflects a more efficient, strategically aligned investment. Portfolio balance can only be meaningfully assessed through these comparative, multidimensional evaluations. Together, these perspectives underscore the core principle of enterprise portfolio risk management: sustainable value creation depends not solely on opportunity size, but on disciplined alignment of risk exposure, resource allocation, and temporal strategic objectives.

9.6 RESOURCE MANAGEMENT

Effective resource management is among the most complex functions in biopharma R&D. Although formal systems track FTE allocations, most organizations rely heavily on judgment-based estimates that are fluid, inconsistent, and frequently revised, limiting the precision of forward-looking capacity forecasts. A central challenge arises from the fundamentally different properties of resources:

1. **Budgetary resources**—fungible, monetary, and readily reallocated across projects

2. **Human resources**—specialized and nonfungible; skills such as medicinal chemistry and pharmacology are not directly interchangeable

Robust planning requires a clear baseline understanding of how both resource types are currently deployed across active projects. Only with this transparency can organizations rationally commit incremental resources to new initiatives. Resource planning is further shaped by the mix of internal versus external FTEs, budget flexibility, retention risks, and the timing and sequencing of resource deployment across development phases.

Shared functions—such as manufacturing, process development, and centralized services—add complexity, as partial FTE allocations must be estimated accurately to avoid real or perceived bottlenecks. Portfolio dynamism, including emergent projects and premature terminations driven by uncontrollable risks, necessitates continuous rebalancing. Consequently, marginal portfolio optimization—focusing on the best use of residual resources—outperforms linear or rank-based allocation methods.

For multi-phase development programs, risk-adjusted human resource planning is essential. As with budget dollars, functional FTE requirements should be weighted by phase-specific probabilities of success. Failure to do so implicitly assumes universal phase success, substantially overstating resource demand. WAV Table 9.8 quantifies this effect: an unadjusted estimate of 287 FTEs is reduced to 76.2 FTEs once phase risks are incorporated, underscoring the efficiency gains from risk-adjusted planning.

Because resource demand and supply estimates are inherently approximate, weak planning can generate perceived constraints, operational bottlenecks, misalignment between planned and actual utilization, opaque handling of dependencies, and ultimately poor returns on invested resources. In contrast, disciplined resource management enables efficient capital deployment, early bottleneck detection, dynamic plan updates, and reduced organizational churn. Even rigorous planning cannot eliminate uncertainty. High success rates may strain capacity through continuous reprioritization, while excessive failures can leave unspent budgets due to carryover restrictions. External acquisitions may further displace internal programs, highlighting the limits imposed by nonfungible human resources.

Balanced resource management, therefore, requires an integrated view of financial and human capital. Clear functional definitions, explicit budget limits, and formal optimization approaches (see Chapter 7) are required to align resource allocation with portfolio value maximization.

As described in Chapter 7, optimization frameworks allow portfolio value to be evaluated as a function of incremental FTE availability. Table 9.9 shows how varying functional FTE constraints alter portfolio composition and eNPV. At baseline constraints of 7.0 FTEs for Function A, 8.0 for Function B, and a $1,500M budget in Year X, seven of ten candidate projects are selected, yielding an eNPV of $3,695M. Relaxing FTE constraints increases portfolio value up to $4,820M, beyond which additional FTEs generate no material gains.

Efficient frontier analysis confirms this result, identifying the optimal portfolio value at 7.8 function A FTEs under fixed function B and budget constraints (see WAV Figure 9.21). Simulation tools extend this analysis by estimating the probability of exceeding critical functional constraints under alternative portfolio strategies. As shown in Figure 9.22, a moderate-risk portfolio has a 79% likelihood of exceeding a 10-FTE threshold, compared with 41% for a low-risk portfolio and 100% for a high-risk portfolio. Depending on organizational risk tolerance, a lower-value portfolio may be preferred if it reduces the likelihood of exceeding critical constraints. Alternatively, high-risk portfolios may require external sourcing of FTEs or tighter budget constraints to fund additional capacity.

Table 9.9 Portfolio optimization at different functional FTE levels.

PROJECT	DECISION VARIABLES	DECISION VARIABLES	DECISION VARIABLES	DECISION VARIABLES	DECISION VARIABLES	DECISION VARIABLES	DECISION VARIABLES
A	0	1	0	1	1	1	1
B	0	0	0	0	1	1	1
C	1	1	1	0	0	0	0
D	1	1	1	1	1	1	1
E	1	1	1	1	1	1	1
F	1	0	1	1	1	1	1
G	0	1	1	1	1	0	0
H	0	0	0	0	0	0	0
I	0	0	1	1	0	1	1
J	1	1	1	1	1	1	1
FTE A CONSTRAINT	7.0	8.0	9.0	10.0	11.0	12.0	13.0
FTE B CONSTRAINT	8.0	9.0	10.0	11.0	12.0	13.0	14.0
BUDGET CONSTRAINT ($M)	1,500	1,500	1,500	1,500	1,500	1,500	1,500
FTE A UTILIZATION	4.0	5.6	7.9	7.7	6.9	7.8	7.8
FTE B UTILIZATION	7.6	8.9	9.9	10.7	11.4	12.7	12.7
BUDGET UTILIZATION ($M)	1,195	1,370	1,480	1,430	1,445	1,480	1,480
eNPV ($M)	3,695	4,245	4,670	4,720	4,745	4,820	4,820

As projects approach phase completion, teams submit forecasts for next-phase resources during stage-gate reviews (see Chapter 1). Under constrained budgets, continuation requests must be prioritized. WAV Table 9.10 and Figure 9.23 illustrate the timing and accumulation of monthly budget requests. In the example shown, $420M is required to fully fund all projects in year X, but a $250M budget constraint forces deferrals. Four projects are nondiscretionary, three of which request funding after the budget ceiling is reached, creating a structural allocation challenge.

Two allocation strategies are possible. A reactive approach funds projects as requests arise, relying on attrition to resolve later shortages. A proactive approach reserves funding for nondiscretionary projects and simulates discretionary investments against the residual budget. As shown in WAV Figure 9.24, this simulation indicates a maximum required investment of $230M, a mode of $120M, and an 81% probability of exceeding the discretionary $70M constraint.

These results frame the core managerial question: whether additional funding (e.g., $50M) can be secured to support the pipeline on a risk-adjusted basis. Absent such flexibility, organizations may need to proactively defer early investments and reallocate resources later in the year as uncertainty resolves.

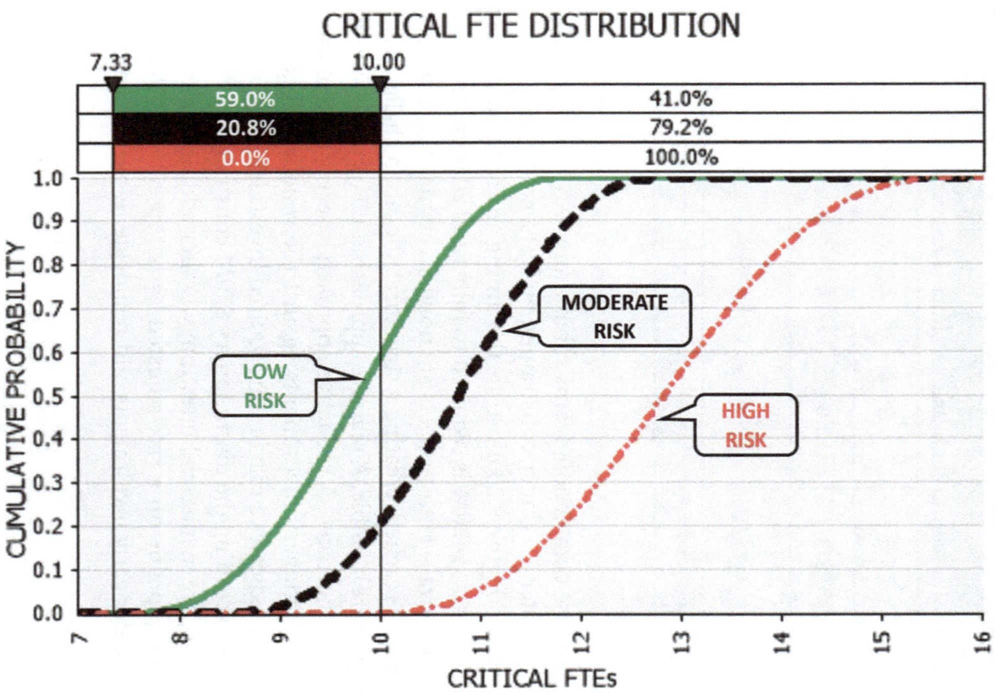

Statistics

	LOW RISK	MODERATE ...	HIGH RISK
Cell	F2	F3	F4
Minimum	7.3619	8.3959	9.8578
Maximum	11.9127	12.8549	15.7522
Mean	9.7633	10.7567	12.8267
90% CI	± 0.00451	± 0.00445	± 0.00588
Mode	9.8647	10.8584	12.8434
Median	9.7791	10.7740	12.8274
Std Dev	0.8679	0.8562	1.1301
Skewness	-0.0768	-0.0857	-0.0030
Kurtosis	2.3412	2.3431	2.3334
Values	100000	100000	100000
Errors	0	0	0
Filtered	0	0	0
Left X	7.33	7.33	7.33
Left P	0.0%	0.0%	0.0%
Right X	10.00	10.00	10.00
Right P	59.0%	20.8%	0.0%

Figure 9.22 Using simulation to estimate the likelihood of exceeding a critical functional FTE constraint.

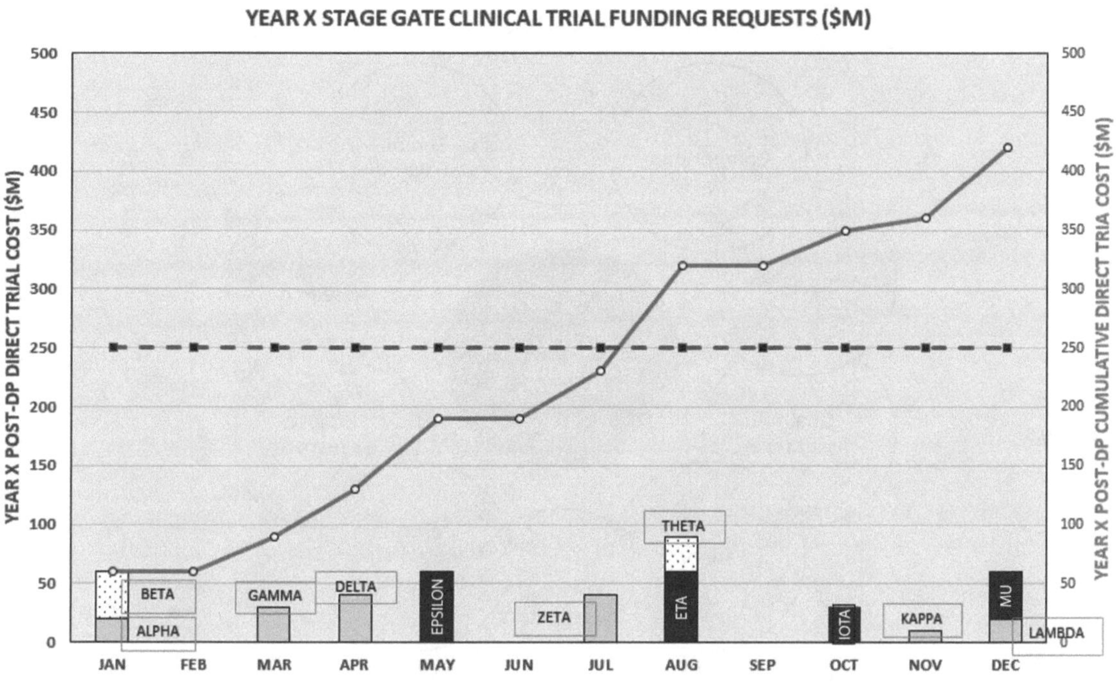

Figure 9.23 Investment inventory—calendarized representation of monthly and cumulative resource requests.

9.7 STAKEHOLDER MANAGEMENT

Clinical trial execution relies on a complex network of internal and external stakeholders with distinct responsibilities, incentives, and decision rights (see Figure 9.25). For portfolio management, understanding these roles is essential to aligning activities with enterprise priorities. Key considerations include stakeholder responsibilities and impact, interests and incentives, conflicting objectives, engagement drivers, information needs, perceptions and influence, and control over critical resources.

While it is commonly termed *stakeholder management*, a more precise characterization is *stakeholder relationship management*, emphasizing the strategic and relational nature of engagement. Effective practice begins with identifying current and anticipated primary stakeholders (see Figure 9.26), defined as any individuals or groups who influence or are affected by portfolio management outcomes.

Once identified, the portfolio management function must clearly articulate its purpose and intended outcomes, including the current status versus the desired state of portfolio management, expected organizational and functional benefits with defined metrics, near-term challenges associated with change, and the long-term value of disciplined portfolio decision making. Early clarity on these elements is critical to securing initial endorsement.

Stakeholder analysis should evaluate needs, expectations, and relative influence, recognizing that political and organizational sensitivities can materially affect success. This work should be guided closely by primary sponsor(s), particularly to identify stakeholders whose authority or domains may be perceived as threatened by effective portfolio discipline. Sustained relationship building is essential because each constructive engagement preserves future support in the event that the value of portfolio management is later questioned.

Figure 9.25 Key stakeholders in clinical trial management.

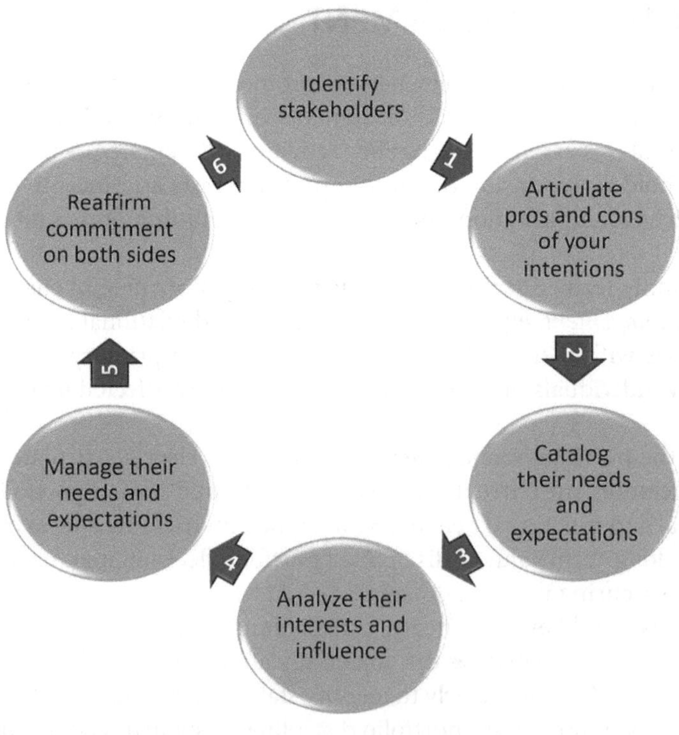

Figure 9.26 Stakeholder management process.

Stakeholder commitment is dynamic and must be reinforced through visible value creation and continuous dialogue. Engagement intensity should be calibrated to interest and influence: low-impact stakeholders should be kept informed, high-impact stakeholders require frequent reaffirmation of alignment and benefit, and intermediate stakeholders benefit from a balanced mix of consultation and communication.

Without a structured stakeholder management approach, portfolio management risks losing organizational traction. Uniform communication strategies are ineffective; excessive detail can overwhelm stakeholders and signal unnecessary complexity. High-performing portfolio functions deliberately migrate stakeholders toward deeper engagement while minimizing informational burden. Resistance to change must be monitored throughout. Highly influential but resistant stakeholders require focused engagement, while less influential but resistant stakeholders should be observed closely due to their potential to shape broader sentiment.

Stakeholder influence evolves across the portfolio life cycle. The steering or executive committee provides direction during initiation, broader stakeholders contribute during execution, and the steering committee synthesizes insights and approves recommendations during decision making. Clear success criteria from the steering committee are critical to prevent portfolio reviews from devolving into non-decision-relevant debate.

Data gathering is the most resource-intensive phase and requires close collaboration with finance and resource management to produce defensible inputs under uncertainty. Stakeholders must understand the value of their contributions to maintain credibility. While analysis is centralized, continued engagement from key functions—especially finance—is essential, with finance endorsement of budget implications required prior to final recommendations. Mature portfolio processes depend on strong collaboration among portfolio management, finance, and the steering committee.

As portfolio management maturity increases, so does the sophistication required for stakeholder engagement (see Figure 9.27). Early maturity levels are project-driven and focus on data aggregation,

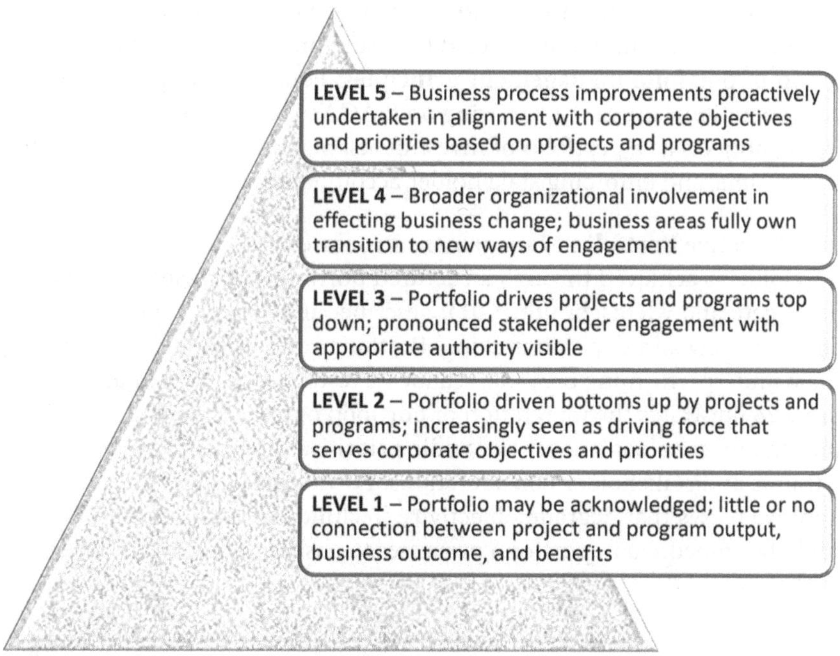

Figure 9.27 Stakeholder management maturity model.

while advanced levels are business-driven, with greater executive sponsorship and shared ownership of outcomes. Continuous assessment of maturity and proactive management of evolving stakeholder expectations are therefore essential to sustaining portfolio value creation.

9.8 A PRACTICAL APPROACH TO IMPROVING ENTERPRISE PORTFOLIO MANAGEMENT COMPETENCIES

Achieving higher competency in EPPM requires deliberate planning, sustained leadership commitment, and a pragmatic implementation strategy. Although the core competencies of portfolio management are well defined, progression toward maturity is rarely linear. Organizations benefit from a phased approach that balances aspiration with operational feasibility:

Step 1: Implement Organizational Change Management to Support the Portfolio Vision
Enterprise-wide adoption of portfolio management frequently encounters resistance rooted in legacy processes and functional silos. A formal organizational change management (OCM) strategy is therefore essential. Key elements include:

1. **Executive sponsorship:** Active endorsement by senior leaders across finance, operations, IT, HR, and BUs is the single most critical success factor. Effective sponsorship accelerates cross-functional alignment and reduces political friction.
2. **Governance and organizational design:** Clear governance structures establish decision rights, accountability, and role clarity. While organizational designs vary, core portfolio roles are consistent across enterprises. A representative allocation of stakeholder responsibilities is shown in WAV Table 9.11 (Merkhofer 2008).
3. **Training and capability development:** Portfolio management competencies must be developed systematically. Training should address decision analysis, portfolio optimization, scenario analysis, data elicitation, and risk management. Broader awareness programs are needed for personnel indirectly affected to ensure understanding of portfolio management.
4. **Adoption of portfolio management principles:** Foundational principles provide conceptual and operational guidance. Drawing on ISACA's Val IT framework (2006), these include managing projects as a portfolio of investments, maximizing value subject to resource and risk constraints, and enforcing stakeholder accountability.

Step 2: Institute Effective Portfolio Management Processes
Operational discipline is achieved through structured portfolio processes. The Stage-Gate™ framework (Cooper, Edgett, and Kleinschmidt 2001), described in Chapter 1, provides a proven foundation for systematic review and decision making. Portfolio reviews should occur multiple times per year to terminate underperforming projects, reallocate resources, and maintain strategic alignment.

WAV Figure 9.28 illustrates the integration of portfolio optimization and prioritization. Optimization identifies the most valuable asset set, while prioritization supports execution by addressing resource constraints and stakeholder considerations. Portfolio decisions are susceptible to motivational and cognitive biases that degrade objectivity. Techniques discussed in Chapters 3 and WAV Table 9.11 should be embedded in portfolio processes to ensure decisions are evidence-based rather than intuition-driven.

Step 3: Address Portfolio Risks Proactively
Risk management must extend beyond individual projects to the portfolio level. Portfolio risks include failure to meet strategic or economic objectives, budget overruns, and over-concentration in specific platforms or products.

A proactive risk culture requires systematic risk identification and categorization, quantification of likelihood and impact, use of decision trees, and incorporation of risk-adjusted metrics into portfolio optimization. Training in risk-adjusted NPV and explicit articulation of organizational risk tolerance are essential.

Step 4: Build Decision-Making Competencies
Objective, transparent decision making differentiates high-performing portfolio organizations. Quantitative and qualitative frameworks such as CREOPM™ and the multiple-objective prioritization process, described in Chapter 8, reduce subjectivity and enhance credibility. When combined with portfolio optimization (see Chapter 7) and disciplined decision framing (see Chapter 3), these approaches strengthen enterprise-level value maximization.

Step 5: Institutionalize Portfolio Management Tools
A portfolio management information system (PMIS) strengthens analytical rigor, data integrity, and reporting efficiency. Core functionality includes decision support, optimization, prioritization, simulation, financial and risk analysis, valuation, resource management, workflow, and integration with financial systems.

Key implementation principles (Merkhofer 2008) include accuracy, logical soundness, completeness, effectiveness in supporting decision timing, and stakeholder acceptability. A phased roadmap for PMIS adoption is shown in Figure 9.29.

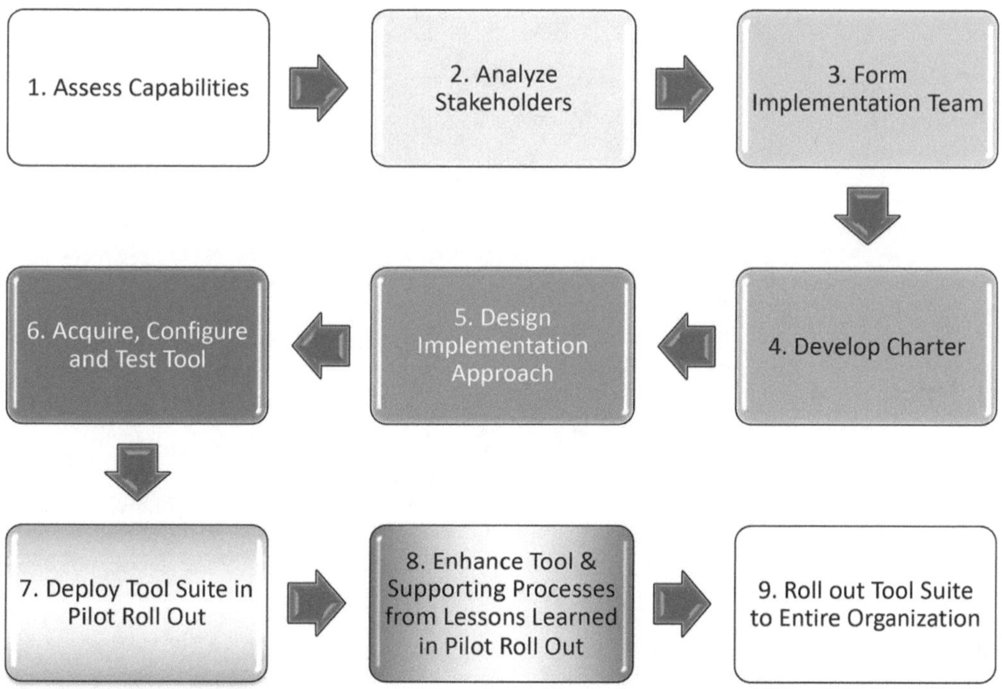

Figure 9.29 Roadmap for implementing a portfolio management information system (Merkhofer 2008).

By systematically addressing organizational change, process discipline, cognitive bias, portfolio risk, decision-making capability, and enabling tools, organizations can materially advance portfolio management maturity—leading to superior investment selection, improved enterprise value creation, and sustainable competitive advantage.

10

EPILOGUE

PORTFOLIO MANAGEMENT IN THE MODERN BIOPHARMA ENTERPRISE

Across virtually every sector, organizations have experienced profound disruption due to globalization, mergers and acquisitions, restructuring, downsizing, cost-optimization programs, outsourcing, and change management initiatives. More recently, the rapid emergence of artificial intelligence (AI) technologies has added a further layer of complexity, requiring portfolio management functions to deliver tangible, sustainable enterprise value under conditions of increasing uncertainty. Large-scale organizations—where economies of scale and scope are critical to maintaining competitive global margins—tend to feel these pressures most acutely.

In the biopharma industry, these dynamics are particularly pronounced. Following decades of robust, often double-digit top-line growth, the combination of intensifying competitive pressures, patent expirations for blockbuster therapeutics, and persistent pricing scrutiny has compelled many organizations to rethink and reshape their operational and strategic models. Drug discovery and preclinical research are increasingly outsourced to specialized entities whose business model is focused exclusively on early-stage development. Clinical development and manufacturing functions are similarly being shifted to contract research organizations (CROs) and contract manufacturing organizations (CMOs), which can deliver superior cost efficiencies and time-to-market advantages that are challenging for traditional internal capabilities to replicate.

It is critical to recognize that operational efficiencies alone—while enhancing productivity per unit of resource—do not intrinsically generate enterprise value. Value creation occurs only when such efficiencies either increase shareholder value or enable additional reinvestment into high-impact research and development (R&D) initiatives. Portfolio management methodologies, when applied with rigor, serve as the mechanism through which these efficiencies translate into strategic value. Among these, the *CREOPM™ Framework* has emerged as a particularly robust approach, capable of maximizing value creation at the enterprise level while minimizing losses associated with suboptimal portfolio decisions.

CREOPM™ should not be misconstrued as a mechanistic algorithm or a deterministic substitute for judgment. Rather, it is an integrated decision framework designed to maximize enterprise value—or, failing that, to minimize value destruction—by making trade-offs explicit, transparent, and defensible (see Figure 10.1). When applied rigorously and consistently, CREOPM™ enables senior leaders to understand not only which portfolio is optimal under a given set of constraints, but also how much value is foregone when organizational realities necessitate the selection of a suboptimal alternative.

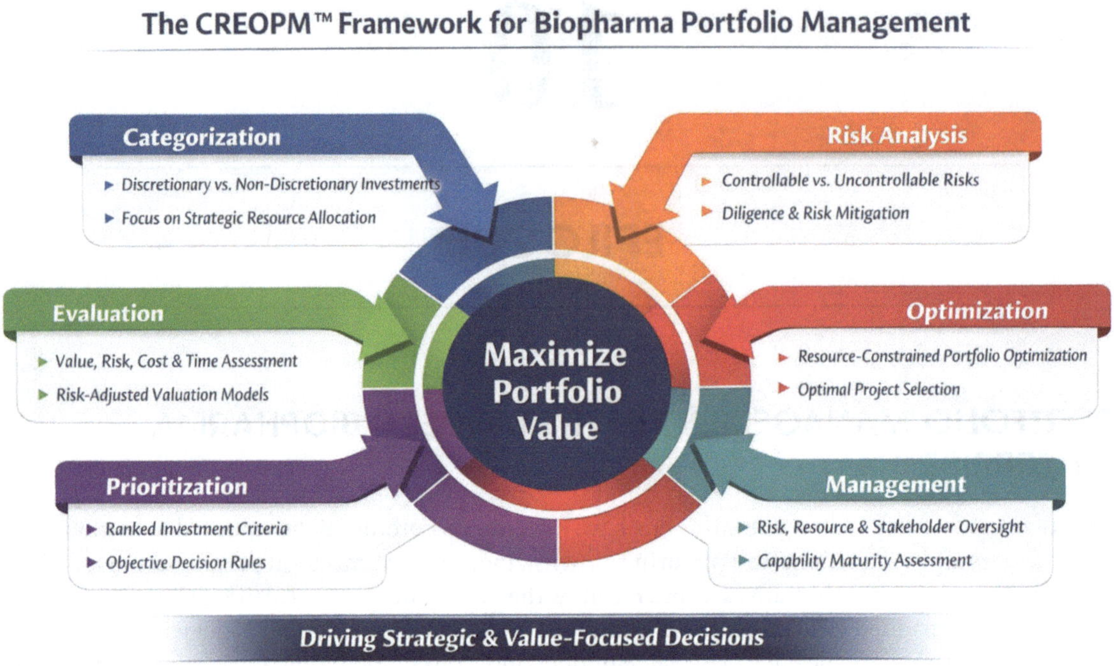

Figure 10.1 CREOPM™: Six pillars of value-driven portfolio management.

At its core, CREOPM™ comprises six tightly coupled facets: **c**ategorization, **r**isk analysis, **e**valuation, **o**ptimization, **p**rioritization, and **m**anagement. While each element has intrinsic merit, the true power emerges from their coherent integration into a single decision architecture. The following sections examine each of the six pillars.

1. Categorization: Establishing Decision Relevance and Focus

The first and most foundational step in CREOPM™ is rigorous categorization of investments. This process explicitly distinguishes between non-discretionary and discretionary projects and programs. Non-discretionary investments—such as regulatory commitments, safety obligations, and contractual requirements—consume resources but are not subject to meaningful choice. Discretionary investments, by contrast, constitute the true decision space of the portfolio and therefore deserve disproportionate analytical attention.

Organizations that fail to enforce this distinction routinely dilute their decision processes, expending analytical effort on investments that cannot realistically be altered. The result is protracted deliberation, delayed decisions, and erosion of value due to opportunity costs.

Importantly, categorization within CREOPM™ is not synonymous with the widespread practice of *bucketing* assets into vaguely labeled groups such as *invest to win* or *invest to compete*. Such constructs often obscure rather than clarify decision logic. In reality, for phase-specific R&D investments, the choice is binary: either the organization commits resources until a success or failure signal is reached, or it does not invest at all. CREOPM™ enforces this discipline by grounding categorization in the development phase, decision rights, and the presence—or absence—of true managerial discretion.

2. Risk Analysis: Decomposing Uncertainty to Improve Decisions

Risk analysis within CREOPM™ is explicitly designed to separate what can be influenced from what cannot. Technical feasibility, development execution, regulatory interaction quality, and commercial preparedness are, to varying degrees, controllable through management action; market dynamics, competitive entry timing, and certain biological uncertainties are not.

By decomposing risk in this manner, CREOPM™ ensures that analytical rigor and managerial attention are applied where they can have the greatest impact. While critics often point to the subjectivity inherent in judgment-based risk assessments, decades of published experience—most notably from organizations such as Eli Lilly—demonstrate that disciplined, well-calibrated processes can yield decision quality superior to intuition alone.

AI is frequently proposed as a solution to the limitations of human judgment in risk assessment. In practice, its role is more nuanced. At best, AI can augment subject-matter expertise by identifying patterns across large datasets. At worst, it may foster false confidence by implying determinism in inherently idiosyncratic and biology-driven R&D processes. CREOPM™ therefore treats AI as an enabling tool, not a replacement for accountable human judgment.

3. Evaluation: Integrating Value, Risk, Cost, and Time

Evaluation represents the analytical core of CREOPM™. Here, the four fundamental value drivers—potential value, probability of success, cost, and time—are integrated into coherent, risk-adjusted metrics that enable explicit trade-off analysis across competing investments.

Decision trees remain the workhorse of this process. Despite frequent criticism regarding imperfect data and layered assumptions, decision trees provide a structured, transparent means of decomposing complex, multi-stage R&D investments into decision-relevant components. The alternative—unstructured debate, political negotiation, and narrative persuasion—offers neither greater accuracy nor greater accountability.

Within CREOPM™, evaluation is not intended to produce spurious precision. Rather, it establishes a common quantitative language that allows decision makers to compare fundamentally different investments on a consistent, phase-appropriate basis.

4. Optimization: Identifying the Value-Maximizing Portfolio

Optimization is the defining feature that distinguishes CREOPM™ from most portfolio practices in the biopharma industry. While prioritization ranks projects individually, optimization determines which combination of projects and programs maximizes enterprise value subject to explicit resource, timing, and policy constraints.

Mathematically, this involves selecting a portfolio that maximizes a defined objective function—typically expected financial value or a closely related metric—while respecting constraints on capital, headcount, development capacity, and temporal dependencies. The result is a reproducible, defensible answer to a question that is otherwise addressed through informal negotiation.

Despite its conceptual clarity, optimization remains underutilized. One plausible explanation is cultural rather than technical: prioritization readily accommodates managerial judgment, whereas optimization constrains it. CREOPM™ deliberately exposes this tension, enabling organizations to see—often uncomfortably—how far intuition-driven selections diverge from value-maximizing portfolios.

5. Prioritization: Ordering Decisions Within an Optimized Context

Prioritization retains an important role within CREOPM™, but it is explicitly subordinated to optimization. Once an optimized portfolio has been identified, prioritization rules can be applied to sequence investments, allocate incremental resources, and manage execution risk without reintroducing subjective bias.

Executives are understandably drawn to rank-ordered lists. However, when prioritization is used as a substitute for portfolio selection rather than a complement to it, value-destructive decisions are often rationalized through re-bucketing and reframing. CREOPM™ makes such behavior visible by contrasting prioritized portfolios with optimized alternatives and quantifying the associated value loss.

6. Management: Institutionalizing Portfolio Discipline

Management is the final facet of CREOPM™. It addresses the organizational conditions required to sustain portfolio discipline over time. This includes proactive management of risk, resources, and stakeholders, as well as the assessment and maturation of portfolio management capability.

Experience suggests that the greatest barrier to institutionalization is not analytical sophistication but ineffective executive sponsorship. Analytical models, when poorly contextualized or weakly sponsored, can alienate rather than align decision makers. CREOPM™ therefore emphasizes the integration of analytical rigor with influential leadership skills, recognizing that successful portfolio management is as much a social process as a technical one.

In the contemporary biopharma landscape, where external pressures and internal complexities intersect, CREOPM™ provides a structured methodology for translating strategic intent into tangible enterprise value. By integrating categorization, risk analysis, evaluation, optimization, prioritization, and management into a coherent framework, organizations can achieve higher confidence in investment decisions, maximize returns on R&D portfolios, and institutionalize a culture of disciplined, value-focused portfolio management.

THE ROLE OF CREOPM™ IN ENTERPRISE VALUE MAXIMIZATION

It is essential to underscore that *CREOPM™ is not a prescriptive algorithm or a universal panacea* for portfolio value creation. Its power lies not in eliminating judgment but in providing a disciplined, transparent framework for evaluating choices, thereby enabling organizations to distinguish between optimal and suboptimal portfolios under specific constraints and obligations.

When consistently applied in organizations with the cultural alignment and operational discipline to adhere to its principles, CREOPM™ facilitates transparent decision making regarding resource allocation across multi-year drug discovery and development pipelines. In the context of biopharmaceutical R&D, this shifts the organizational mindset away from treating projects as *random shots on goal* toward a structured approach in which budgetary and human resources are deployed strategically to maximize expected enterprise value over time.

Unlike ad hoc prioritization based solely on next year's budget or immediate tactical considerations, CREOPM™ offers a longitudinal and policy-driven framework. It enables clear articulation of which projects and programs to pursue, why they are pursued, and how they align with the organization's strategic and financial objectives. By quantifying trade-offs in expected value between optimal and suboptimal portfolios, decision makers at all levels can make defensible, transparent choices without abdicating the need for human judgment.

A common question arises: "*How can one demonstrate that implementing a formal portfolio management framework, such as CREOPM™, adds measurable enterprise value?*" While definitive empirical proof requires comparison against a controlled baseline—portfolios selected in the absence of a structured methodology—conceptual and theoretical analysis already illustrates the value differential. Organizations that rely primarily on heuristics, judgment, or prioritization methods are, by definition, exposed to suboptimal portfolio selection. CREOPM™ mitigates this exposure by providing quantitative insight into the value foregone when less optimal decisions are made.

It is important to note that portfolio management frameworks should be judged on selection quality, not project outcomes. Not every project within a portfolio will result in a successful drug or therapeutic milestone; however, robust frameworks ensure that the portfolio itself is configured to maximize expected value based on available information. In the absence of such frameworks, even high-performing organizations risk leaving value on the table—or, worse, eroding potential enterprise value.

This distinction reinforces a central tenet of contemporary biopharma portfolio management: *good enough is no longer good enough.* Suboptimal portfolio selection, regardless of isolated successes, is functionally equivalent to value destruction when assessed against the potential of optimized decision making. Consequently, enterprise and shareholder value must not merely be created—they must be maximized.

From a practical standpoint, there are two primary pathways to achieve this:

1. **Optimal portfolio selection:** Choosing a portfolio configuration that maximizes risk-adjusted or expected value across all potential investments
2. **Best-in-class suboptimal selection:** When the truly optimal portfolio is unattainable due to constraints, selecting the most advantageous alternative among available suboptimal options

Empirical experience across a spectrum of organizations consistently demonstrates that adherence to formal portfolio management principles, including those embodied in CREOPM™, significantly enhances enterprise value creation. By systematically addressing gaps in decision making, resource allocation, and risk-adjusted valuation, organizations can convert otherwise latent opportunities into tangible strategic advantage.

In summary, CREOPM™ is both a framework and a discipline: it provides the structure to make informed, defensible portfolio choices and the rigor to continuously improve the organization's capacity to maximize value from complex, resource-constrained R&D portfolios. This methodology is not optional in today's biopharma landscape; it is foundational to sustainable competitive performance and shareholder value maximization.

BIBLIOGRAPHY

Abraham, S.C. *Strategic Planning: A Practical Guide for Competitive Success.* Thompson South-Western, Mason, OH. 2006.

Biotechnology Innovation Organization; Informa Pharma Intelligence; Quantitative Life Sciences, 2021. Clinical development success rates 2006–2015, *Bio industry analysis report*, June, 2016.

Celona, J. and McNamee, P. *Decision Analysis for the Professional.* Strategic Decisions Group (Editions 1 and 2) and SmartOrg (Editions 3 and 4), 2005.

Centre for Medicines Research (CMR) International. www.clarivate.com/life-sciences-healthcare/consulting-services/benchmarking-and-rd-consulting-services.

Cook, D., Brown, D., Alexander, R., March, R., Morgan, P., Satterthwaite, G., and Pangalos, M.N. *Lessons learned from the fate of AstraZeneca's drug pipeline: A five-dimensional framework. Nature Reviews | Drug Discovery.* Vol. 13, pp. 419–431, 2014.

Cooper, R.G., Edgett, S.J., and Kleinschmidt, E.J. *Portfolio Management for New Products.* Basic Books. 2001.

DiMasi, J.A., Grabowski, H.G., & Hansen, R.W. *Innovation in the pharmaceutical industry: New estimates of R&D costs.* Journal of Health Economics, Vol. 47, pp. 20–33, 2016.

DuBrin, A.J. *Fundamentals of Organizational Behavior.* Thomson South-Western, Mason, OH. 2007.

Duran, C.O., Bonam, M., Bjork, E., Hughes, R., Ghiorghiu, S., Massacesi, C., Campbell, A., Hutchinson, E., Pangalos, M.N., and Galbraith, S. *Implementation of digital health technology in clinical trials: The 6R framework.* Nature Medicine. Vol. 29, 2023.

Evaluate Pharma. www.evaluate.com.

Fernando, K, Menon, S., Jansen, K., Naik, P., Nucci, G., Roberts, J., Wu, S.S., and Dolsten, M. *Achieving end-to-end success in the clinic: Pfizer's learnings on R&D productivity.* Drug Discovery Today. Vol. 27, pp. 697–704, 2022.

Gido, J. and Clements, J.P. *Successful Project Management.* Thomson South-Western, Mason, OH. 2009.

Goodwin, P. and Wright, G. *Decision Analysis for Management Judgment.* John Wiley & Sons Ltd., Hoboken, NJ. 2010.

Hammond, J.S., Keeney, R. L., and Raiffa, H. *Smart Choices—A practical guide to making better life decisions.* Broadway. 2002.

Hill, C.W. and Jones, G.R. *Strategic Management: An integrated approach.* South-Western Cengage Learning, Mason, OH. 2010.

ISACA. VAL IT Overview. Retrieved from Information Systems Audit and Control Association (ISACA): http://www.isaca.org/Content/ContentGroups?Val_IT1/Val_IT.htm.

Iqvia Institute for Human Data Science. Global Trends in R&D 2025: Progress in recapturing momentum in Biopharma innovation. 2025.

Keeney, R.L. *Decision Analysis: An overview.* Operations Research. Vol. 30, pp. 803–838, 1982.

Kirkwood, C.W. *Strategic Decision Making: Multi-objective Decision Analysis with spreadsheets.* Duxbury Press. 1996.

KMR Group. www.kmrgroup.com

Kola, I. and Landis, J. *Can the pharmaceutical industry reduce attrition rates?* Nature Reviews | Drug Discovery, Vol. 3, pp. 711–715, 2004.

Letavec, C. *The Program Management Office: Establishing, Managing, and Growing the Value of a PMO.* J. Ross Publishing, Ft. Lauderdale, FL. 2006.

Lewis, J.P. *Project Leadership.* McGraw-Hill, NY. 2003.

McKay, M.D., Beckman, R.J., and Conover, W.J. *A Comparison of Three Methods for Selecting Values of Input Variables in the Analysis of Output from a Computer Code.* Technometrics (American Statistical Association), pp. 239–245, 1979.

Merkhofer, L. *Keys to implementing project portfolio management.* Retrieved from http://www.prioritysystem.com. 2008.

Mintzberg, M., Ahkstrand, B., and Lampel, J. *Strategy Safari: A guided tour through the wilds of strategic management.* The Free Press, NY. 1998.

Morgan, P., Brown, D.G., Lennard, S., Anderton, M.J., Barrett, J.C., Eriksson, U., Fidock, M., Hamren, B., Johnson, A., March, R.E., Matcham, J., Mettetal, J., Nicholls, D.J., Platz, S., Rees, S., Snowden, M.A., and Pangalos, M.N. *Impact of a five-dimensional framework on R&D productivity at AstraZeneca.* Nature Reviews | Drug Discovery. Vol. 17, pp. 167–181, 2018.

Morgan, P., Van Der Graaf, P.H., Arrowsmith, J., Feltner, D.E., Drummond, K.S., Wegner, C.D., and Street, S.D.A. *Can the flow of medicines be improved? Fundamental pharmacokinetic and pharmacological principles toward improving Phase II survival.* Drug Discovery Today. Vol. 17, pp. 419–424, 2012.

Office of Government Commerce (OGC). Portfolio, Programme, and Project Offices. The Stationery Office.

Owens, P.K., Raddad, E., Miller, J.W., Ştille, J.R., Olovich, K.G., Smith, N.V., Jones, R.S., and Scherer, J.C. *A decade of innovation in pharmaceutical R&D: The Chorus model.* Nature Reviews | Drug Discovery, Vol. 3, pp. 711–715, 2015.

Paul, S.M., Mytelka, D.S., Dunwiddie, C.T., Persinger, C.C., Munos, B.H., Linborg, S.R., and Schacht, A.L. *How to improve R&D productivity: The pharmaceutical industry's grand challenge.* Nature Reviews | Drug Discovery. Vol. 9, pp. 203–214, 2010.

Persinger, C. and Andersen, J. Probability Assessments to Support Portfolio Management at Eli Lilly, *Foresight: The International Journal of Applied Forecasting*, no. 76, pp. 6–14, 2025.

Ragsdale, C.T. *Spreadsheet Modeling & Decision Analysis: A practical introduction to business analytics*. 8th edition. Cengage Learning. 2017.

Ramirez, R. and Wilkinson, A. *Strategic Reframing: The Oxford Scenario Planning Approach*. Oxford University Press. 2018.

Rang, H.P. *Drug Discovery and Development*. Churchill Livingstone Elsevier. 2006.

Ringel, M., Tollman, P., Hersch, G., and Schulze, U. *Does size matter in R&D productivity? If not, what does?* Nature Reviews | Drug Discovery. Vol. 12, pp. 901–902, 2013.

Saaty, T.L. *The Analytic Hierarchy Process*. McGraw-Hill, NY. 1980.

Sertkaya, A., Beleche, T., Jessup, A., and Sommers, B.D. *Cost of drug development and research and development intensity in the U.S. 2000–2018*. JAMA Network Open. Vol. 7, pp. 1–13, 2024.

Shapiro, A., Dentcheva, D., and Ruszczyski, A. *Lectures on stochastic programming: Modeling and theory*. 2009.

Skinner, D.C. *Introduction to Decision Analysis: A practitioner's guide to improving decision quality*. Probabilistic Publishing. 1999.

Strategic Decisions Group. See Celona and McNamee. 2005.

Taleb, N.N. *The Black Swan: The impact of the highly improbable*. Random House. 2007.

Tebbey, P.W. and Rink, C. *Target Product Profile: A renaissance for its definition and use*. J. Medical Marketing. Vol. 9, 4, pp. 301–307, 2009.

Tufte, R.E. *The Visual Display of Quantitative Information*. Graphic Press. 2001.

Wehling, M. *Nature Reviews | Drug Discovery*, Vol. 8, pp. 541–546, 2009.

Wendler, A. and Wehling, M. *Journal of Translational Medicine*, 15:226, 2017.

Winston, W. and Albright, C. *Practical Management Science*. Thomson South-Western, Mason, OH. 2018.

Wu, S.S., Fernando, K., Allerton, C., Jansen, K.U., Vincent, M.S., and Dolsten, M. *Reviving an R&D pipeline: A step change in the Phase II success rate*. Drug Discovery Today. Vol. 26, pp. 308–314, 2020.

Wyss, G.D. and Jorgensen, K.H. *A User's Guide to LHS: Sandia's Latin Hypercube Sampling Software*. 1998.

INDEX

Note: Page numbers followed by "*f*" and "*t*" refer to figures and tables, respectively.